Automotive Computer Controlled Systems

Automotive Computer Controlled Systems

Diagnostic tools and techniques

Allan W. M. Bonnick

MPhil CEng MIMechE MIRTE

OXFORD AMSTERDAM BOSTON LONDON NEW YORK PARIS
SAN DIEGO SAN FRANCISCO SINGAPORE SYDNEY TOKYO

Butterworth-Heinemann
An imprint of Elsevier Science
Linacre House, Jordan Hill, Oxford OX2 8DP
200 Wheeler Road, Burlington, MA 01803

First published 2001
Transferred to digital printing 2003

British Library Cataloguing in Publication Data
A catalogue record for this book is available from the British Library

Library of Congress Cataloguing in Publication Data
A catalogue record for this book is available from the Library of Congress

ISBN 0 7506 5089 3

For information on all Butterworth-Heinemann publications
visit our website at www.bh.com

Contents

Preface

Improvements in design, materials and manufacturing techniques have combined to produce vehicles that are, in general, very reliable. Many servicing and repair tasks, such as rebores, big-end repairs, gearbox overhauls etc., are no longer commonplace and this sometimes gives the impression that today's vehicle technicians do not need the range of skills that once were necessary.

It may be the case that the so called 'traditional' skills are less important, but the change in automotive technology that has resulted from the introduction of many computer controlled systems has meant that technicians require additional skills. These additional skills are discussed. However, it remains the case that technicians need to have a thorough understanding of technical and scientific principles that lie behind the operation of vehicle systems. For example, an exhaust emission system may be malfunctioning and a first reaction might be that the exhaust catalyst has failed. But what about other factors, such as air filter, fuel pressure, condition of the injectors, condition of the ignition system, engine valves, cylinder compression etc.? I have assumed that most readers of this book will be engaged in vehicle service work, in training or education and that they will have knowledge of the basic technology and science that enables them to 'think through' the connections between defects in computer controlled systems and the factors that may be contributing to them.

The text concentrates on areas of technology that are common to a range of systems. For example, air flow meters are a common feature on most petrol engines and they are of two types: volumetric flow (the flap), and mass flow such as the hot wire and the hot film. The outputs from these sensors are broadly similar and they can be measured accurately with the type of equipment that is described. Most exhaust gas oxygen sensors are of the zirconia type and the output signals, on almost all vehicles to which they are fitted, will be broadly identical.

There are families or groups of sensors and actuators that operate on broadly similar principles and this makes them amenable to testing by means that are widely available. When an object, such as a sensor, bears similar properties to other objects it may be referred to as belonging to a genus and the term 'generic testing' is sometimes used since the tests can be applied to most, if not all, of the same type of sensor. Many diagnostic equipment manufacturers are now making equipment that enables technicians to perform a wide range of tests on computer controlled systems. The aim of this book is to show how, with the aid of equipment, suitable training and personal endeavour, service technicians and trainees may equip themselves with the knowledge and skill that will permit them to perform accurate diagnosis and repair.

Chapters 5, 6 and 7 show how knowledge of the technology that is common to many of the systems can be used to perform effective diagnosis on a range of computer controlled systems. Also covered is a range of modern computer controlled systems, computer technology and features such as CAN and OBD II.

This book has been designed to meet the needs of students and trainees who are working for NVQ level 3, BTEC National Certificate and Diploma, Higher National and similar vocational qualifications. However, the treatment of topics is sufficiently broad as to provide useful background knowledge for students of design and technology, and those on computing courses who are studying in schools and colleges. DIY motorists, particularly those with an interest in computing, may also find the book helpful in obtaining a better understanding about their own vehicles, particularly in relation to features such as the European OBD, which is likely to cause widespread attention when it becomes more widely used in the UK.

Allan Bonnick

Acknowledgements

Thanks are due to the following companies who supplied information and in many cases permission to reproduce illustrations.
Crypton Technology Group
Ford Motor Company
Fluke (UK) Ltd
Gunson Limited
Lucas Aftermarket Operations Ltd
Lucas Diesel Systems
Lucas Varity Ltd
Motorola
Renault Ltd
Robert Bosch Ltd (Mr Richard Clayton - Garage equipment dept.)
Rover Car Company
Society of Automotive Engineers, Inc. (Reprinted with permission from SAE J 2012 MAR99 © 1999)
Toyota Motor Company
Volvo Cars
Wabco

Special thanks also to Shirley and to Phil Handley who persuaded me that the effort was worthwhile, especially when the going was hard.

1
Common technology

The aim of this chapter is to review a number of computer controlled vehicle systems that are in current use and to make an assessment of the technology involved that is common to a range of systems. It is this knowledge that is 'common' to many systems that enables a vehicle technician to develop a 'platform' of skills that will assist in diagnostic work across the spectrum of vehicles, from small cars to heavy trucks.

Subsequent chapters concentrate on aspects of the technology that enable garage technicians to perform diagnostic and other tasks related to the maintenance and repair of modern vehicles. In order to achieve this aim a representative range of systems is examined in outline, so as to give a broad understanding of their construction and mode of operation, as opposed to an 'in depth' study of each system. Later chapters look at the individual aspects of each system, such as sensors and the computer (ECM), and provide detailed explanations since the evidence suggests that more detailed knowledge assists in the diagnostic process.

1.1 Common technology

Changes in electronics technology and manufacturing methods take place rapidly and for some years now, microcontrollers (mini-computers) have formed the heart of many of the control systems found on motor vehicles.

Microcontrollers, in common with other computers, contain a control unit and presumably in order to avoid any possible confusion, the 'black box' that used to be known as the Electronic Control Unit (ECU) is now commonly referred to as the Electronic Control Module (ECM). In this book, the term electronic control module (ECM) is used when referring to the control module that was formerly known as the ECU.

As vehicle systems have developed it has become evident that there is a good deal of electronic and computing technology that is common to many vehicle electronic systems and this suggests that there is good reason for technicians to learn this 'common technology' because it should enable them to tackle diagnosis and repair on a range of vehicles. Indeed, many manufacturers of automotive test equipment are now producing equipment which, when supported by information and data about diagnostic trouble codes (fault codes), provides the knowledgeable technician with the support that should enable him/her to go forward in to the 2000s with a degree of confidence in their ability to maintain and repair modern systems.

We will now look at a representative selection of commonly used modern systems in order to enable us to 'tease out' the common elements that it will be useful to learn more about.

1.2 Engine-related systems

The engine systems that are surveyed are those that are most commonly used, namely ignition and fuelling, plus emission control. A major purpose of these system surveys is to identify common ground in order to focus on the components of the systems that can realistically be tested with the aid of reasonably priced tools, rather than the more exotic systems that require specialized test equipment.

By examining three ignition systems it should be possible to pick out certain elements that are commonly used. In the process of examining a number of other systems we shall see that certain basic principles are common to several types of systems that are used on vehicles. In effect, there is a good deal of knowledge that can be transferred across a considerable range of technology.

1.3 Ignition systems

Electronic ignition systems make use of some form of electrical/electronic device to produce the electrical pulse that switches the ignition coil primary current 'on and off', so that a high voltage is induced in the coil secondary winding in order to produce a spark in the required cylinder at the correct time.

There are several methods of producing the basic 'triggering' pulse for the ignition, but three of these methods are more widely used than the others. It is the ignition systems that are based on the use of these three methods that are now dealt with in some detail.

1.3.1 THE CONSTANT ENERGY IGNITION SYSTEM

Figure 1.1 shows a type of electronic ignition distributor that has been in use for many years. The distributor shaft is driven from the engine camshaft and thus rotates at half engine speed.

Each time a lobe on the rotor (reluctor) passes the pick-up probe a pulse of electrical energy is induced in the pick-up winding. The pick-up winding is connected to the electronic ignition module and when the pulse generator voltage has reached a certain level (approximately 1 V) the electronic circuit of the module will switch on the current to the ignition coil primary winding.

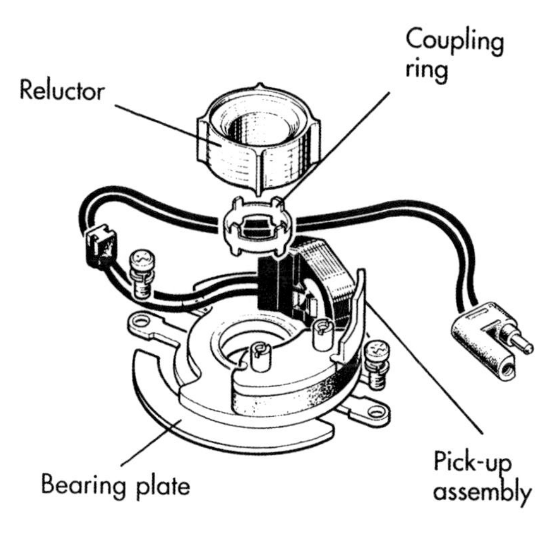

Fig. 1.1 Reluctor and pick-up assembly

As the reluctor continues to rotate, the voltage in the pick-up winding begins to drop and this causes the ignition module to 'switch of' the ignition coil primary current; the high voltage for the ignition spark is then induced in the ignition coil secondary winding. The period between switching on and switching off the ignition coil primary current is called the dwell period. The effective increase in dwell angle as the speed increases means that the coil current can build up to its optimum value at all engine speeds. Figure 1.2 shows how the pulse generator voltage varies due to the passage of one lobe of the reluctor past the pick-up probe. From the graphs in this figure may be seen that the ignition coil primary current is switched on when the pulse generator voltage is approximately 1 V and is switched off again when the voltage falls back to the same level. At higher engine speeds the pulse generator produces a higher voltage and the switching-on voltage (approximately 1 V) is reached earlier, in terms of crank position, as shown in the second part of Fig. 1.2. However, the 'switching-of' point is not affected by speed and this means that the angle (dwell) between switching the coil primary current on and off increases as the engine speed increases. This means that the build-up time for the current in the coil primary winding, which is the important factor affecting the spark energy, remains virtually constant at all speeds. It is for this reason that ignition systems of this type are known as 'constant energy systems'. It should be noted that this 'early' type of electronic ignition still incorporates the centrifugal and vacuum devices for automatic variation of the ignition timing.

1.3.2 DIGITAL (PROGRAMMED) IGNITION SYSTEM

Programmed ignition makes use of computer technology and permits the mechanical, pneumatic and other elements of the conventional distributor to be dispensed with. Figure 1.3 shows an early form of a digital ignition system.

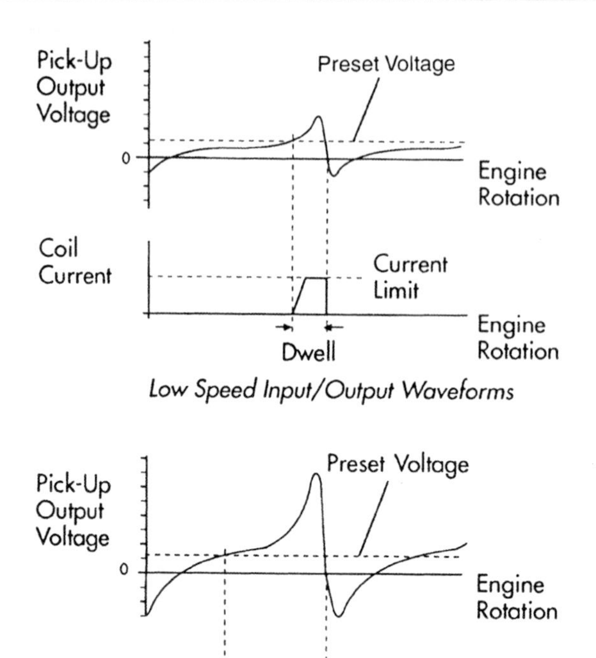

Fig. 1.2 Pick-up output voltage at low and high speeds

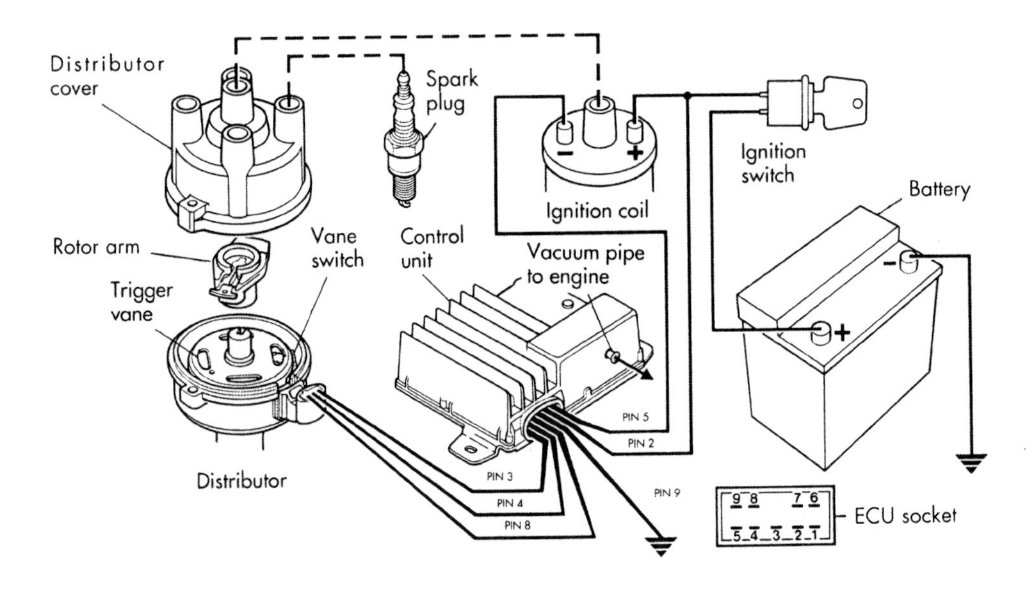

Fig. 1.3 A digital ignition system

The control unit (ECU or ECM) is a small, dedicated computer which has the ability to read input signals from the engine, such as speed, crank position, and load. These readings are compared with data stored in the computer memory and the computer then sends outputs to the ignition system. It is traditional to represent the data, which is obtained from engine tests, in the form of a three-dimensional map as shown in Fig. 1.4.

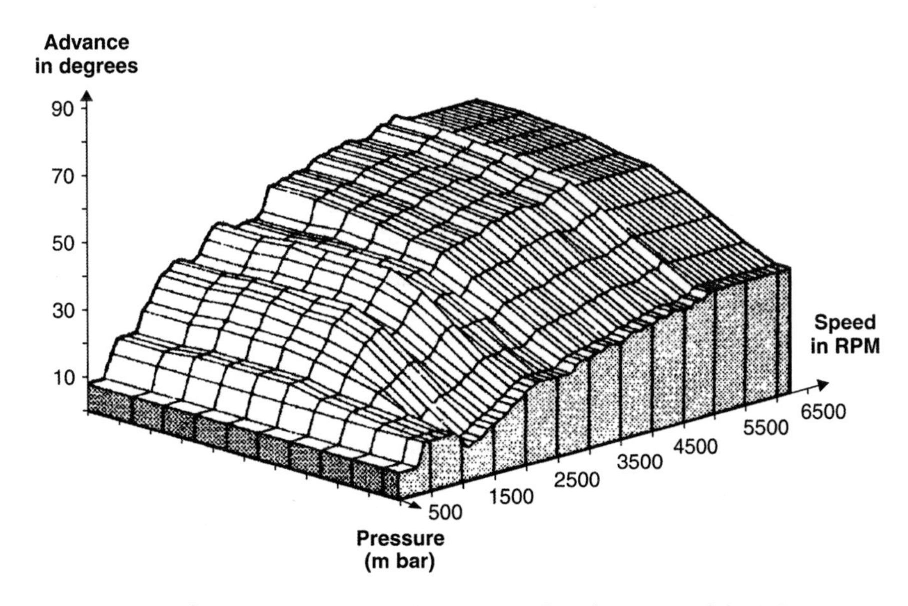

Fig. 1.4 An ignition map that is stored in the ROM of the ECM

Any point on this map can be represented by a number reference: e.g. engine speed, 1000 rpm; manifold pressure (engine load), 0.5 bar; ignition advance angle, 5°. These numbers can be converted into computer (binary) code words, made up from 0s and 1s (this is why it is known as digital ignition). The map is then stored in the computer memory where the processor of the control unit can use it to provide the correct ignition setting for all engine operating conditions.

In this early type of digital electronic ignition system the 'triggering' signal is produced by a Hall effect sensor of the type shown in Fig. 1.5.

When the metal part of the rotating vane is between the magnet and the Hall element the sensor output is zero. When the gaps in the vane expose the Hall element to the magnetic field, a voltage pulse is produced. In this way, a voltage pulse is produced by the Hall sensor each time a spark is required. Whilst the adapted form of the older type ignition distributor is widely used for electronic ignition systems, it is probable that the trigger pulse generator driven by the crankshaft and flywheel is more commonly used on modern systems. This is a convenient point at which to examine the type of system that does not use a distributor of the conventional form but uses a flywheel-driven pulse generator.

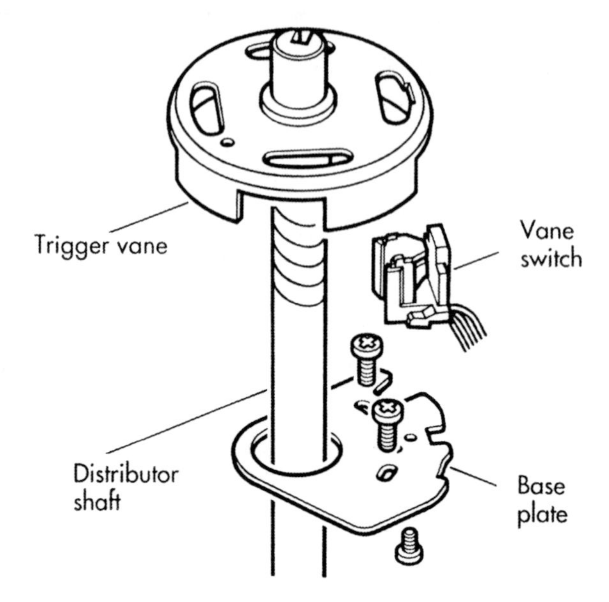

Fig. 1.5 A Hall type sensor

1.3.3 DISTRIBUTORLESS IGNITION SYSTEM

Figure 1.6 shows an ignition system for a four-cylinder engine. There are two ignition coils, one for cylinders 1 and 4, and another for cylinders 2 and 3. A spark is produced each time a pair of cylinders reaches the firing point which is near top dead center (TDC). This means that a spark occurs on the exhaust stroke as well as on the power stroke. For this reason, this type of ignition system is sometimes known as the 'lost spark' system.

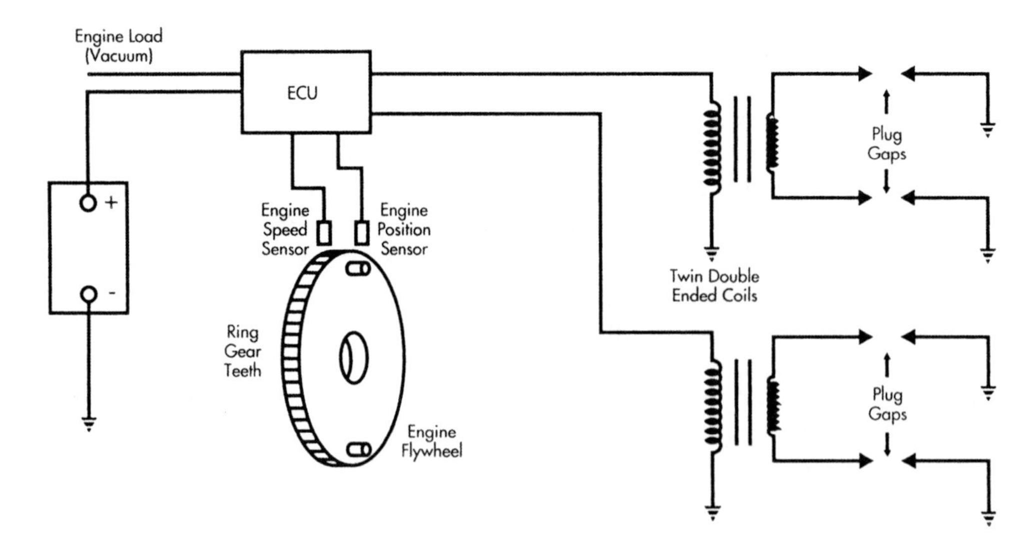

Fig. 1.6 A distributorless ignition system

Figure 1.6 shows that there are two sensors at the flywheel: one of these sensors registers engine speed and the other is the trigger for the ignition. They are shown in greater detail in Fig. 1.7 and they both rely on the variable reluctance principle for their operation.

An alternative method of indicating the TDC position is to use a toothed ring, attached to the flywheel, which has a tooth missing at the TDC positions, as shown in Fig. 1.8. With this type of sensor, the TDC position is marked by the absence of an electrical pulse. This is also a variable reluctance sensor. The other teeth on the reluctor ring, which are often spaced at 10° intervals, are used to provide pulses for engine speed sensing.

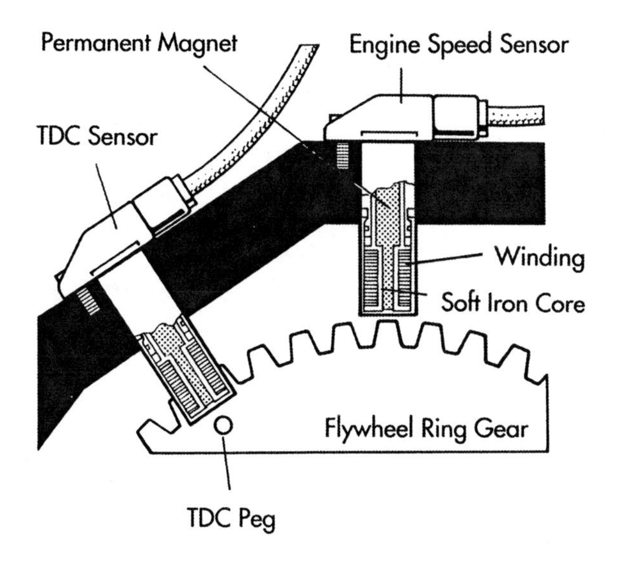

Fig. 1.7 Details of engine speed and crank position sensors

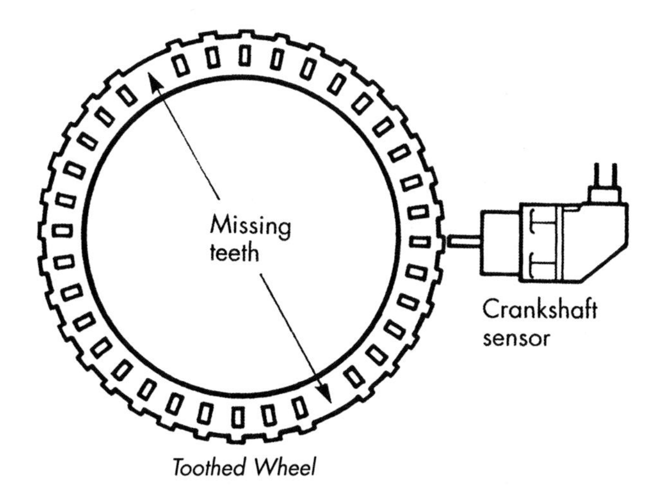

Fig. 1.8 Engine speed and position sensor that uses a detachable reluctor ring

1.3.4 OPTOELECTRONIC SENSING FOR THE IGNITION SYSTEM

Figure 1.9 shows the electronic ignition photoelectronic distributor sensor used on a Kia. There are two electronic devices involved in the operation of the basic device. One is a light-emitting diode (LED), which converts electricity into light, and the other is a photodiode that can be 'switched on' when the light from the LED falls on it.

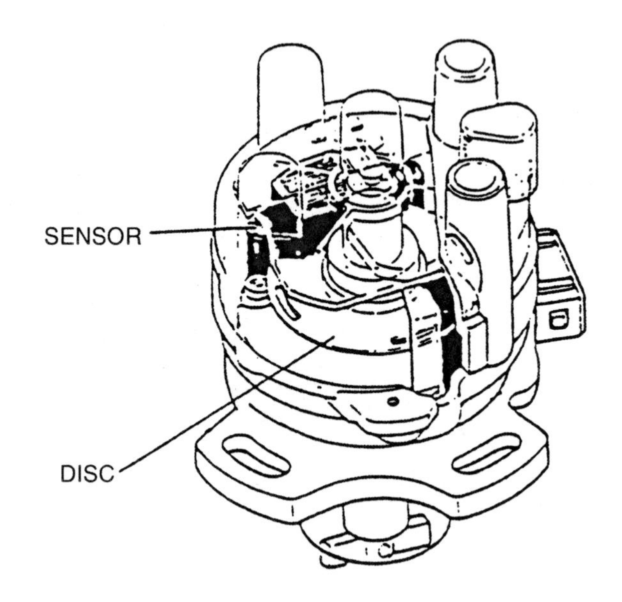

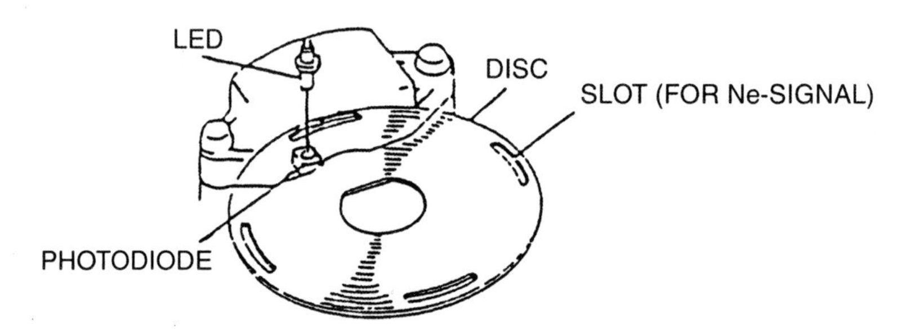

Fig. 1.9 An optoelectronic sensor

Another version of this type of sensor is shown in Fig. 1.10. Here the rotor plate has 360 slits placed at 1° intervals, for engine speed sensing, and a series of larger holes for TDC indication that are placed nearer the center of the rotor plate. One of these larger slits is wider than the others and it is used to indicate TDC for number 1 cylinder.

As the processing power of microprocessors has increased it is natural to expect that system designers will use the increased power to provide further features such as combustion knock sensing and adaptive ignition control.

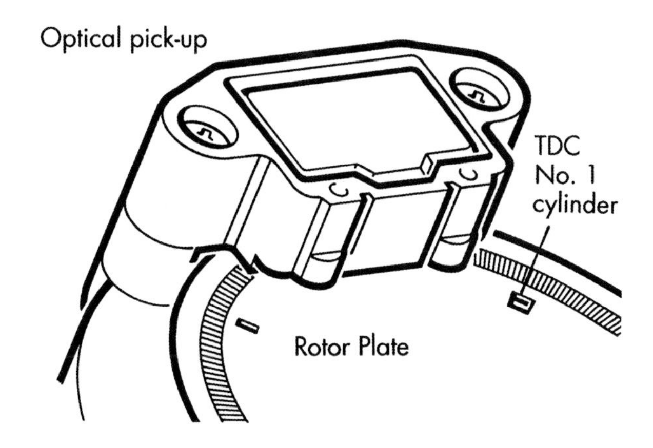

Fig. 1.10 An alternative form of optoelectronic sensor

1.3.5 KNOCK SENSING

Combustion knock is a problem that is associated with engine operation. Early motor vehicles were equipped with a hand control that enabled the driver to retard the ignition when the characteristic 'pinking' sound was heard. After the pinking had ceased the driver could move the control lever back to the advanced position. Electronic controls permit this process to be done automatically and a knock sensor is often included in the make-up of an electronic ignition system. Figure 1.11 shows a knock sensor as fitted to the cylinder block of an in-line engine.

The piezoelectric effect is often made use of in knock sensors and the tuning of the piezoelectric element coupled with the design of the sensor's electronic circuit permits combustion knock to be selected out from other mechanical noise. The combustion knock is represented by a voltage signal which is transmitted to the ECM and the processor responds by retarding the ignition to prevent knocking. The ECM retards the ignition in steps, approximately 2° at a time, until knocking ceases. When knocking ceases the ECM will again advance the ignition, in small steps, until the correct setting is reached.

Figure 1.11 shows the knock sensor in position. The operating principles and test procedure are described in more detail in Chapters 5 and 7.

1.3.6 ADAPTIVE IGNITION

The computing power of modern ECMs permits ignition systems to be designed so that the ECM can alter settings to take account of changes in the condition of components, such as petrol injectors, as the engine wears. The general principle is that the best engine torque is achieved when combustion produces maximum cylinder pressure just after TDC. The ECM monitors engine acceleration by means of the crank sensor, to see if changes to the ignition setting produce a better result, as indicated by increased engine speed as a particular cylinder fires. If a better result is achieved then the ignition memory map can be reset so that

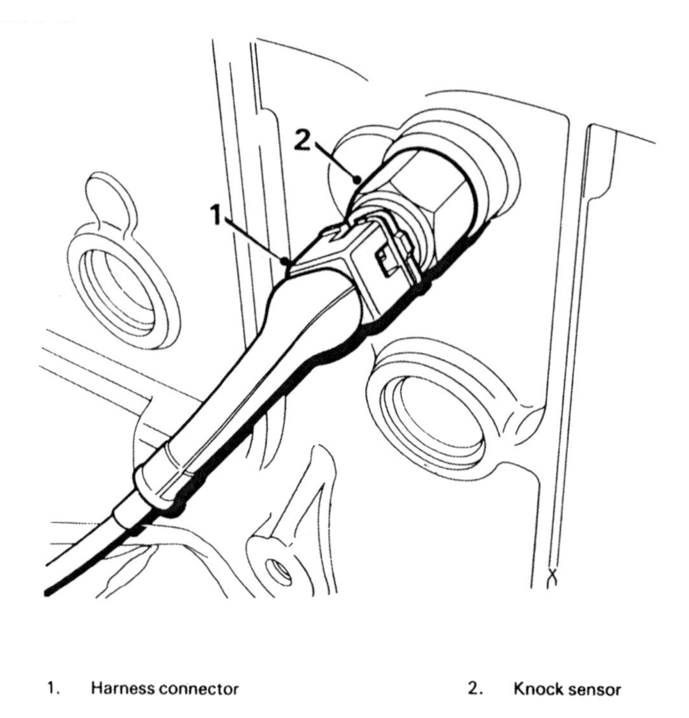

1. Harness connector 2. Knock sensor

Fig. 1.11 The knock sensor on the engine

the revised setting becomes the one that the ECM uses. This 'adaptive learning strategy' is now used quite extensively on computer controlled systems and it requires technicians to run vehicles under normal driving conditions for several minutes after replacement parts and adjustments have been made to a vehicle. This review of ignition systems gives a broad indication of the technology involved and, more importantly, it highlights certain features that can reasonably be said to be common to all ignition systems. These are: crank position and speed sensors, an ignition coil, a knock sensor, and a manifold pressure sensor for indicating engine load. In the next section, computer controlled fuelling systems are examined and it will be seen that quite a lot of the technology is similar to that used in electronic ignition systems.

1.4 Computer controlled petrol fuelling systems

Computer controlled petrol injection is now the normal method of supplying fuel - in a combustible mixture form - to the engine's combustion chambers. Although it is possible to inject petrol directly into the engine cylinder in a similar way to that in diesel engines, the practical problems are quite difficult to solve and it is still common practice to inject (spray) petrol into the induction manifold. There are, broadly speaking, two ways in which injection into the induction manifold is performed. One way is to use a single injector that sprays fuel into the region of the throttle butterfly and the other way is to use an injector for each cylinder, each injector being placed near to the inlet valve. The two systems are known as single-point injection (throttle body injection), and multi-point injection. The principle is illustrated in Fig. 1.12.

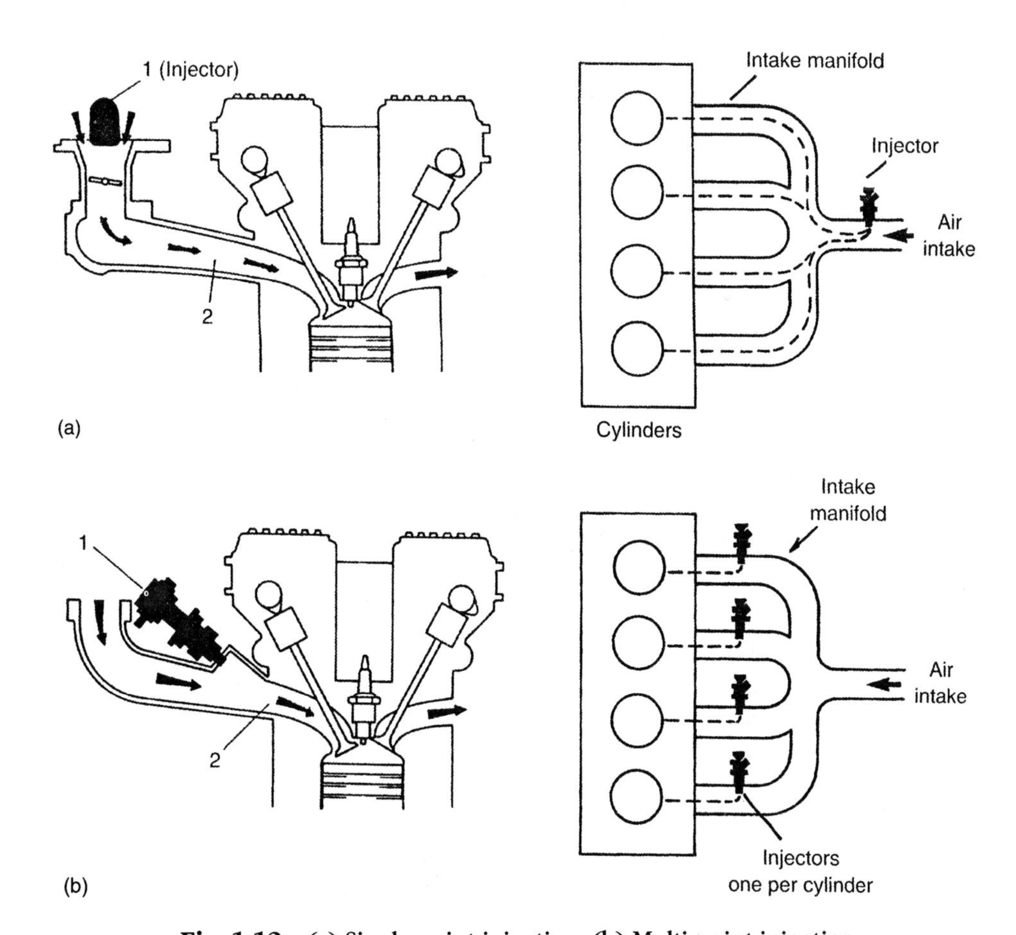

Fig. 1.12 (a) Single-point injection. (b) Multi-point injection

1.4.1 SINGLE-POINT INJECTION

In its simplest form, petrol injection consists of a single injector that sprays petrol into the induction manifold in the region of the throttle butterfly valve, as shown at (4) in Fig. 1.13.

Finely atomized fuel is sprayed into the throttle body, in accordance with controlling actions from the engine computer (EEC, ECM), and this ensures that the correct air–fuel ratio is supplied to the combustion chambers to suit all conditions. The particular system shown here uses the speed density method of determining the mass of air that is entering the engine, rather than the air flow meter that is used in some other applications. In order for the computer to work out (compute) the amount of fuel that is needed for a given set of conditions it is

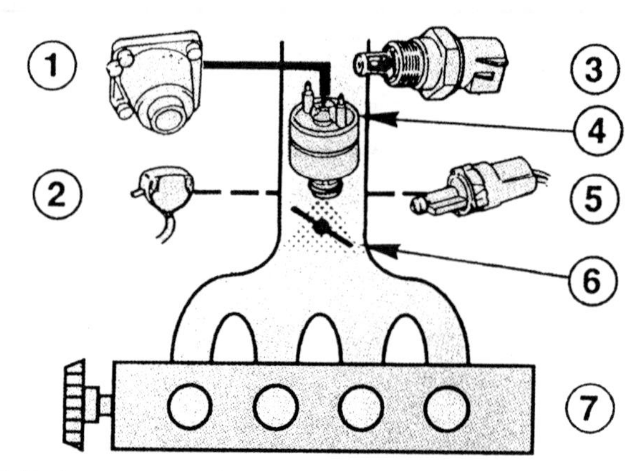

1. Fuel pressure regulator
2. Throttle position sensor
3. Air charge temperature sensor
4. Injector valve
5. Throttle plate control motor (actuator)
6. Throttle plate
7. Engine

Fig. 1.13 Single-point injection details

necessary for it to have an accurate measure of the air entering the engine. The speed density method provides this information from the readings taken from the manifold absolute pressure (MAP) sensor, the air charge temperature sensor, and the engine speed sensor.

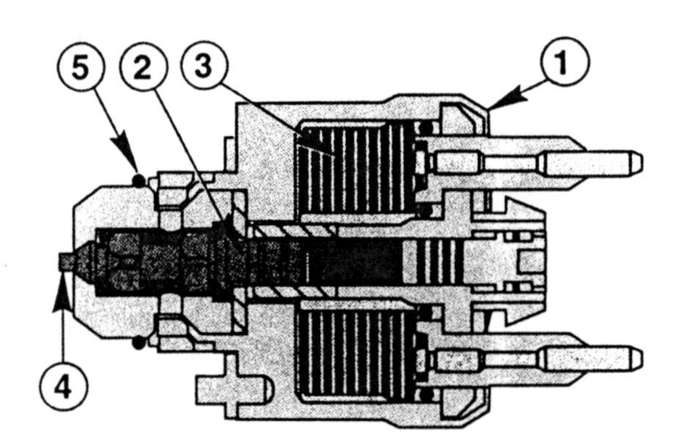

1 Housing
2 Fuel duct
3 Solenoid
4 Jet needle
5 O-ring

Fig. 1.14 The single point CFI (central fuel injection) unit

The actual central injector unit is shown in Fig. 1.14. The injector valve is operated by the solenoid (3) which receives electric current in accordance with signals from the engine control computer. When the engine is operating at full or part load the injector sprays fuel during each induction stroke. When the engine is idling the injector operates once per revolution of the crankshaft. Because the fuel pressure regulator maintains a constant fuel pressure at the injector valve, the amount of fuel injected is determined by the length of time for which the solenoid holds the valve in the open position.

The throttle plate (butterfly valve) motor is operational during starting, coasting, when shutting down the engine, and when the engine is idling.

1.4.2 MULTI-POINT INJECTION

In petrol injection systems it is often the practice to supply the injectors with petrol, under pressure, through a fuel gallery or 'rail'. Each injector is connected to this gallery by a separate pipe, as shown in Fig. 1.15.

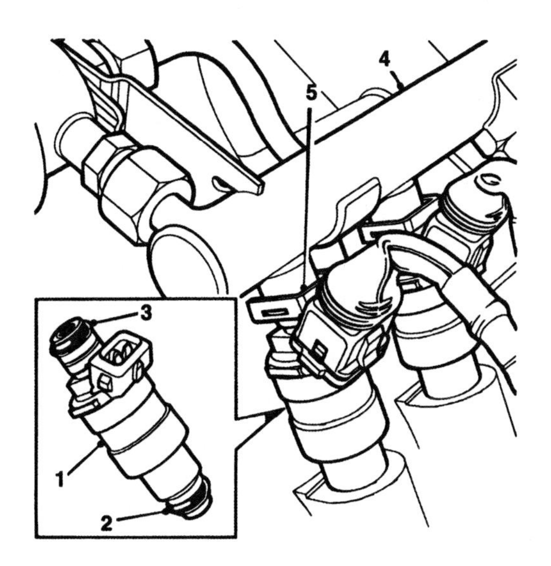

1. Injector
2. 'O' ring seal (manifold)
3. 'O' ring seal (fuelrail)
4. Fuel gallery
5. Retaining clip

Fig. 1.15 The fuel gallery

The pressure of the fuel in the gallery is controlled by a regulator of the type shown in Fig. 1.16. This particular pressure regulator is set, during manufacture, to give a maximum fuel pressure of 2.5 bar.

In operation the petrol pump delivers more fuel than is required for injection and the excess pressure lifts the regulator valve (5) off its seat to allow the excess fuel to return to the fuel tank via the return connection (6). The

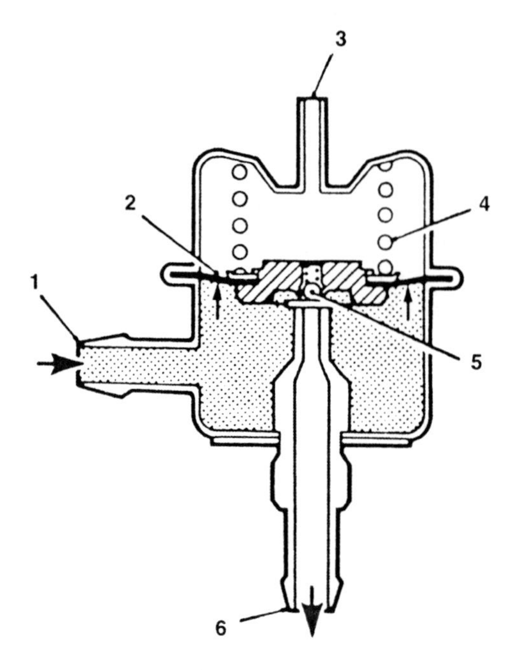

1. Fuel inlet
2. Diaphragm
3. Vacuum pipe connection
4. Pressure control spring
5. Valve
6. Fuel return connection

Fig. 1.16 The fuel pressure regulator

internal diaphragm (2) of the regulator is subject to inlet manifold pressure (vacuum) and this permits the diaphragm and spring to regulate fuel pressure to suit a range of operating conditions. Raising the diaphragm, against the spring, lowers the fuel pressure and this permits a low pressure of approximately 1.8 bar. Lowering the diaphragm, in response to higher pressure in the inlet manifold (wider throttle opening), gives a high fuel pressure of approximately 2.5 bar.

With this arrangement, the amount of fuel that each injector sprays into the inlet manifold is determined by the length of time for which the injector valve is opened by its operating solenoid. By varying the length of time for which the injector valve remains open, the amount of fuel injected is made to suit a range of requirements.

Fuelling requirements for a particular engine are known to the designer and they are placed in the ECM memory (ROM). In operation, the ECM receives information (data) from all of the sensors connected with the engine's fuel needs. The ECM computer compares the input data from the sensors with the data stored in the computer memory. From this comparison of data the ECM computer provides some output data which appears on the injector cables as an electrical pulse that lasts for a set period. This injector electrical pulse time varies from approximately 2 milliseconds (ms), to around 10 ms. The 'duty cycle' concept is based on the

A typical square waveform is shown in the figure, a single cycle is indicated by 'C', which consists of an ON time 'A', and an OFF time 'B'.

Duty cycle is the length of the ON time 'A' compared to the whole cycle 'C', expressed as a percentage. (Please note, On time can be High and Low on certain systems.)

Using the figure and the time periods, the duty cycle is 25%.

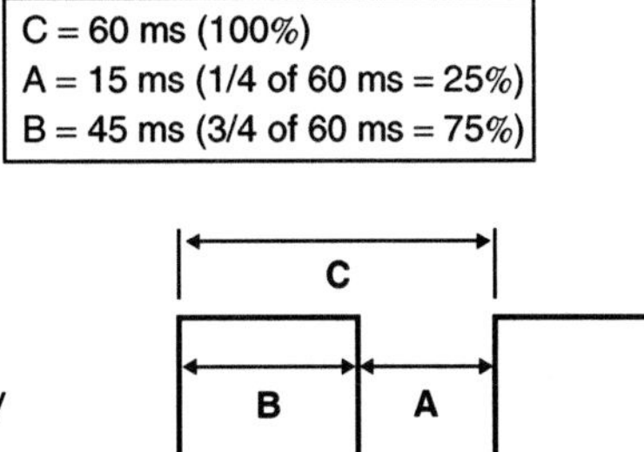

Fig. 1.17 Duty cycle

percentage of the available time for which the device is energized, as shown in Fig. 1.17.

The performance characteristics of the engine and the driveability of the vehicle are determined by the 'quality' of the input that is put into the design, and part of this design is the computer program that is held in the read only memory (ROM) of the ECM. A fuelling map is similar to the ignition map shown in Fig. 1.4. The differences being that engine load is represented by throttle position and spark advance is replaced by air/fuel ratio. Each point of the surface on this map can be represented by a binary code and a range of points from a map is stored in the ROM of the ECM. The values stored in the ROM are compared with input signals from sensors in order that the computer can determine the duration of the fuel injection pulse.

Some computerized systems are designed so that the franchised dealership can alter the computer program to match customer requirements. A re-programmable ROM is necessary for this to be done and the work can only be done by qualified personnel acting under the control of the vehicle manufacturer.

Multi-point injection systems commonly use one of two techniques.

1. Injection of half the amount of fuel required to all inlet ports, each time the piston is near top dead center.
2. Sequential injection, whereby injection occurs only on the induction stroke.

In the multi-point injection system shown in Fig. 1.18 there is one petrol injector (number 12 in the diagram) for each cylinder of the engine. Each of the injectors is designed so that it sprays fuel on to the back of the inlet valve. The actual position and angle at which injection takes place varies for different types of engines.

In the system shown in Fig. 1.18 the air flow is measured by the hot-wire type of mass air flow meter. The control computer receives the signal from the air flow meter and uses this signal together with those from other sensors, such as engine speed, engine coolant temperature, throttle position etc., to determine the length of time of the injection pulse.

Sequential multi-point injection is a term that is used to describe the type of petrol injection system that provides one injection of fuel for each cylinder during each cycle of operation. To assist in providing the extra controlling input that

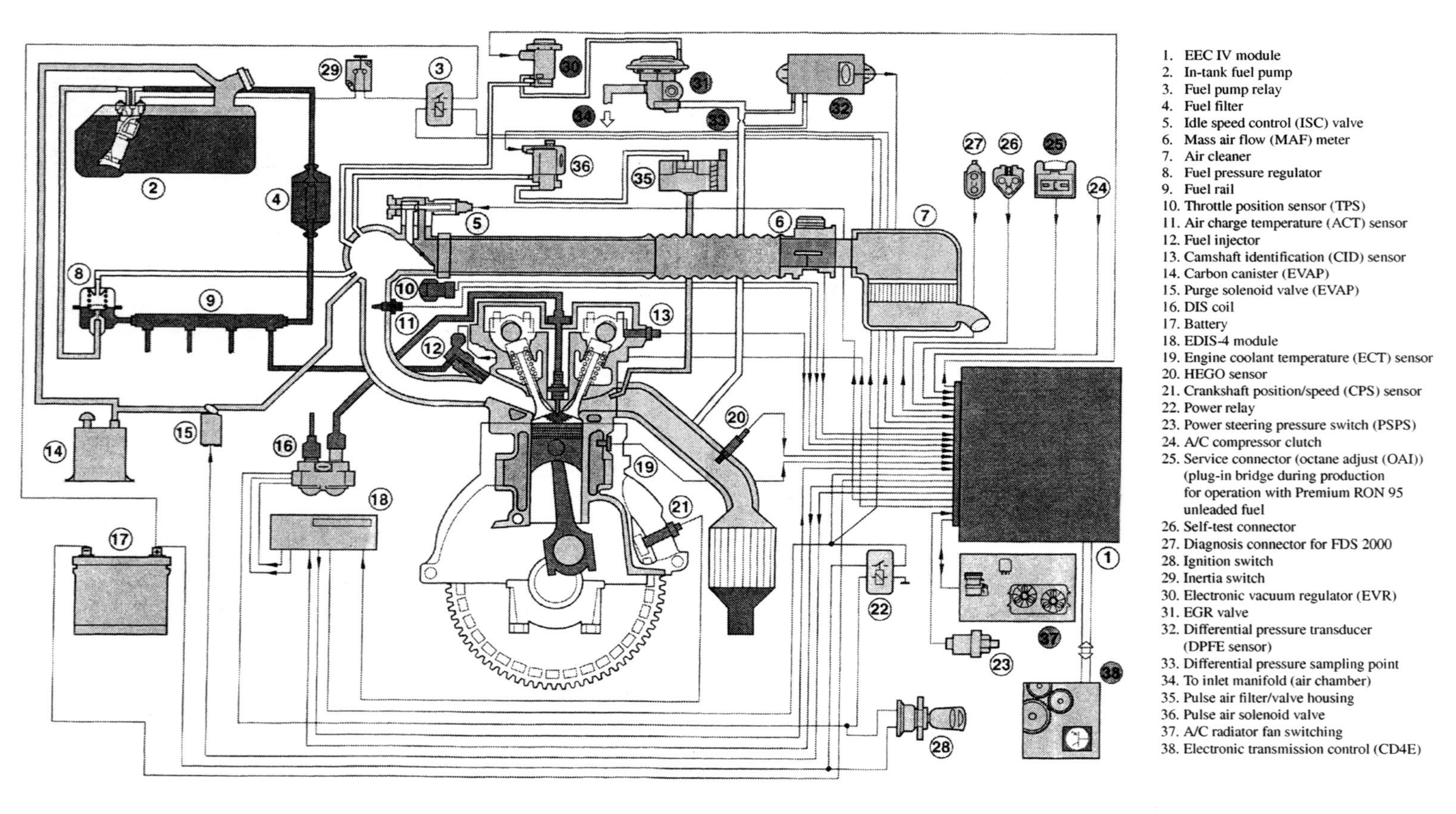

1. EEC IV module
2. In-tank fuel pump
3. Fuel pump relay
4. Fuel filter
5. Idle speed control (ISC) valve
6. Mass air flow (MAF) meter
7. Air cleaner
8. Fuel pressure regulator
9. Fuel rail
10. Throttle position sensor (TPS)
11. Air charge temperature (ACT) sensor
12. Fuel injector
13. Camshaft identification (CID) sensor
14. Carbon canister (EVAP)
15. Purge solenoid valve (EVAP)
16. DIS coil
17. Battery
18. EDIS-4 module
19. Engine coolant temperature (ECT) sensor
20. HEGO sensor
21. Crankshaft position/speed (CPS) sensor
22. Power relay
23. Power steering pressure switch (PSPS)
24. A/C compressor clutch
25. Service connector (octane adjust (OAI)) (plug-in bridge during production for operation with Premium RON 95 unleaded fuel
26. Self-test connector
27. Diagnosis connector for FDS 2000
28. Ignition switch
29. Inertia switch
30. Electronic vacuum regulator (EVR)
31. EGR valve
32. Differential pressure transducer (DPFE sensor)
33. Differential pressure sampling point
34. To inlet manifold (air chamber)
35. Pulse air filter/valve housing
36. Pulse air solenoid valve
37. A/C radiator fan switching
38. Electronic transmission control (CD4E)

Fig. 1.18 A multi-point petrol injection system

is required for sequential injection the engine is often fitted with an additional sensor which is driven by the engine camshaft. Hall type sensors and variable reluctance sensors driven by the camshaft are often used for this purpose to assist the computer to determine TDC on number 1 cylinder. Figure 1.19 shows one of these sensors which is fitted to an overhead camshaft engine.

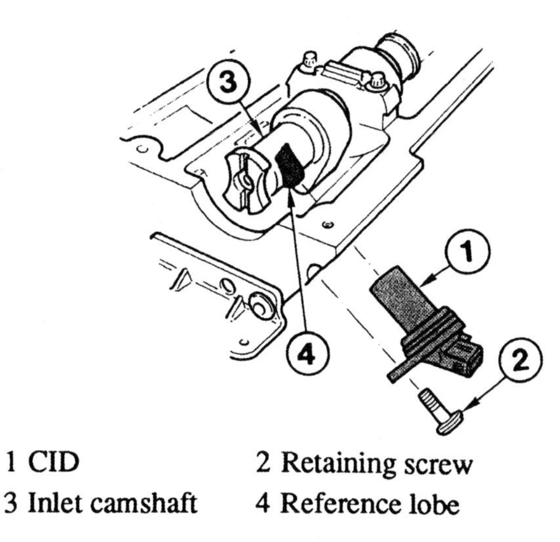

1 CID 2 Retaining screw
3 Inlet camshaft 4 Reference lobe

Fig. 1.19 A cylinder identification sensor

Some of the sensors used for fuelling are the same as those used for ignition systems, e.g. crank speed and top dead center sensors, manifold pressure to indicate engine load etc. Because some of the sensor signals can be used for both ignition and fuelling it has become common practice to place them under the control of a single computer and the resulting system is known as an engine management system.

1.5 Engine management systems (EMS)

Engine management systems are designed to ensure that the vehicle complies with emissions regulations as well as to provide improved performance. This means that the number of sensors and actuators is considerably greater than for a simple fuelling or ignition system. The system shown in Fig. 1.18 is fairly typical of modern engine management systems and selected items of technology are now given closer attention. The aim here is to concentrate on the aspects of engine control that were not covered in the sections on fuel and ignition systems.

The first component to note is the oxygen sensor at number 20. This is a heated sensor (HEGO) and the purpose of the heating element is to bring the sensor to its working temperature as quickly as possible. The HEGO provides a feedback signal that enables the ECM to control the fuelling so that the air-fuel ratio is kept very close to the chemically correct value where lambda = 1, since this is the value that enables the catalytic converter to function at its best. Oxygen sensors are common to virtually all modern petrol engine vehicles and this is obviously an area of technology that technicians need to know about. The zirconia type oxygen sensor is most commonly used and it produces a voltage signal that represents oxygen levels in the exhaust gas and is thus a reliable indicator of the air-fuel ratio that is entering the combustion chamber. The voltage signal from this sensor is fed back to the control computer to enable it to hold lambda close to 1.

1.5.1 EXHAUST GAS RECIRCULATION

Two items in Fig. 1.18, the electronic vacuum regulator at (30) and the exhaust gas recirculation (EGR) valve at (31), play an important part in this and many other engine management systems and they warrant some attention. In order to reduce emissions of NO_x it is helpful if combustion chamber temperatures do not rise above approximately 1800°C because this is the temperature at which NO_x can be produced. Exhaust gas recirculation helps to keep combustion temperatures below this figure by recirculating a limited amount of exhaust gas from the exhaust system back to the induction system, on the engine side of the throttle valve. Figure 1.20 shows the principle of an EGR system.

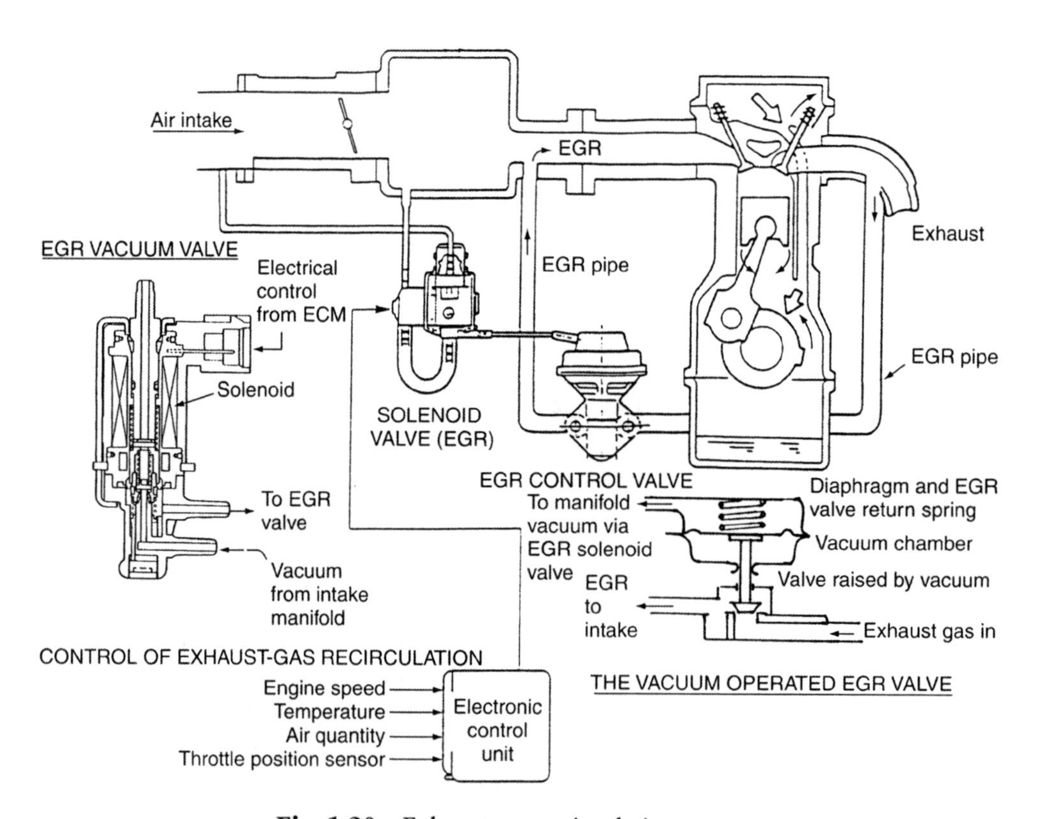

Fig. 1.20 Exhaust gas recirculation system

In order to provide a good performance, EGR does not operate when the engine is cold or when the engine is operating at full load. The inset shows the solenoid valve that controls the EGR valve and this type of valve is operated on the duty cycle principle. Under reasonable operating conditions it is estimated that EGR will reduce NO_x emissions by approximately 30%.

1.5.2 COMPUTER CONTROL OF EVAPORATIVE EMISSIONS

Motor fuels give off vapors that contain harmful hydrocarbons, such as benzene. In order to restrict emissions of hydrocarbons from the fuel tank, vehicle systems are equipped with a carbon canister. This canister contains activated charcoal which has the ability to bind toxic substances into hydrocarbon molecules. In the evaporative emission control system the carbon canister is connected by valve and pipe to the fuel tank, as shown in Fig. 1.21.

The evaporative purge solenoid valve connects the carbon canister to the induction system, under the control of the ECM, so that the hydrocarbon vapors can be drawn into the combustion chambers to be burnt with the main fuel-air mixture. The control valve is operated by duty cycle electrical signals from the computer which determine the period of time for which the valve is open. When the engine is not running the vapor from the fuel in the tank passes into the carbon canister. When the engine is started up the ECM switches on the solenoid valve so that the vapor can pass into the induction system. The frequency of operation of the solenoid valve after this is dependent on operating conditions.

Evaporative emissions control is part of the emissions control system of the vehicle and it must be maintained in good order.

1.6 Anti-lock braking (ABS)

Anti-lock braking is another form of a computer controlled system that is commonly used. Figure 1.22 shows a relatively modern system that uses individual wheel control for ABS and is known as a four-channel system. The braking system shown here uses a diagonal split of the hydraulic circuits: the brakes on the front left and rear right are fed by one part of the tandem master cylinder, and the brakes on the front right and rear left are fed from the other part of the tandem master cylinder.

The wheel sensors operate on the Hall principle and give an electric current output which is considered to have advantages over the more usual voltage signal from wheel sensors. The ABS control computer is incorporated into the ABS modulator and, with the aid of sensor inputs, provides the controlling actions that are designed to allow safe braking in emergency stops.

Starting at the top left corner of Fig. 1.23 there are two hydraulic accumulators (A1 and A2) which act as pressure reservoirs for hydraulic fluid. Below these is the modulator pump which is under computer control. At the bottom of the diagram are the four wheel brakes and above these are the inlet and outlet valves (labelled

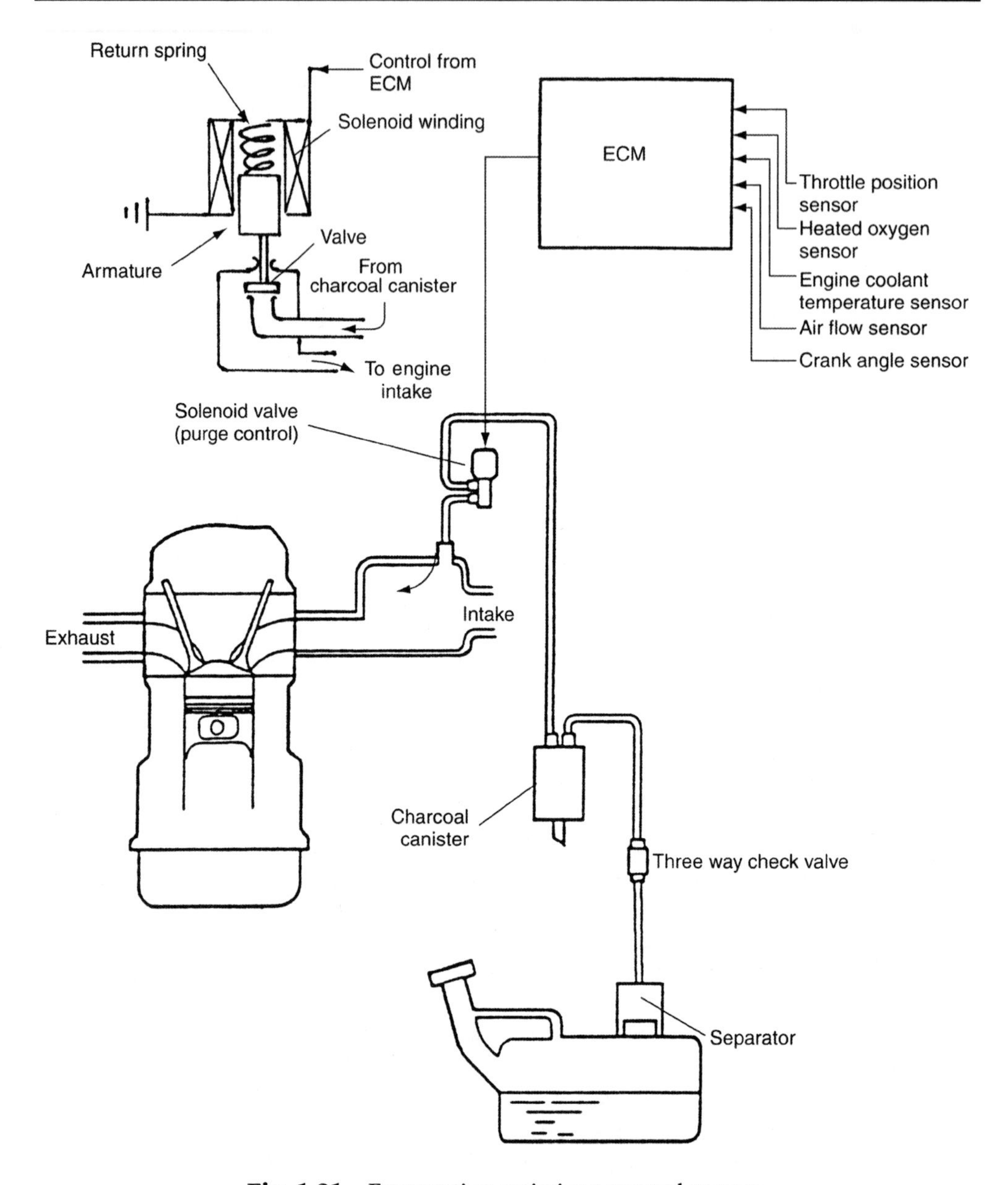

Fig. 1.21 Evaporative emissions control system

C and D, respectfully) which, under computer control, determine how braking is applied when the ABS system is in operation.

ABS is not active below 7 km/h and normal braking only is available at lower speeds. When ABS is not operating, the inlet valves rest in the open position (to permit normal braking) and the outlet valves rest in the closed position. At each inlet valve there is a pressure sensitive return valve that permits rapid release of pressure when the brake pedal is released and this prevents any dragging of the brakes.

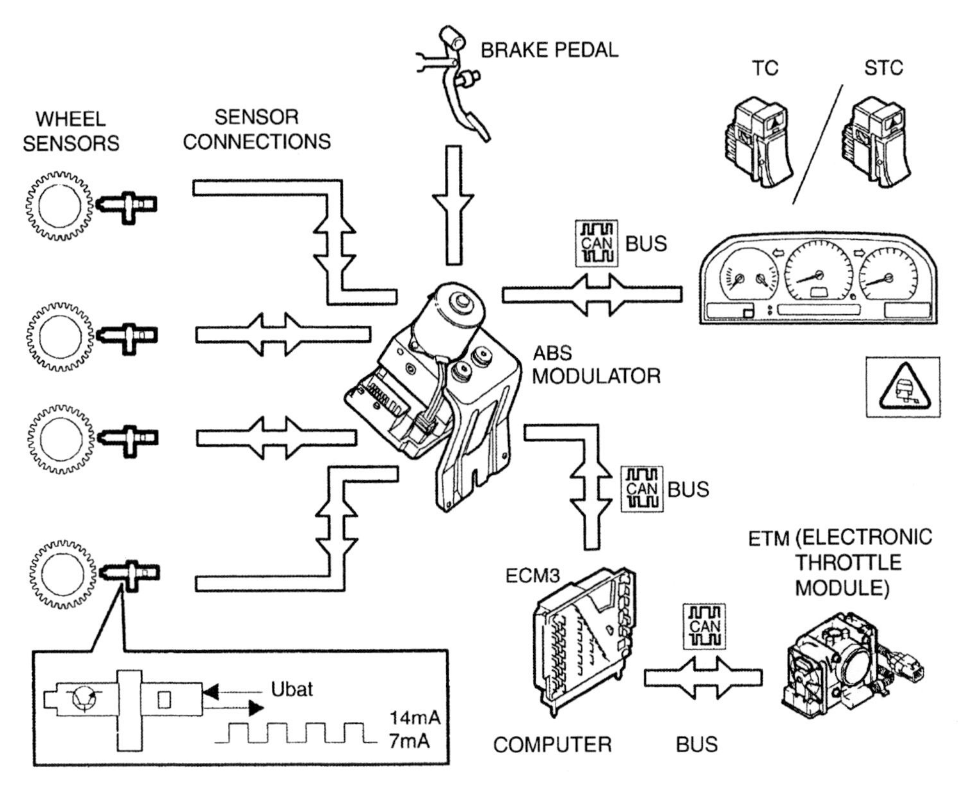

Fig. 1.22 Elements of a modern ABS system

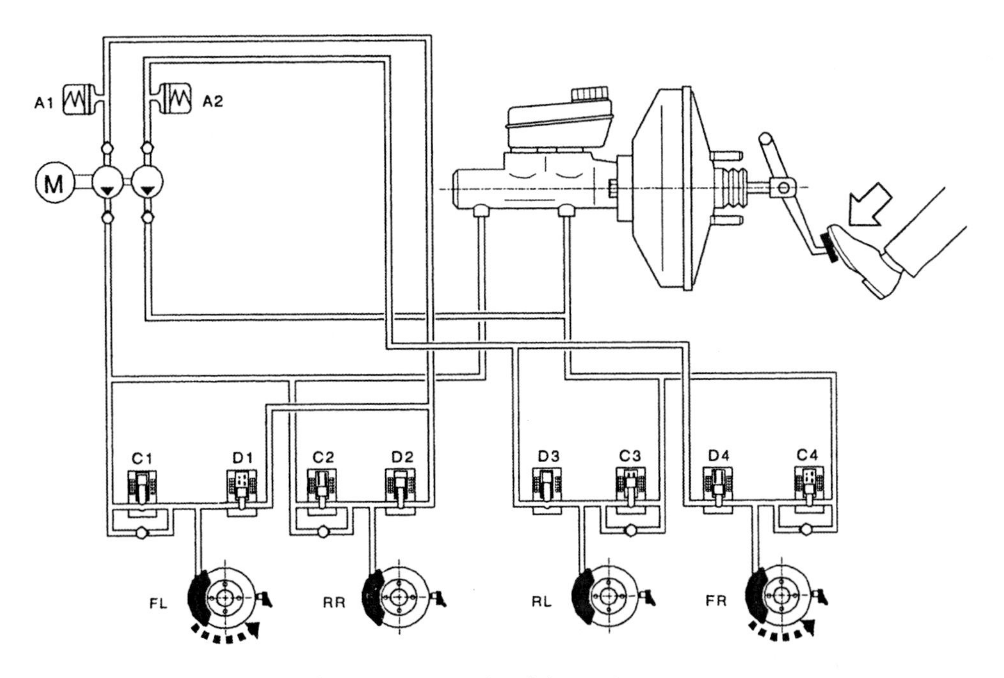

Fig. 1.23 Details of the ABS system

1.6.1 OPERATION OF ABS

Depressing the brake pedal operates the brakes in the normal way. For example, should the wheel sensors indicate to the computer that the front right wheel is about to lock, the computer will start up the modulator pump and close the inlet valve C4. This prevents any further pressure from reaching the right front brake. This is known as the 'pressure retention phase'. If the wheel locks up, the computer will register the fact and send a signal that will open the outlet valve D4 so that pressure is released. This will result in some rotation of the right front wheel. This is known as the 'pressure reduction phase'. If the sensors indicate that the wheel is accelerating, the computer will signal the outlet valve D4 to close and the inlet valve C4 to open and further hydraulic pressure will be applied. This is known as the 'pressure increase phase'. These three phases of ABS braking, i.e. pressure retention, pressure release and pressure increase, will continue until the threat of wheel lock has ceased or until the brake pedal is released.

1.6.2 SOME GENERAL POINTS ABOUT ABS

The system shown in Fig. 1.23 illustrates one mode of ABS operation. The front right and rear right brakes are in the pressure retention phase, the front left brake is in the pressure increase phase, and the rear left brake is in the pressure reduction phase. This is indicated by the open and closed positions of the inlet valves C1 - C4 and the outlet valves D1 - D4.

During ABS operation the brake fluid returns to the master cylinder and the driver will feel pulsations at the brake pedal which help to indicate that ABS is in operation. When ABS operation stops the modulator pump continues to run for approximately 1 s in order to ensure that the hydraulic accumulators are empty.

1.7 Traction control

The differential gear in the driving axles of a vehicle permits the wheel on the inside of a corner to rotate more slowly than the wheel on the outside of the corner. For example, when the vehicle is turning sharply to the right, the right-hand wheel of the driving axle will rotate very slowly and the wheel on the left-hand side of the same axle will rotate faster. Figure 1.24 illustrates the need for the differential gear.

However, this same differential action can lead to loss of traction (wheel spin). If for some reason one driving wheel is on a slippery surface when an attempt is made to drive the vehicle away, this wheel will spin whilst the wheel on the other side of the axle will stand still. This will prevent the vehicle from moving. The loss of traction (propelling force) arises from the fact that the differential gear only permits transmission of torque equal to that on the weakest side of the axle. It takes very little torque to make a wheel spin on a slippery surface, so the small amount of torque that does reach the non-spinning wheel is not enough to cause the vehicle to move.

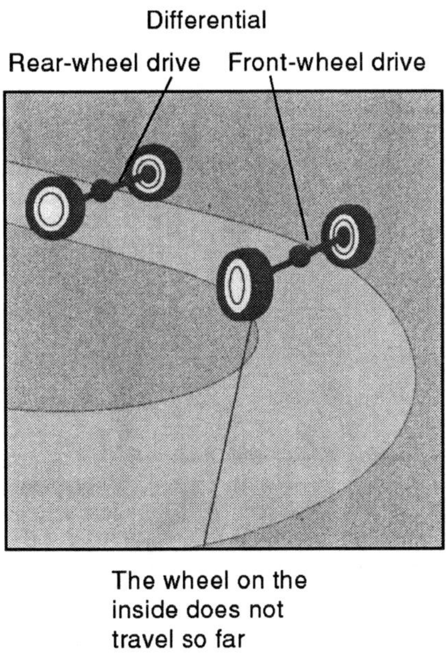

Fig. 1.24 The need for a differential gear

Traction control enables the brake to be applied to the wheel on the slippery surface. This prevents the wheel from spinning and allows the drive to be transmitted to the other wheel. As soon as motion is achieved, the brake can be released and normal driving can be continued. The traction control system may also include a facility to close down a secondary throttle to reduce engine power and thus eliminate wheel spin. This action is normally achieved by the use of a secondary throttle which is operated electrically. This requires the engine management system computer and the ABS computer to communicate with each other, and a controller area network (CAN) system may be used to achieve this.

Figure 1.25 gives an indication of the method of operation of the throttle.

The ABS system described in section 1.6 contains most of the elements necessary for automatic application of the brakes, but it is necessary to provide additional valves and other components to permit individual wheel brakes to be applied. Figure 1.26 shows the layout of a traction control system that is used on some Volvo vehicles.

In the traction control system, shown in Fig. 1.26, the ABS modulator contains extra hydraulic valves (labelled 1), solenoid valves (labelled 2) and by-pass valves (labelled 3). The figure relates to a front-wheel drive vehicle and for this reason we need to concentrate on the front right (FR) brake and the front left (FL) brake. In this instance wheel spin is detected at the FR wheel which means that application of the FR brake is required.

The solenoid valves (2) are closed and this blocks the connection between the pressure side of the pump (M) and the brake master cylinder. The inlet valve (C1) for the FL brake is closed to prevent that brake from being applied.

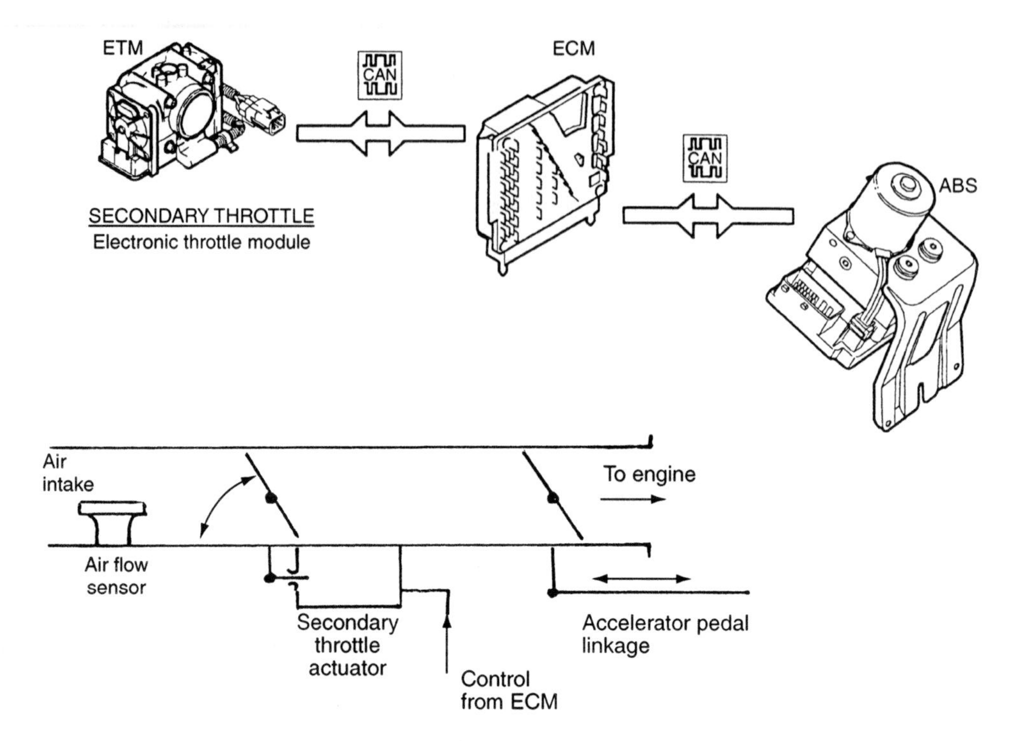

Fig. 1.25 The electrically-operated throttle used with the traction control system

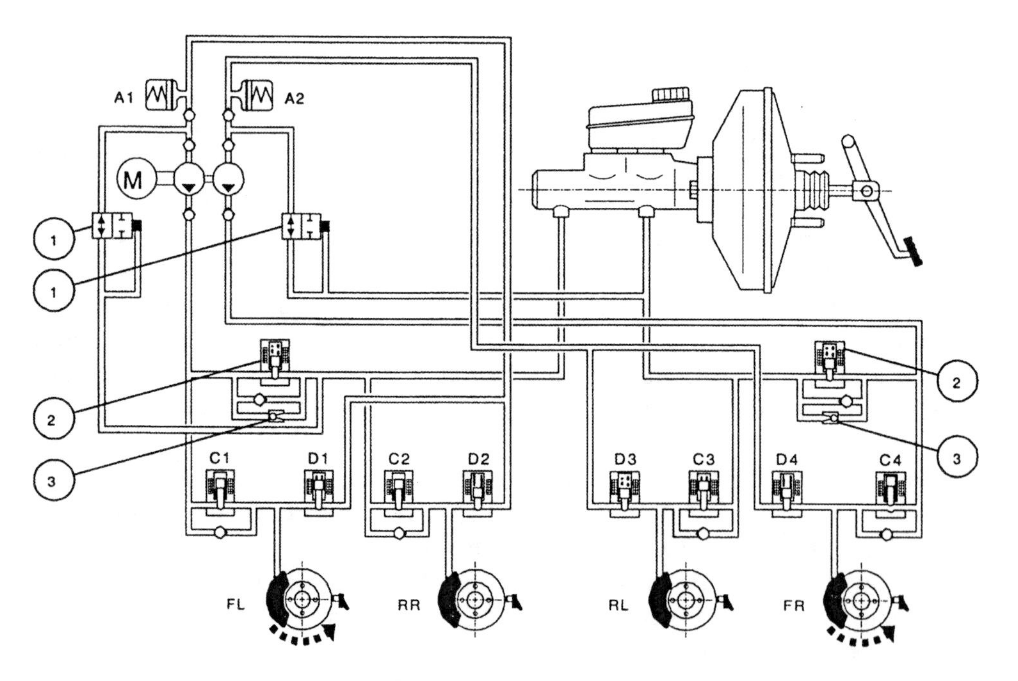

Fig. 1.26 A traction control system

The modulator pump starts and runs continuously during transmission control operation and takes fluid from the master cylinder, through the hydraulic valve 1, and pumps it to the FR brake through the inlet valve (C4).

When the speed of the FR wheel is equal to that of the FL wheel, the FR brake can be released, by computer operation of the valves, and then re-applied until such time as the vehicle is proceeding normally without wheel spin. In the case here, of spin at the FR wheel, the controlling action takes place by opening and closing the inlet valve (C4) and the outlet valve (D4).

When the computer detects that wheel spin has ceased and normal drive is taking place, the modulator pump is switched off, the solenoid valves (2) open and the valves (C4) and (D4) return to their positions for normal brake operation. Because the modulator pump is designed to provide more brake fluid than is normally required for operation of the brakes, the by-pass valves (3) are designed to open at a certain pressure so that excess brake fluid can be released back through the master cylinder to the brake fluid reservoir.

The system is designed so that traction control is stopped if:

1. the wheel spin stops;
2. there is a risk of brakes overheating;
3. the brakes are applied for any reason;
4. traction control is not selected.

1.8 Stability control

The capabilities of traction control can be extended to include actions that improve the handling characteristics of a vehicle, particularly when a vehicle is being driven round a corner. The resulting system is often referred to as 'stability control'.

Figure 1.27 shows two scenarios. In Fig. 1.27(a) the vehicle is understeering. In effect it is trying to continue straight ahead and the driver needs to apply more steering effect in order to get round the bend. Stability control can assist here by applying some braking at the rear of the vehicle, to the wheel on the inside of the bend. This produces a correcting action that assists in 'swinging' the vehicle, in a smooth action, back to the intended direction of travel.

In Fig. 1.27(b) oversteer is occurring. The rear of the vehicle tends to move outwards and effectively reduce the radius of turn. It is a condition that worsens as oversteer continues. In order to counter oversteer, the wheel brakes on the outside of the turn can be applied and/or the engine power reduced, via the secondary throttle, by the computer. In order to achieve the additional actions required for stability control it is necessary to equip the vehicle with additional sensors, such as a steering wheel angle sensor, and a lateral acceleration sensor that has the ability to provide the control computer with information about the amount of understeer or oversteer.

To achieve stability control it is necessary for the engine control computer, the ABS computer and the traction control computer to communicate, and

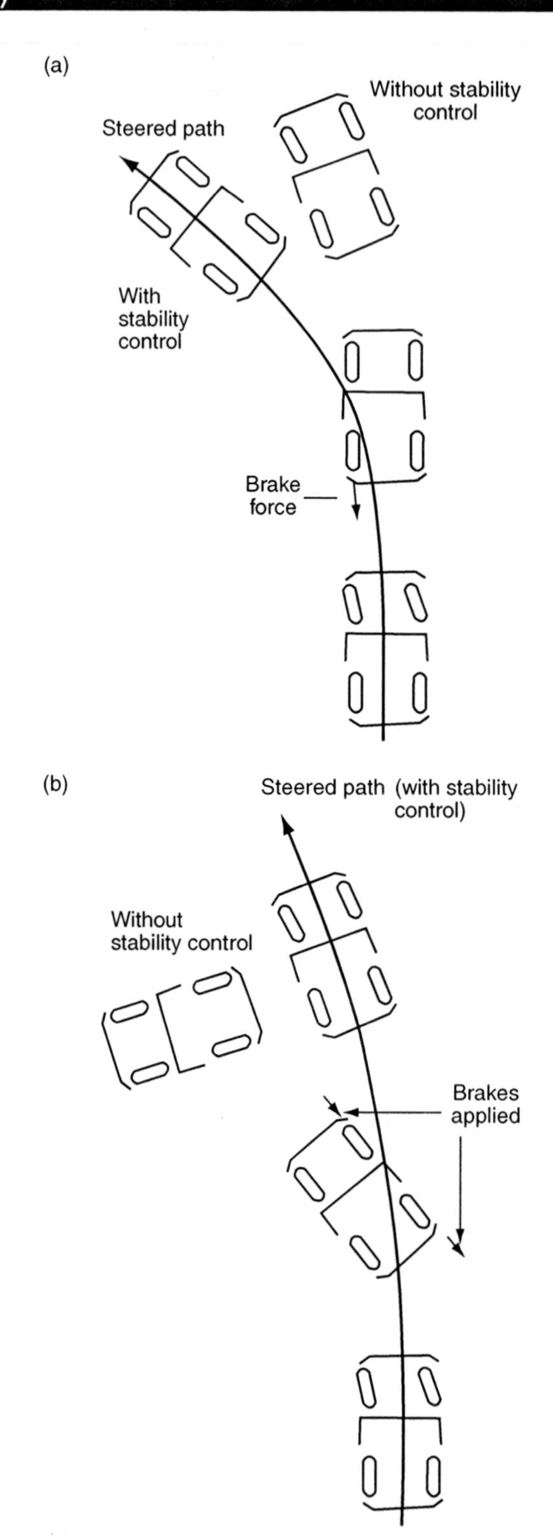

Fig. 1.27 Stability control; (a) understeer, (b) oversteer

this they do via the CAN network as shown in Fig. 1.25. This figure also illustrates the form of output from the Hall type wheel sensors. CAN networking is covered in Chapter 2 and more details about Hall type sensors are explained in Chapter 5.

1.9 Air conditioning

Maintaining a comfortable temperature inside the passenger/driving compartment of a vehicle is a function that is normally performed by a computer controlled system. Providing heat to the vehicle interior is usually achieved by redirecting heat from the engine via directional ducts and fans. However, cooling down the interior of the vehicle normally requires the use of an extra machine-driven cooling system that will take heat from the interior and transfer it to the atmosphere surrounding the exterior of the vehicle. It is the air conditioning system that performs this function. Figure 1.28 illustrates the outline principle of a vehicle air conditioning system.

The liquid (refrigerant) that is used to carry heat away from the vehicle interior and transfer it to the outside is circulated around the closed system by means of a compressor that is driven by the engine of the vehicle. Inside the system the refrigerant constantly changes state between liquid and a vapor as it circulates.

The reducing valve is an important agent in the operation of the system. The 'throttling' process that takes place at the reducing valve causes the refrigerant to vaporize and its pressure and temperature to fall. After leaving the reducing valve, the refrigerant passes into a heat exchanger called the evaporator where it collects heat from the vehicle interior and thus cools the interior in the process. The heat collected causes the refrigerant to vaporize still further and it is returned to the compressor where its pressure and temperature are raised.

From the compressor, the refrigerant passes into another heat exchanger where it gives up heat to the atmosphere. This heat exchanger is known as a condenser because the loss of heat from the refrigerant causes it to become wet. After the condenser, the refrigerant passes through the accumulator, which serves to separate liquid from the vapor. The refrigerant then returns to the reducing valve and evaporator, thus completing the cycle.

Because the compressor takes a considerable amount of power from the engine it is necessary for the air conditioning computer to be aware of the operational state of the engine. For example, the idling speed of the engine will be affected if the air conditioning compressor is operating, and the engine ECM will normally cause an increase in idle speed to prevent the engine from stalling. To allow the air conditioning compressor to be taken in and out of operation it is driven through an electromagnetic clutch which is shown in Fig. 1.29.

This clutch permits the compressor to be taken out of operation at a speed just above idling speed and, in order to protect the compressor, it is also disconnected at high engine speed. In some cases where rapid acceleration is called for, temporary disengagement of the compressor may also occur.

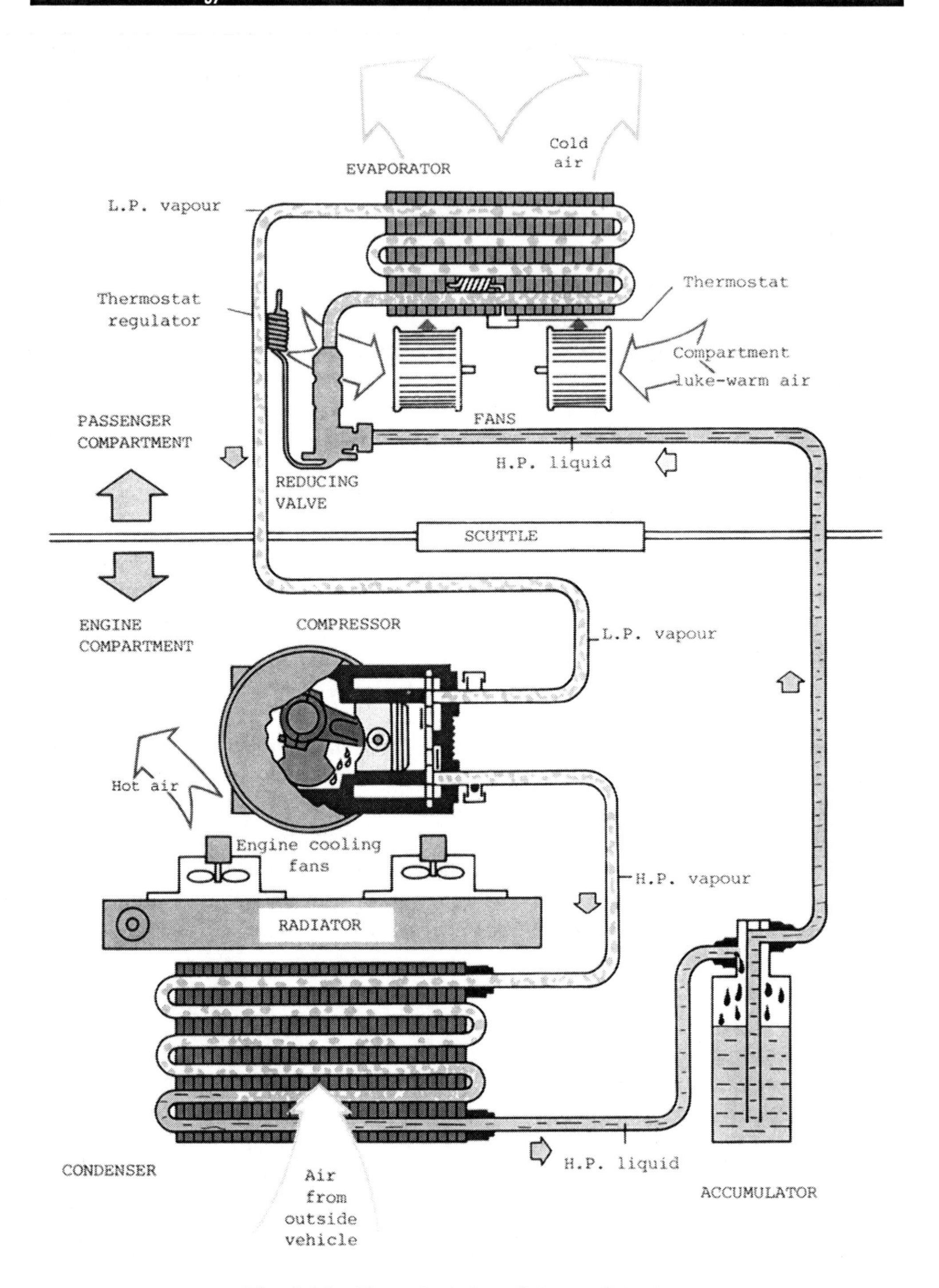

Fig. 1.28 The principles of air conditioning

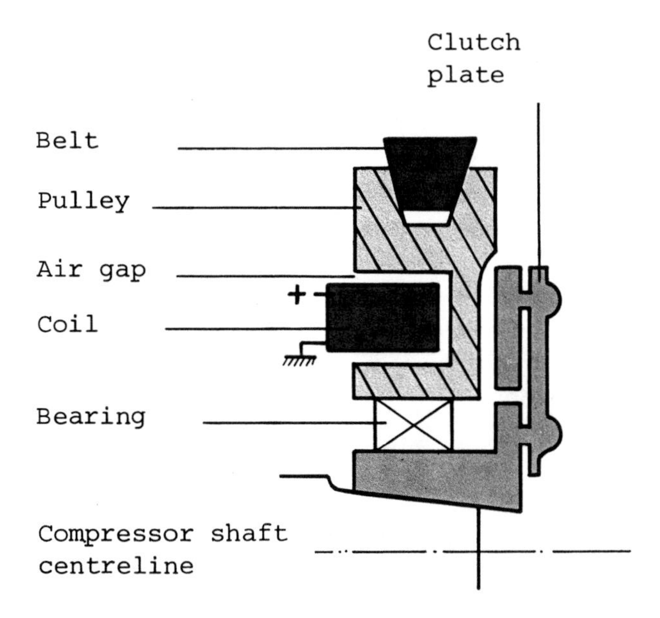

Fig. 1.29 The electromagnetic clutch

In addition to engine operating considerations, the interior temperature of the vehicle must constantly be compared with the required setting and the exterior temperature, and this is achieved by temperature sensors which are similar to those used for engine coolant temperature sensing. The following is a list of the controlling functions of an air conditioning ECM.

- Calculation of required outlet air temperature
- Temperature control
- Blower control
- Air inlet control
- Air outlet control
- Compressor control
- Electric fans control
- Rear defogger control
- Self diagnosis

1.9.1 DEALING WITH AIR CONDITIONING REFRIGERANT

Refrigerants that are used in air conditioning systems can be harmful to persons who come into contact with them and they are also considered to be harmful to the environment. For these reasons the servicing of air conditioning systems requires the use of specialized equipment, and technicians must be trained for the specific application that they are working on. Most garage equipment manufacturers market air conditioning service equipment and the Bosch Tronic R134 kit is an example. Equipment suppliers and vehicle manufacturers provide training for air conditioning systems and I strongly recommend that all garage technicians receive such training as air conditioning is now found in many vehicles that are used in Europe.

Some points of general application are as follows.

- There are strict rules about releasing refrigerant into the atmosphere. Technicians must familiarize themselves with the local rules and abide by them.
- The refrigerant is held in the system under pressure. Any small leak must be repaired.
- Some refrigerants produce poisonous gas when a flame is introduced near them. This eventuality must be avoided.
- If refrigerant gets on to the body it can cause cold burns and damage to the eyes - this must be avoided.

These are some of the reasons why special training is so important.

1.10 Computer controlled damping rate

Forcing oil through an orifice is a commonly used method of providing the damping in vehicle suspension systems. The amount of damping force that is applied is dependent, among other factors, on the size of the orifice through which the damping fluid is forced by the action of the suspension damper. The damping force can thus be changed by altering the size of the damping orifice. In practice this can be achieved by means of a valve which, under the control of the ECM, varies the size of the damping orifice to provide softer or stiffer damping, as required. Figure 1.30 shows the location of a solenoid-operated damping valve that is used on some Ford systems. The solenoid is controlled by the adaptive damping computer and provides two damping rates, a soft one and a stiff one.

The suspension damping rate is varied to suit a range of driving conditions, such as acceleration mode, braking (deceleration), bumpy roads and cornering etc. In order to provide the required damping for the various conditions the computer is fed information from a number of sensors. The input data is then compared with the design values in the computer ROM and the processor then makes decisions that determine the required damping rate. Figure 1.31 gives an indication of the types of sensors involved for adaptive damping.

The speed sensor can be the one that is used for other systems on the vehicle and it will probably be of the electromagnetic type. The steering position sensor is frequently of the opto-electronic type. This utilizes an infrared beam which is interrupted by the perforated disc as shown in Fig. 1.32. The wheel speed sensor signal can be derived from the ABS computer, and the brake light signal is derived from the stop light switch.

1.11 Computer controlled diesel engine management systems

Diesel engines rely on the compression pressure being high enough to ignite the fuel when it is injected into the combustion chamber. In order to achieve the

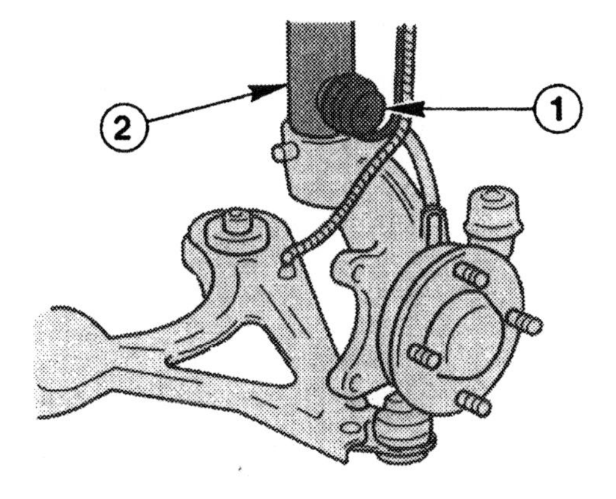

Location of solenoid valve on front axle

1 Solenoid valve
2 Damper

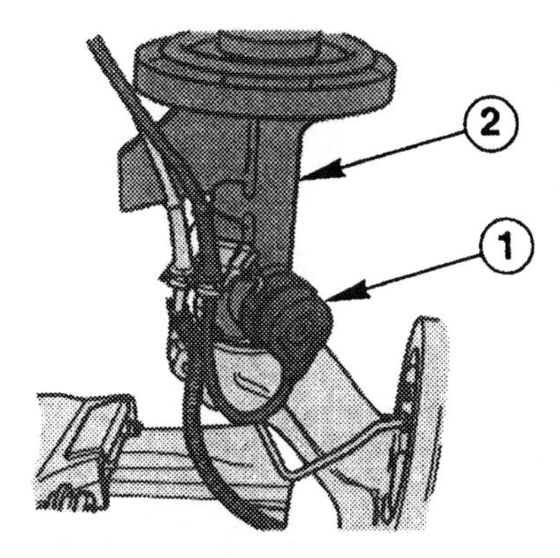

Location of solenoid valve on rear axle

1 Solenoid valve
2 Damper

Fig. 1.30 The adaptive damping solenoid

required pressure and temperature the mass of air that is compressed in each cylinder remains approximately constant throughout the engine's operating range. The power output is regulated by varying the amount of fuel injected. This means that diesel engines have a very weak mixture at idling speeds and a richer mixture for maximum power.

The operating principles of diesel engines therefore preclude the use of the exhaust oxygen sensor feedback principle that is used to help control emissions from petrol engines. However, only relatively small amounts of HC and CO appear in diesel engine exhaust gas and these can be reduced further by the use of an oxidizing catalyst. The reduction catalyst that is normally used to

● **Float and Pitch Control**

Electronic control in which the front and rear wheels are controlled independently to constantly obtain an optimal damping force in response to the bumps and dips on the road.

● **Vehicle Speed Sensitive Control**

Control to ensure high-speed stability by increasing the minimum dampening force as the vehicle speed increases so that the vehicle's riding comfort and road-holding performance reach an optimal balance.

● **Anti-Dive Control**

Control to ensure driving stability by restraining the diving of the vehicle body, which occurs when the brakes are applied.

● **Thumping Sensitive Control**

Control to ensure riding comfort by making it more difficult for the dampening force to increase on a flat road surface.

● **Anti-Roll Control**

Control to ensure driving stability by restraining the vehicle body roll that occurs when the steering wheel is turned.

● **Anti-Squat Control**

Control to ensure driving stability by restraining the vehicle body squat that occurs during acceleration by switching to a higher dampening force.

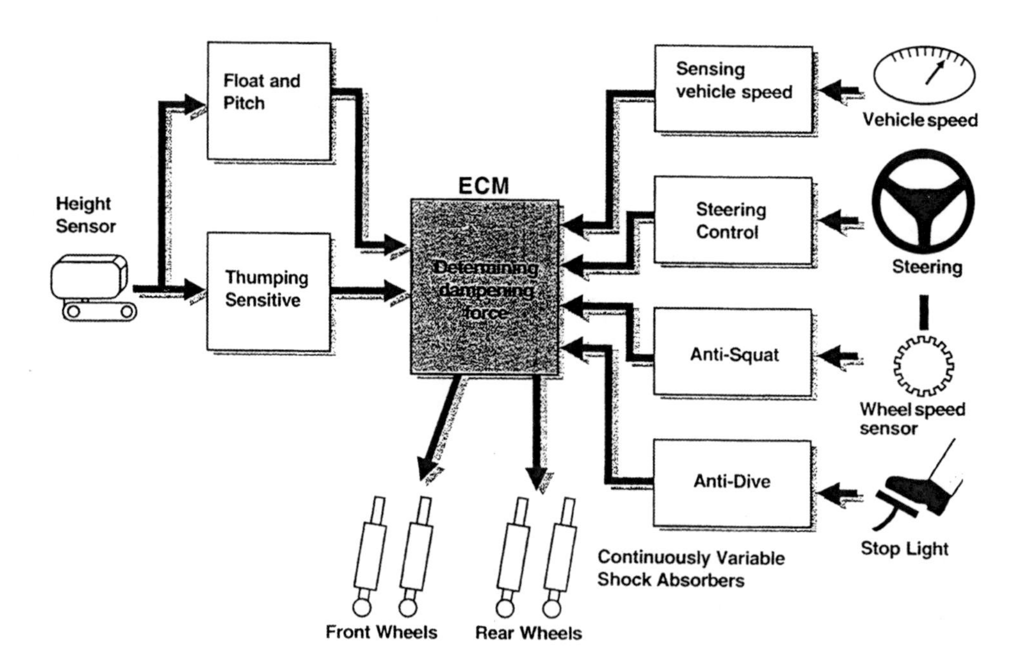

Fig. 1.31 Computer controlled variable rate damping - inputs and outputs

reduce NO_x emissions, however, cannot be used. A commonly used alternative method of NO_x reduction on diesel engines is electronically controlled exhaust gas recirculation.

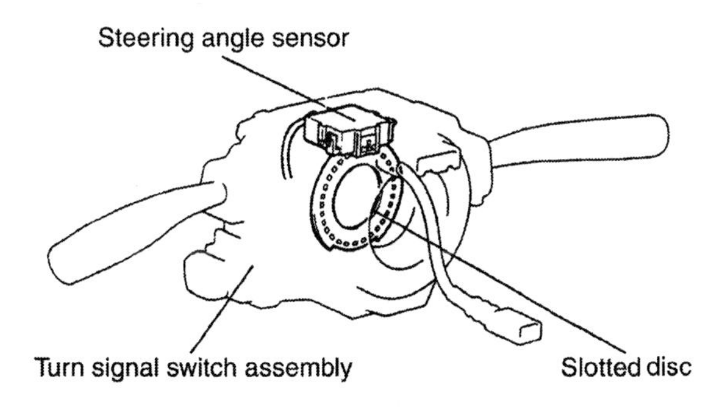

Fig. 1.32 The steering position sensor

Another emission that causes concern is soot. Soot emission can be reduced by electronic control of the mass of fuel injected, injection timing and turbocharging (exhaust system particulate traps also have a beneficial effect).

Figure 1.33 shows a computer controlled diesel engine management system. The exhaust system is equipped with an oxidizing catalyst that converts CO into carbon dioxide and HC into carbon dioxide and steam. There is no exhaust oxygen sensor because diesel engines, when operating correctly, have a certain amount of excess oxygen in the exhaust and this aids the operation of the catalytic converter. NO_x is kept within the required limits by careful control of fuelling and exhaust gas recirculation. The effectiveness of the emissions system is, in the UK, checked by means of a smoke meter, and the emissions test is part of the annual inspection. The emissions are also subject to spot checks by the enforcement authorities at any time.

The power output of a diesel engine is controlled by the quantity of fuel that is injected into each cylinder, whilst the quantity of air that is drawn into the cylinder on each induction stroke remains approximately the same. The main aim of computer control is to ensure that the engine receives the precise amount of fuel that is required, at the correct time and under all operating conditions. There are three areas of computer control. If we examine these in detail we shall find that for their operation they rely on well tried devices, such as solenoids and valves. The three items are:

1. fuel quantity (spill control)
2. injection timing control
3. idle speed control.

1.11.1 SPILL CONTROL

Figure 1.34 shows a cross-section of a rotary-type fuel injection pump. The high pressure pump chamber that produces the several hundred bars of pressure

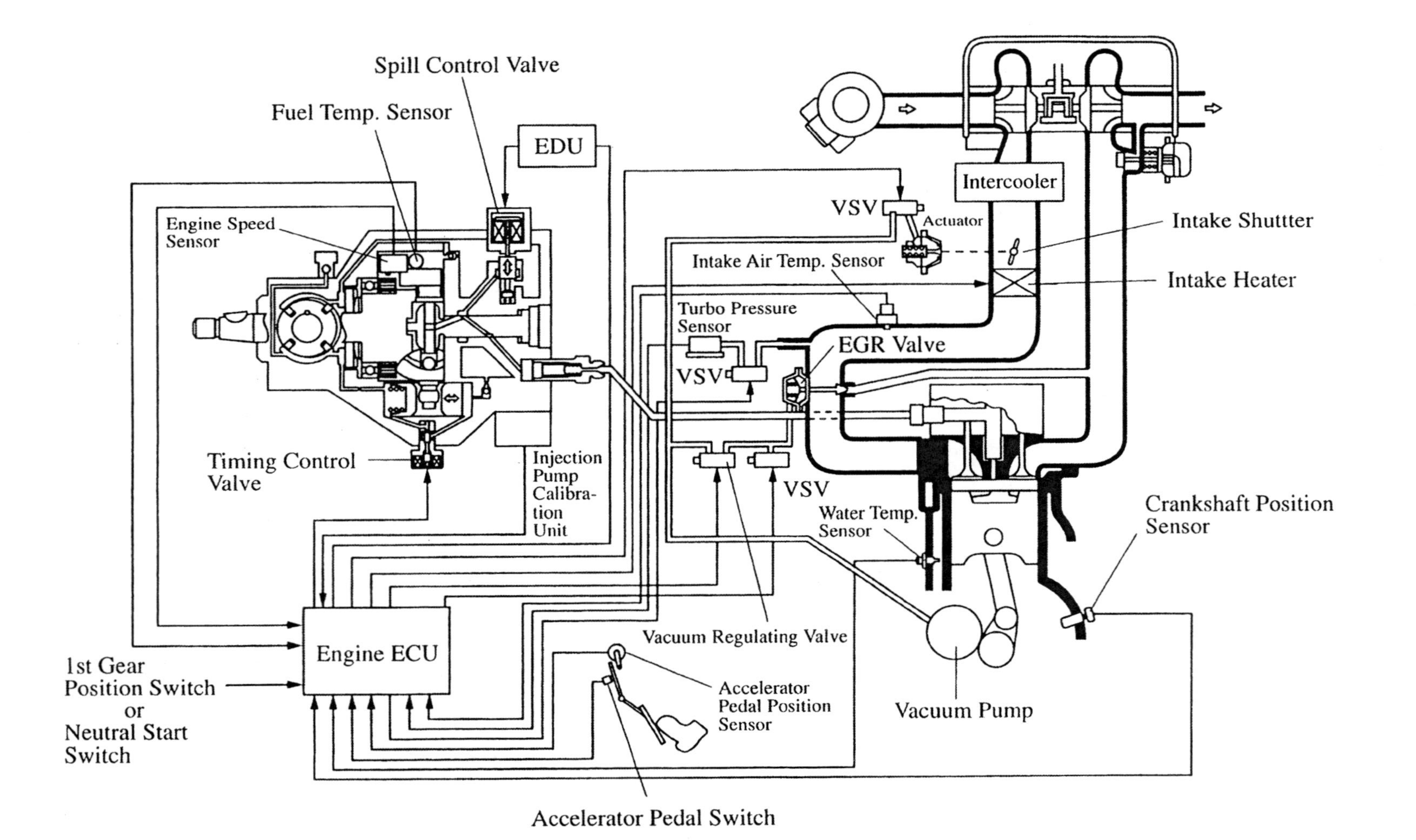

Fig. 1.33 Computer controlled diesel engine system

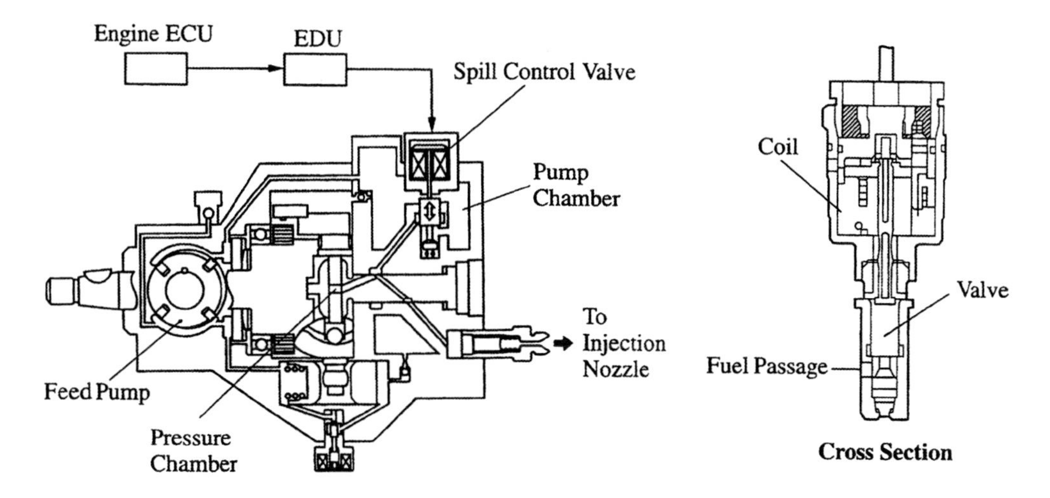

Fig. 1.34 Rotary-type fuel injection pump

that operate the fuel injectors, has two outlet ports. One of these outlet ports connects to the solenoid-operated spill valve, and the other one connects to the port and pipe that supplies the injector. When the spill valve is opened by ECM signals to the solenoid, fuel from the feed pump enters the pressure chamber at a pressure of 15–20 bar, thus charging the high pressure pump element. When signals from the ECM to the solenoid cause the spill valve to close, the high pressure pump plungers will force the fuel through the outlet to operate the injector. When injection is completed, the ECM will again signal the solenoid to open the spill valve, ready for the next sequence. The electronic driving unit (EDU) contains a device that amplifies the 5 V computer pulse into a 150 V supply to operate the spill valve at high speed.

1.11.2 TIMING CONTROL

The timing control valve is a solenoid-operated hydraulic valve that directs a regulated supply of fuel to the plungers that rotate the high pressure pump cam ring, clockwise or anticlockwise, in order to advance or retard the injection point as required. The sensor inputs that are required for this operation are shown on Fig. 1.35.

1.11.3 IDLE SPEED CONTROL

The idling speed of a diesel engine is controlled by the amount of fuel that is injected into the cylinders. As the conditions under which the engine is required to idle vary, the computer program must be arranged to provide the correct fuelling to ensure a steady idling speed under all conditions. The inputs to the ECM that are shown in Fig. 1.36 give an indication of the sensor inputs that are

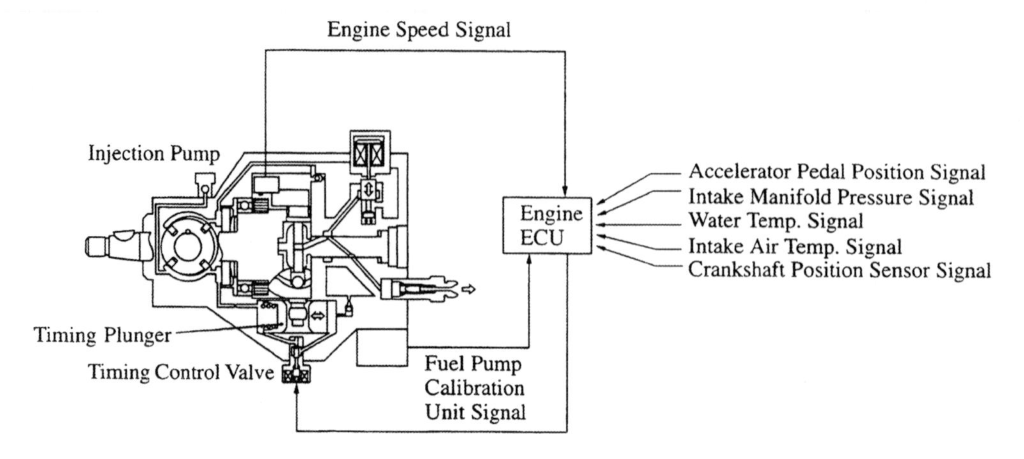

Fig. 1.35 Sensor inputs for the timing control valve

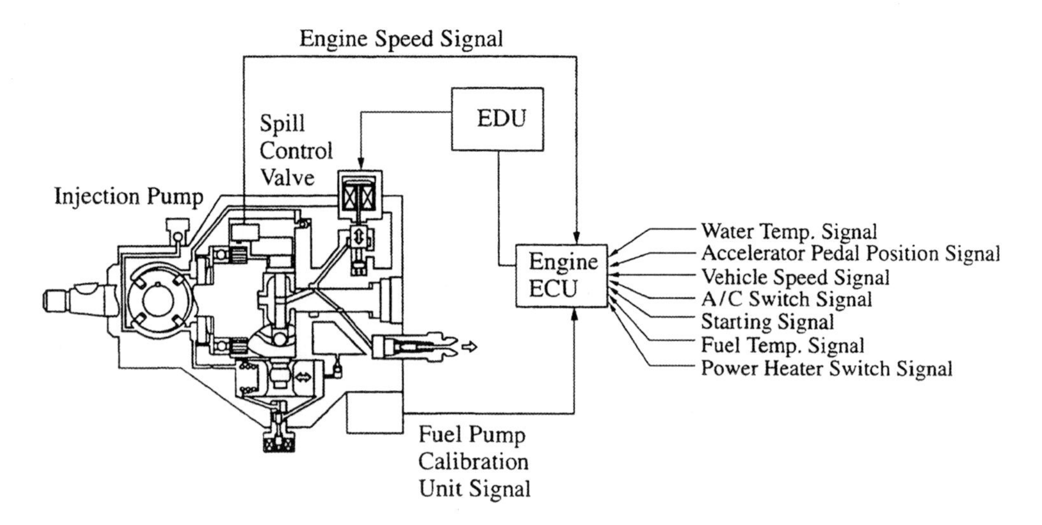

Fig. 1.36 Diesel engine idle speed control

required in order that the ECM can provide the correct signals to the spill control valve.

Another recent development in computer controlled diesel systems is the common rail system shown in Fig. 1.37. In this common rail system, the fuel in the common rail (gallery) is maintained at a constant high pressure. A solenoid-operated control valve that is incorporated into the head of each injector is operated by the ECM. The point of opening and closing of the injector control valve is determined by the ROM program and the sensor inputs. The injection timing is thus controlled by the injector control valve and the ECM. The quantity of fuel injected is determined by the length of time for which the injector remains open and this is also determined by the ECM.

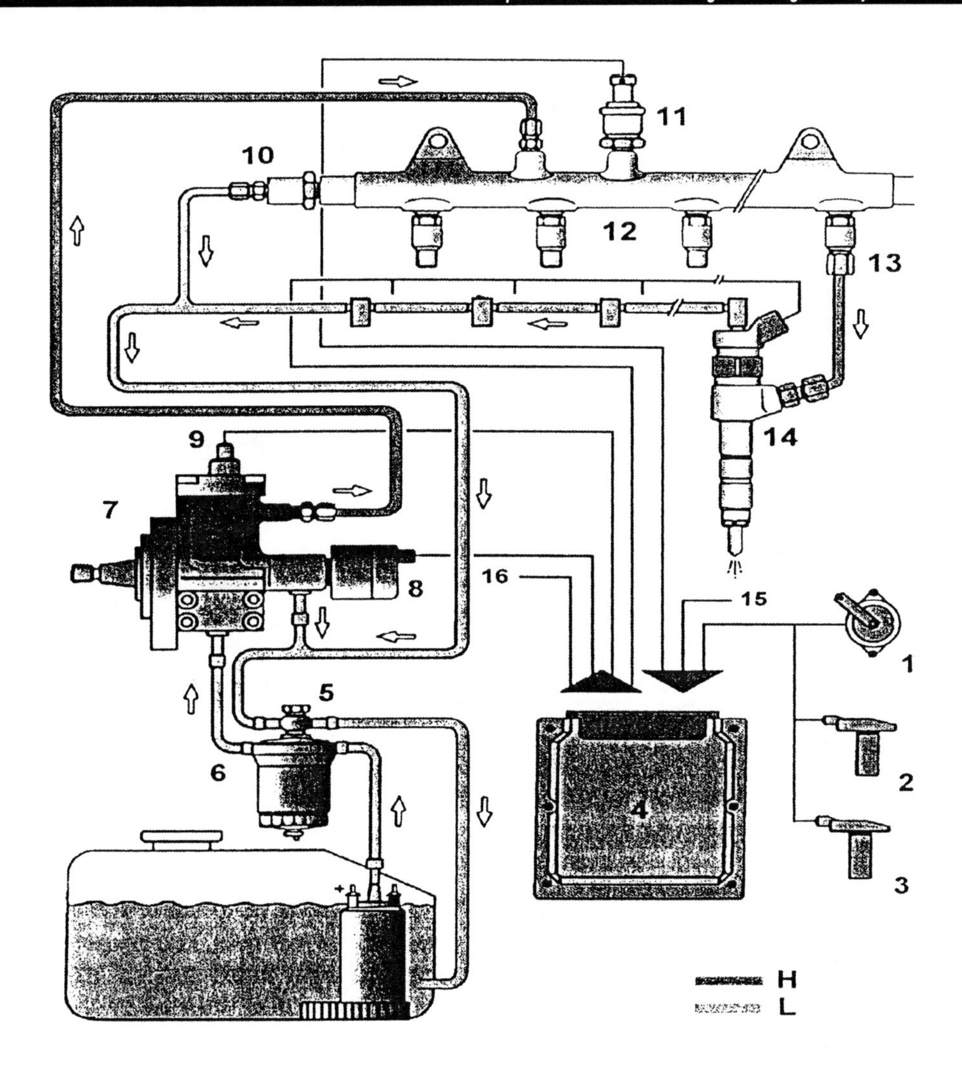

1. Accelerator
2. Engine speed (crank)
3. Engine speed (cam)
4. Engine control module
5. Overflow valve
6. Fuel filter
7. High pressure pump
8. Pressure regulating valve
9. Plunger shut-off
10. Pressure limiting valve
11. Rail pressure sensor
12. Common rail
13. Flow limiter
14. Injector
15. Sensor inputs
16. Actuator outputs

Fig. 1.37 The Rover 75 common rail diesel fuel system

1.12 Summary

As a result of this survey of computer controlled systems it is possible to see that electromagnetism, semiconductors, variable resistance, circuits and computers all figure prominently in the systems reviewed. This gives a lead to the types of background technology that is common to a range of systems. In the following chapters some of this background technology is examined in greater detail and in Chapter 7 there are examples of many tests that can be applied to aid fault diagnosis on computer controlled vehicle systems.

An important element in diagnostics on computer controlled systems is the self-diagnostic power of the computer and the fault codes that are created. This aspect is discussed in Chapters 2 and 3. However, fault codes are often just the beginning of a fault-finding task, and reading of the codes is often followed by a process of testing the performance of sensors and actuators and their interconnecting circuits. The range of systems reviewed in this chapter indicates that there is a body of knowledge about sensors and actuators that is applicable across a range of systems. The operating principles of sensors and actuators are examined in greater depth in Chapters 5 and 6 and once this basic knowledge has been acquired it can be used to advantage in the testing of many systems, as will be seen in Chapter 7.

1.13 Review questions (see Appendix 2 for answers)

1. The purpose of exhaust gas recirculation is:
   (a) to reburn the exhaust gas?
   (b) to reduce combustion temperature and reduce NO_x emissions?
   (c) to increase power output?
   (d) to give better fuel economy?
2. A Hall effect sensor:
   (a) generates electricity?
   (b) shuts off current in the Hall element so that the signal voltage is zero when the magnetic field is blocked?
   (c) gives an increase in signal current as the speed increases?
   (d) is only used in ignition systems?
3. In an ABS system:
   (a) the computer uses the peak-to-peak voltage from the wheel sensor to control braking?
   (b) the computer compares frequencies from wheel sensors to help control braking?
   (c) the warning light will go out when the vehicle speed reaches 50 km/h?
   (d) the braking distance is greatly reduced in all conditions?
4. In sequential multi-point petrol injection systems there is one injection of fuel to each cylinder:
   (a) on each stroke of the piston?
   (b) each time a piston approaches TDC on the exhaust stroke?

(c) whenever hard acceleration takes place?
(d) when the knock sensor transmits a signal?

5. The manifold absolute pressure sensor is used in speed density fuel injection systems to:
   (a) provide a signal that enables the ECM to calculate the amount of fuel entering the engine?
   (b) provide a signal that enables the ECM to calculate the amount of air entering the engine?
   (c) control the fuel pressure at the injectors?
   (d) control the EGR valve?
6. An adaptive strategy:
   (a) is a procedure that allows the ECM to set new values for certain operating variables as the system wears?
   (b) is a limited operating strategy that allows the ECM to set values that will get a vehicle back to the workshop for repair?
   (c) alters map values in the ROM?
   (d) is a procedure for fault tracing?
7. In diesel engines:
   (a) the fuel and air are mixed in the intake manifold?
   (b) ignition is caused by glow plugs?
   (c) the heat generated by compression causes combustion to take place?
   (d) a mixture of fuel and air is forced into the cylinder by the injector?
8. In adaptive suspension systems:
   (a) the ECM changes the damping rate to suit driving conditions?
   (b) the steering angle sensor is fitted to the front wheel drive shafts?
   (c) the system must not operate at speeds greater than 25 km/h?
   (d) the ECM learns a new set of values if a suspension spring breaks?

2
The Computer ECM

Whilst vehicle computers (ECMs) are not made to be repaired in garage workshops, there are certain factors that require technicians to have an appreciation of computer technology. For example, diagnostic trouble codes (DTCs) are an important part of fault finding and DTCs are stored in the computer memory. The means by which these codes are read out varies from vehicle to vehicle and it is helpful for technicians to understand why a procedure for reading DTCs on one vehicle may not work on another vehicle. It is also the case that technicians in some main dealer workshops are required to use special equipment to amend the computer operating program. Increasingly, use is being made of 'freeze frame' data. This is 'live' data that is captured whilst the system is in operation and it is useful in helping to determine the causes of a system fault. Whilst these operations are normally performed through the use of 'user friendly' diagnostic equipment, it is still the case that an understanding of what can and what cannot be done via the ECM is useful.

2.1 The fundamental parts of a computer

Figure 2.1 shows the general form of a computer that consists of the following parts:

- a central processing unit (CPU)
- input and output devices (I/O)
- memory
- a program
- a clock for timing purposes.

Data processing is one of the main functions that computers perform. Data, in computer terms, is the representation of facts or ideas in a special way that allows it to be used by the computer. In the case of digital computers this usually means binary data where numbers and letters are represented by codes made up from 0s and 1s. The input and output interfaces enable the computer to read inputs and to make the required outputs. Processing is the manipulation and movement of data

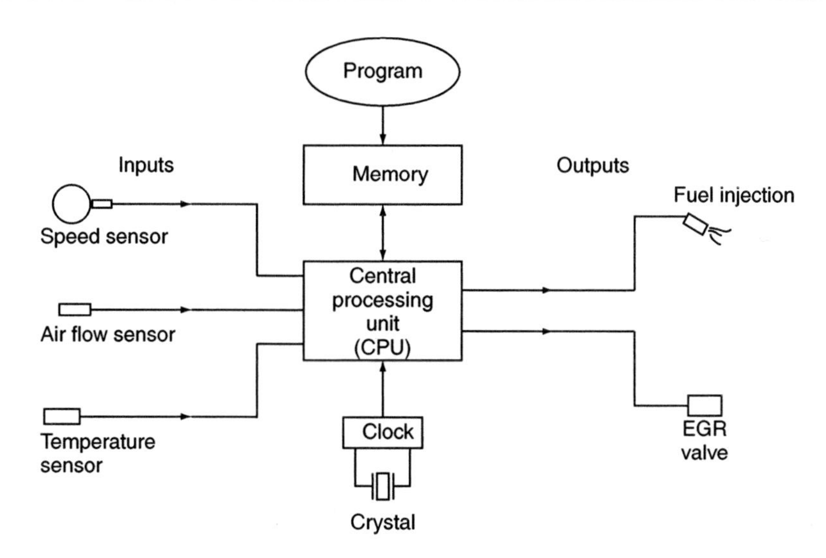

Fig. 2.1 The basic components of a computer system

and this is controlled by the clock. Memory is required to hold the main operating program and to hold data temporarily while it is being worked on.

2.1.1 COMPUTER MEMORY

Read only memory (ROM) is where the operating program for the computer is placed. It consists of an electronic circuit which gives certain outputs for predetermined input values. ROMs have large storage capacity.

Read and write, or random access memory (RAM), is where data is held temporarily while it is being worked on by the processing unit. Placing data in memory is referred to as 'writing' and the process of using this data is called 'reading'.

2.1.2 THE CLOCK

The clock is an electronic circuit that utilizes the piezoelectric effect of a quartz crystal to produce accurately timed electrical pulses that are used to control the actions of the computer. Clock speeds are measured in the number of electrical pulses generated in one second. One pulse per second is 1 Hertz and most computer clocks operate in millions of pulses per second. One million pulses per second is 1 megahertz (1 MHz).

2.2 A practical automotive computer system

Figure 2.2 shows a computer controlled transmission system. At the heart of the system is an electronic module. This particular module is a self-contained

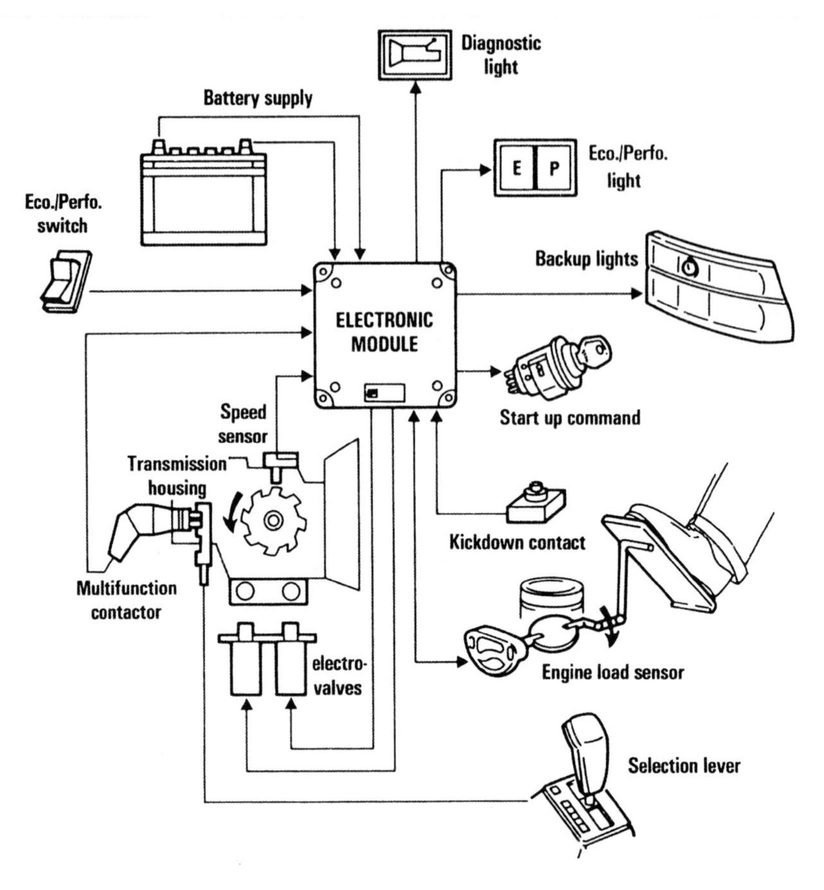

Fig. 2.2 A computer controlled transmission system

computer which is also known as a microcontroller. Microcontrollers are available in many sizes, e.g. 4, 8, 16 and 32 bit, which refers to the length of the binary code words that they work on. In this system it is an 8-bit microcontroller.

Figure 2.3 shows some of the internal details of the computer and the following description gives an insight into the way that it operates.

The microcomputer (1)

This is an 8-bit microcontroller. In computer language a bit is a 0 or a 1. The 0 normally represents zero, or low voltage, and the 1 normally represents a higher voltage, probably 1.8 V.

The microcontroller integrated circuit (chip) has a ROM capacity of 2048 bytes (there are 8 bits to one byte) and a RAM that holds 64 bytes. The microcontroller also has an on-chip capacity to convert four analogue inputs into 8-bit digital codes.

The power supply (2)

The power supply is a circuit that takes its supply from the vehicle battery then provides a regulated d.c. supply of 5 V to the microcontroller, and this is its

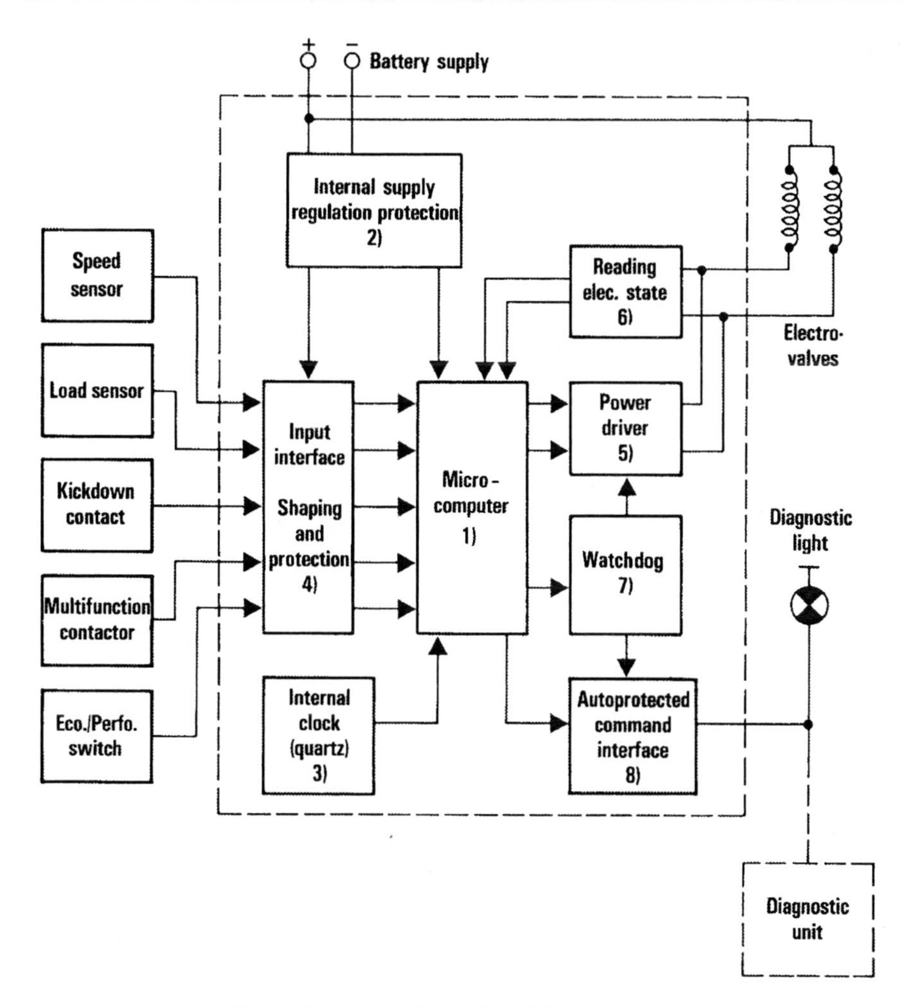

Fig. 2.3 Internal details of the computer

working voltage. The power supply also includes protection against over voltage and low voltage. The low voltage protection is required if battery voltage is low and it often takes the form of a capacitor.

The clock circuit (3)

In this particular application the clock operates at 4 MHz. The clock controls the actions of the computer, such as counting sensor pulses to determine speed and timing the output pulses to the electrovalves so that gear changes take place smoothly and at the required time.

The input interface (4)

The input interface contains the electronic circuits that provide the electrical power for the sensors and switches that are connected to it. Some of these inputs are in an electrical form (analogue) that cannot be read directly into the computer and these inputs must be converted into computer (digital) form at the interface.

The output (power) interface (5)
The power driver consists of power transistors that are switched electronically to operate electrovalves that operate the gear change hydraulics.

Feedback (6)
At (6) on the diagram the inscription reads ' Reading electrical state'. This means that the computer is being made aware of the positions (on or off) of the electrovalves.

The watchdog (7)
The watchdog circuit is a timer circuit that prevents the computer from going into an endless loop that can sometimes happen if false readings occur.

The diagnostic interface (8)
The diagnostic interface is a circuit that causes a warning lamp to be illuminated in case of a system malfunction. It can also be used to connect to the diagnostic kit.

2.3 Principles of operation

As with all automotive computer controlled systems, this one relies on inputs from sensors. The computer compares these input values with values that are held in the program memory (ROM) and then determines what signals are to be delivered to the actuators (electrovalves) in order to cause gear changes.

In this particular application the computer program makes use of a concept known as the vehicle operating point. The operating point is dependent on two sensor inputs, vehicle speed and load. Speed is determined by an electromagnetic sensor and the load is determined from the throttle position sensor. (Details of both types of sensor are given in Chapter 5.)

A set of vehicle operating points is stored in the ROM and these are used, whilst the vehicle is operating, as references for making gear changes. When the vehicle speed, as measured by the speed sensor, is greater than the speed held in the ROM for a given operating point the computer will call for a change to a higher gear. A lower gear (change down) will be called for when the vehicle speed falls below that held in the ROM.

Should the microcomputer detect a sensor reading that is out of limits for a sequence of readings (probably three readings) then the fault detection system comes into operation. In the case of this transmission control the types of failure that may be detected are:

- electrohydraulic valve failure
- throttle position (load) sensor failure

- speed sensor failure
- power source (battery) voltage failure.

In the event of a failure being detected, the warning lamp(s) are illuminated and the 'limp home' facility is activated and, in all probability, a code will be entered in a section of RAM for later use in diagnosing cause and effect. In this case the 'limp home' mode consists of a computer subroutine that causes the system to operate in high gear. This is intended to permit the driver to drive to a service garage where the required rectification work can be performed.

2.4 Computer data

At the base level inside the computer, the values that the computer processor works on are all expressed in digital form. That is to say that sensor readings, such as engine speed, coolant temperature etc., will be presented in the computer as binary codes made up of 0s and 1s. The 0s and 1s are electrical signals and values often used are binary 0 = 0.0–0.8 V and binary 1 = 2–5 V. The computer thus operates on electrical signals.

2.4.1 DATA TRANSFERS

Coded data is transmitted as electrical pulses, both inside the computer and between the computer and other computers connected to it. Inside the computer a coded number such as 20, which is 0001 0100 in 8-bit binary, is transmitted on a parallel bus which comprises eight wires, side by side. This is known as parallel data transmission and the concept is shown in Fig. 2.4(a).

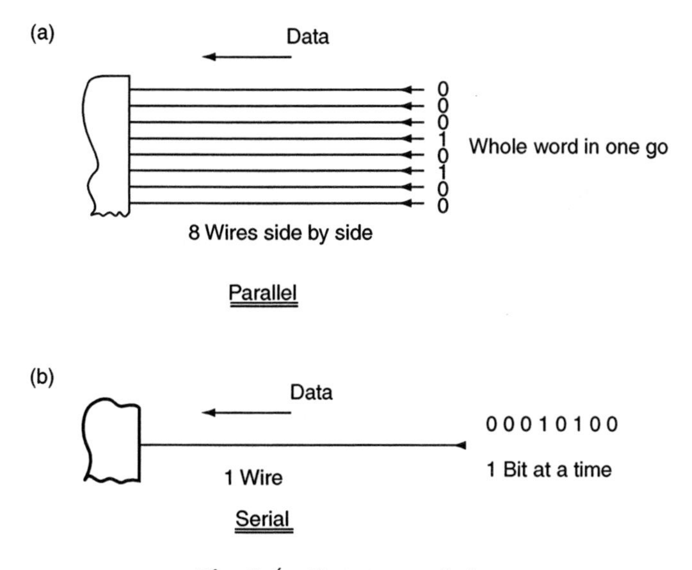

Fig. 2.4 Data transmission

When data is transmitted along a single wire, as happens in networked systems, each bit (0 or 1) is transmitted so that one bit follows another until the whole set of 0s and 1s has been transmitted. This method is known as serial data transmission. The speed of transmission is measured in the number of 0s and 1s that are sent in 1 s. Each 0 or 1 is a 'bit' and the number of bits per second is known as the 'Baud' rate, named after the person who first thought of measuring bits per second. The principle of serial data transmission is shown in Fig. 2.4(b).

2.4.2 DATA TRANSFER REQUIREMENTS

In order for the computer controlled systems to function correctly certain requirements must be met.

1. There must be a method that enables the computer processor to identify a specific device's interface from all other interfaces and memory devices that are attached to the internal buses of the computer.
2. There must be a temporary storage space (buffer) where data can be held (if necessary) when it is being transferred between the computer processor and a peripheral, such as a sensor or an actuator.
3. The peripherals, such as speed sensors and fuel injectors, must supply status information to the computer processor, via the interfaces, to inform the computer processor that they are ready to send data to it, or receive data from it.
4. The computer must generate and receive timing and control signals that are compatible with the computer's processor. These timings and signals must also be compatible with the sending or receiving device, i.e. the sensor or actuator.
5. There must be a means to convert the sensor signals into digital data that the computer can use and also a means to convert digital data into a form that the actuators can use.

2.5 Computer interfaces

Often the interface between the computer and peripherals, such as sensors and actuators, is based on a single chip integrated circuit. Figure 2.5 shows such an interface which is based on the Motorola MC6805 ACIA (asynchronous communications interface adaptor).

The blocks marked R1 and R2 are shift registers. R1 receives a data word as 8 parallel bits and then sends it out as a stream of serial bits. R2 receives a stream of data as serial bits and transfers it to the bus as an 8-bit parallel word. The control logic circuit is operated from the computer and the 8-bit bus at the top left communicates with the control unit and other internal circuits of the microcontroller.

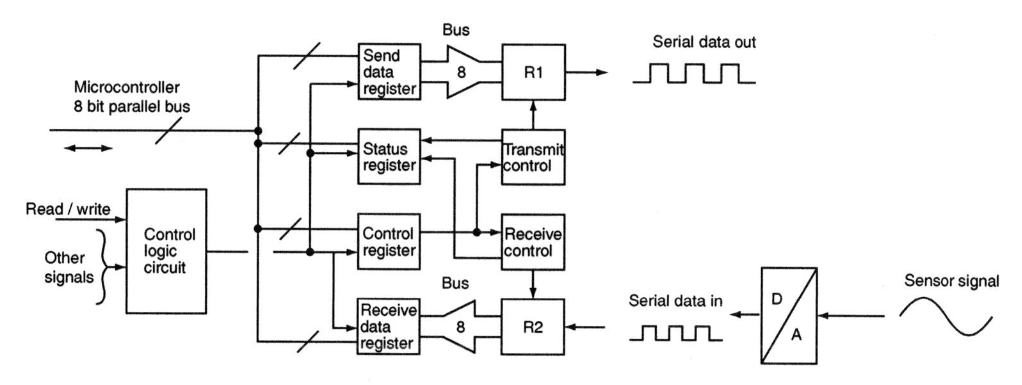

Fig. 2.5 A communications interface

2.6 Control of output devices

In many cases the commands from the computer are used to connect a circuit to earth, via a transistor. Two methods are used; one method is known as 'duty cycle' and the other is known as pulse width modulation. Each of these methods produces a different voltage pattern when the operation of the device is checked with an oscilloscope. The two patterns are shown in Fig. 2.6.

With duty cycle control the transistor that drives the device may be switched on for 100% of the time available, or for a small amount of the time available, for example 30%. The control of the device being operated is thus achieved by the length of the 'on time'. So a duty cycle reading of 50% means that the device, such as a petrol injector or mixture control valve, is switched on for 50% of the available time.

With pulse width control the transistor is switched on and off at fairly high frequency. The use of pulse width modulation (PWM) can reduce the heating effect in the solenoid of the device (injector) that is being operated.

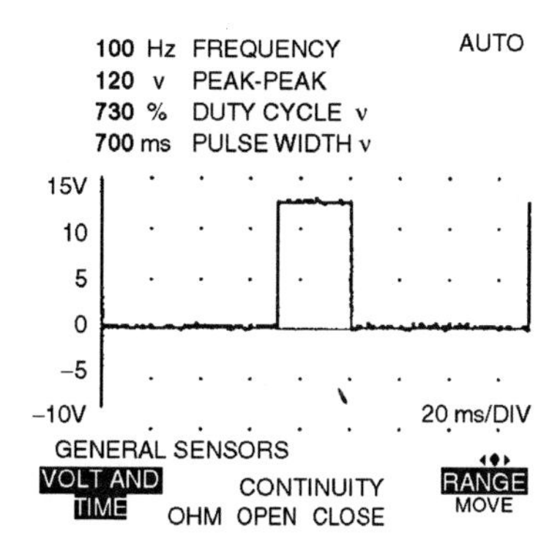

Fig. 2.6(a) Duty cycle control of a mixture control solenoid

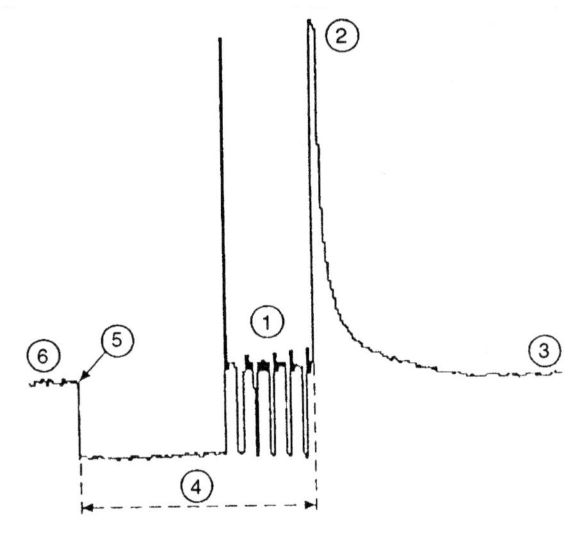

1. Current flow pulsed on and off enough to keep hold in winding activated.
2. Peak voltage caused by collapse of the injector coil, when current is reduced.
3. Return to battery (or source) voltage.
4. Injector ON-time.
5. Driver transistor turns on, pulling the injector pintle away from its seat, starting fuel flow.
6. Battery voltage (or source voltage) supplied to the injector.

Fig. 2.6(b) Pulse width modulation applied to a fuel injector

2.7 Computer memories

The term 'memory chip' derives from the fact that most computer memories are circuits that are made on a silicon chip. Automotive computers use memory chips that are very similar to those used in personal computers. The correct name for a 'chip' is integrated circuit or I/C. A large part of electronic memory is made of transistors and the number of transistors that can be made on a very small piece of silicon runs into millions.

Figure 2.7 shows a memory element known as a D type flip-flop. The D type flip-flop is called a memory device because it has the property that, whatever appears at D, either 0 or 1, will appear at Q when the clock pulse (C) is 1. When

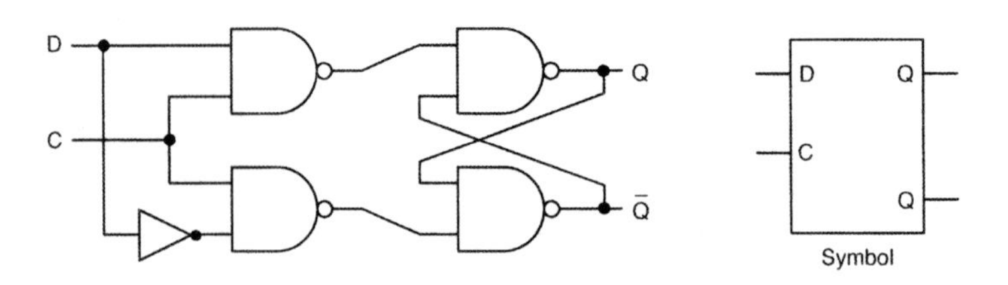

Fig. 2.7 A computer memory element

C goes to 0, Q stops following the D value and holds the value it has when C changed to 0. Q will follow D again when C goes back to 1.

The logic gates shown in this diagram are made up from electronic components such as transistors, and large numbers of them can be made on a single integrated circuit.

2.7.1 *READ ONLY MEMORIES*

The program that controls the system is stored in the ROM, or read only memory. Various types of ROM are used in automotive computers and it is important to understand the differences between them because a service procedure that is used on one make of vehicle may not necessarily work on a similar type of vehicle. Part of the reason for this is to be found in the different types of ROM that are used.

Figure 2.8 shows a block diagram of a small ROM circuit. The truth table shows the data that appears on the output lines (F_0 to F_3). It is read as follows: when the electrical inputs at A, B and C represent logic 0, the output values F_0 to F_3 are 1010, and so on.

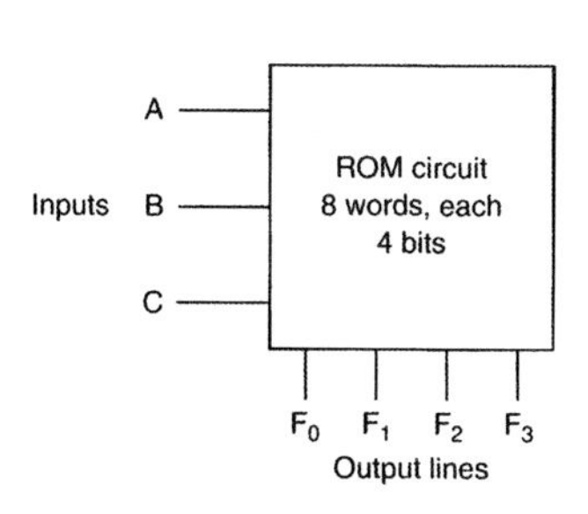

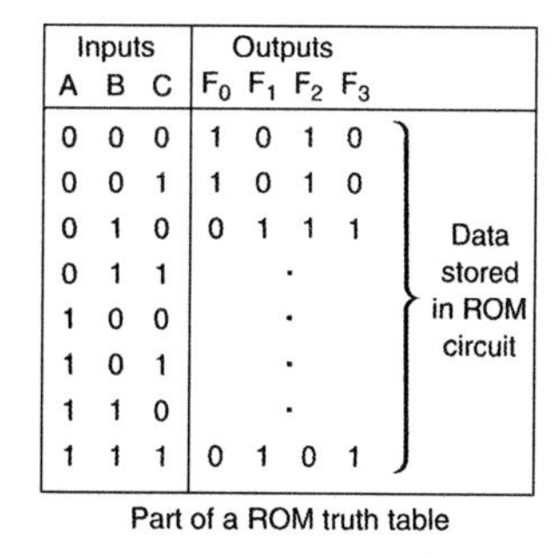

Inputs			Outputs			
A	B	C	F_0	F_1	F_2	F_3
0	0	0	1	0	1	0
0	0	1	1	0	1	0
0	1	0	0	1	1	1
0	1	1		·		
1	0	0		·		
1	0	1		·		
1	1	0		·		
1	1	1	0	1	0	1

Part of a ROM truth table

Fig. 2.8 A ROM circuit and truth table

ROMs are available in various forms. Normally the main program memory cannot be changed once it has been configured, unless the integrated circuit is changed. However, there are other types of ROM and some of these are used in vehicle computers. Some vehicles are equipped with the type of program memory that can be changed in service, but only by approved agents. The ROM is the part of the ECM where the program that controls the working of the system is stored and any attempt to tamper with it could have disastrous consequences.

Mask programmable ROMS

This type of ROM is widely used in large scale manufacture. The term 'mask' refers to the mask that is used during the stage when the ROM's electronic circuit is being constructed. Once a mask programmable ROM has been configured it cannot be altered.

PROM

Programmable ROMs are sometimes referred to as field programmable ROMs because the circuits can be changed 'in the field', that is to say away from the factory. The memory circuits contain fusible links that can be 'blown' selectively by the use of special equipment called a PROM programmer. Once the fusible links are blown (broken) they cannot be 'unblown'.

EPROM

Electrically programmable read only memories (EPROM) are memory circuits that use an electrical charge storage mechanism (like a capacitor) that keeps the memory alive, even when the main source of electrical power is removed. The memory can be erased by exposing the I/C to ultraviolet light. This type of memory chip may be identified by a small window (usually taped over) in the chip housing. After erasure of the memory it can be reset electrically.

EEPROM

An electrically erasable PROM works in a similar way to the EPROM, the main difference is that the memory can be erased and reset by a special charge pump circuit that is controlled by the microcontroller according to the working program of the ECM.

2.7.2 RANDOM ACCESS MEMORY

The RAM, or random access memory, is the section of memory that is used for temporary storage of data while it is being worked on. It must be the type of memory that can be written to (i.e. data is placed in it) and read from (i.e. data is taken from it). This means that the memory contents are constantly changing whenever the computer is operating. It is also known as a read-and-write memory. The contents of RAM are sustained by electricity and when the source of electrical power is removed, the contents of the RAM are lost. This is why the RAM is called volatile memory.

2.7.3 OTHER TYPES OF COMPUTER MEMORY

Hard discs are made of material that can be magnetized in very small localized areas arranged in circular tracks. A magnetized area probably represents a '1' in computer language, and a non-magnetized area represents a '0'. When the disc is rotated through a read-and-write head, the magnetism is converted into an electrical signal and it is these electrical signals that produce the data that operates the computer. Hard discs can hold many millions of bits of data. Floppy discs operate on similar principles, but they have a smaller storage capacity.

Compact discs (CDROMs) also have a large storage capacity. They are similar to an old fashioned gramophone record in that they have grooves which are deeper in some places than they are in others. The depth of the groove is read by a laser beam that is connected to an electronic circuit and this circuit converts the laser readings into voltages that represent the 0s and 1s used by the computer.

2.8 Fault codes

When a microcontroller (computer) is controlling the operation of an automotive system, such as engine management, it is constantly taking readings from a range of sensors. These sensor readings are compared with readings held in the operating program and if the sensor reading agrees with the program value in the ROM, the microcontroller will make decisions about the required output to the actuators, such as injectors.

If the sensor reading is not within the required limits it will be read again and if it continues to be 'out of limits' a fault code will be stored in a section of RAM. It is also likely that the designer will have written the main program so that the microcontroller will cause the system to operate on different criteria until a repair can be made, or until the fault has cleared. The fault codes, or diagnostic trouble codes (DTCs), are of great importance to service technicians and the procedures for gaining access to them need to be understood. It should be clear that if they are held in ordinary RAM, they will be erased when the ECM power is removed. This is why various methods of preserving them are deployed.

The term keep alive memory (KAM) refers to the systems where the ECM has a permanent, fused, supply of electricity. Here the fault codes are preserved, but only while there is battery power. Figure 2.9 shows a circuit for a KAM system.

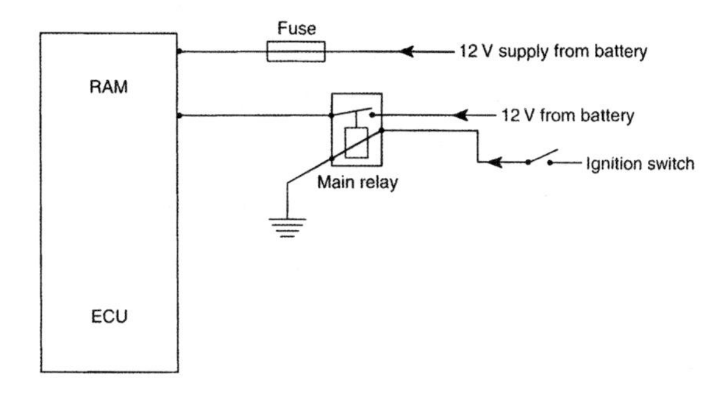

Fig. 2.9 A KAM system

EEPROMs are sometimes used for the storage of fault codes and other data relating to events relating to the vehicle system. This type of memory is sustained even when power is removed. The use of fault codes is discussed in Chapter 3.

2.9 Adaptive operating strategy of the ECM

During the normal lifetime of a vehicle it often happens that compression pressures and other operating factors change. To minimize the effect of these changes, many computer controlled systems are programmed to generate new settings that are used as references, by the computer, when it is controlling the system. These new (learned) settings are stored in a section of memory, normally RAM. This means that such 'temporary' operating settings can be lost if electrical power is removed from the ECM. In general, when a part is replaced or the electrical power is removed for some reason, the vehicle must be test driven for a specified period in order to permit the ECM to 'learn' the new settings. It is always necessary to refer to the repair instructions for the vehicle in question, because the procedures do vary from vehicle to vehicle.

2.9.1 LIMITED OPERATING STRATEGY (LOS)

When a defect occurs that affects the engine, but is not serious enough to prevent the vehicle from being driven, the ROM program will normally contain an alternative loop in the program that will allow the vehicle to be driven to a service point. This mode of operation is often referred to as the 'limp home mode'. It should be noted that any attempt to cure a defect must take account of the fact that the system may be in its limited operating mode.

2.10 Networking of computers

As the use of separate computer controlled systems has increased, the desirability of linking the systems together has become evident and it is now quite common to find systems, such as engine management, traction control, anti-lock braking etc., working together to produce improved vehicle control. When computer controlled systems are linked together they are said to be 'networked'. The networking of computers on vehicles is often referred to as 'multiplexing', but as an introduction to the topic it is helpful to consider some general principles of computer networking as this provides a good insight into the networking that is used on vehicles.

2.10.1 A BUS-BASED SYSTEM

Figure 2.10 shows the basic principle of a number of computers which are linked together by a common wire along which are sent the messages that the computers use to share data.

A principal advantage of this system is that it reduces the number of wires that are needed but, as you will appreciate, there are likely to be problems if more than one message is 'on the data bus' at any one time. The problem is overcome by having strict rules about the way in which data is moved between the computers connected to the bus. These sets of rules are known as 'protocols'.

2.10.2 STAR CONNECTED COMPUTERS

An alternative to the bus system of connecting computers together is the star system shown in Fig. 2.11. An advantage of this system is that a break in the

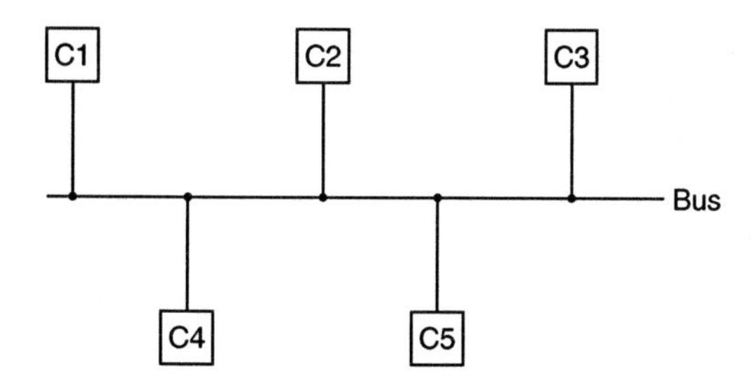

Fig. 2.10 A simple bus-based network of computers

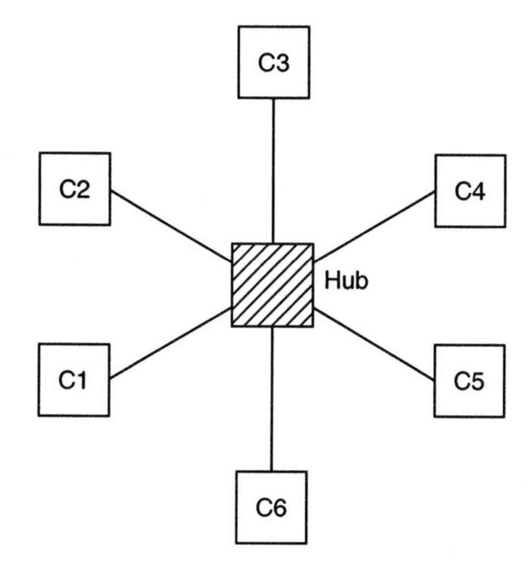

Fig. 2.11 A star-connected computer network

connection between one computer and the hub will not cause a failure of the entire network. The central hub can also be an electronic switch that receives messages from any of the computers. It then determines which of the computers (ECMs) on the network is the intended destination and then sends the message to that computer (ECM) only.

Networked computer systems of both types are often found on the same vehicle. Controller area network (CAN) is a networking system devised by Bosch that is widely used for high speed automotive networks. (Note, the term 'high speed' refers to the speed at which data is moved around the network and not the speed of the vehicle.) When a vehicle is equipped with networks that operate at different speeds (baud rates, or data bits per second), the normal practice is to permit them to communicate with each other via an interface known as a 'gateway'.

2.10.3 MESSAGES

Message is the term used to describe a data item that is sent from one computer (ECM) to another across a network. A message can be long, e.g. the contents of a file made up from many megabytes, or short comprising only a few bytes.

For a given speed of data transmission (baud rate) a long message will take a long time to be transmitted and this can cause problems if another computer needs to send a more urgent message. To prevent this from happening, messages are divided up into smaller parts called packets. Each packet is transmitted on the bus as a separate entity, the receiving computer then reassembles the packets to reconstruct the full message.

(Note, 1 baud = 1 bit, binary 0 or 1, per second.)

The packet is formed into a frame before it is transmitted. The frame consists of:

- the packet itself;
- extra data bits (0s and 1s) to enable errors to be detected;
- data bits to enable the sending computer to be identified;
- data bits which enable the destination computer to be recognized;
- a sequence of data bits to signify the start of the frame;
- a number of data bits to signify the length of the frame and/or the end of the frame.

Each computer on the network has a unique network address.

2.10.4 PROTOCOLS

In order for networks to function there must be rules (protocols) governing the transfer of data. A commonly used protocol for communication between networked computers is that known as carrier-sense multiple access with collision avoidance (CSMA-CD). CSMA-CD works as follows.

A computer on the network must wait for the network to be idle before it can transmit a frame. When the frame has been transmitted all computers on the network check the destination address and, when the checks are completed, the destination computer accepts (reads in) the whole frame. The destination computer then carries out an error check and if an error is detected it will transmit an error message to the transmitting computer. On receipt of an error message, the transmitting computer will re-transmit the entire frame.

If two computers try to transmit a frame simultaneously there will be problems on the bus. This is known as a collision and it causes the frames to be corrupted. If this happens the protocol requires the transmitting computers to stop transmitting their frames immediately and to transmit a 'jamming signal'. The jamming signal warns other computers on the network and they then ignore the parts of frames that have been transmitted. Each computer must then wait a short period of time before attempting to re-transmit.

Each computer, or other device, that is connected to the network must be equipped with a suitable interface. The interface circuit card is equipped with a microprocessor that permits it to receive and check frames without interfering with the tasks that the computer's main processor is dealing with.

To summarize, in order for a local area network (LAN) to function the following must happen:

- the data must be divided into packets;
- error detection bits must be added;
- each packet must be formed into a frame;
- the frame must be transmitted onto the network;
- collision detection must take place, transmission must stop and a jamming signal must be sent out;
- computers must wait for a random period of time before transmitting again.

It should be noted that this all takes place in microseconds under the control of the computer clock.

2.11 Vehicle network systems

A primary purpose of automotive networked systems is to reduce the amount of wire that is used. Some estimates suggest that as much as 15 kg of wire can be eliminated by the use of networking on a single vehicle. Other savings are made in the use of sensors. For example, systems such as traction control and engine management that make use of the engine speed sensor can make use of a single engine speed sensor by placing the reading on the data bus as required (i.e. the sensor is multiplexed), instead of having a separate sensor for each system. Similar economies are possible with a range of systems and this can result in a reduction in the total number of sensors on a vehicle.

There are several areas of vehicle control where data buses can be used to advantage. Some of these, such as lighting and instrumentation systems, can operate at fairly low speeds of data transfer, e.g. 1000 bits per second. Others such as engine and transmission control require much higher speeds, probably 250 000 bits per second, and these are said to operate in 'real time'. To cater for these differing requirements the Society of Automotive Engineers (SAE) recommends three classes known as Class A, Class B and Class C.

- Class A. Low speed data transmission, up to 10 000 bits/s, used for body wiring such as exterior lamps etc.
- Class B. Medium speed data transmission, 10 000 bits/s up to 125 000 bits/s, used for vehicle speed controls, instrumentation, emission control etc.
- Class C. High speed (real time) data transmission, 125 000 bits/s up to 1 000 000 bits/s (or more), used for brake by wire, traction and stability control etc.

2.11.1 THE PRINCIPLE OF A BUS-BASED VEHICLE SYSTEM

The controls (switches) for almost every system on a vehicle must be near the driver's seat. With ordinary wiring this means that there is a cable taking electrical power to the switch and another taking the electricity from the switch to the unit being operated.

As the amount of electrically-operated equipment on vehicles has increased a very large number of electrical cables have become concentrated near the driving position. This causes problems such as finding sufficient space for the wires, extra cable connectors that can cause defects etc. Multiplexed, or data bus-based systems overcome some of these problems.

Figure 2.12 shows the basic concept of multiplexed vehicle wiring. In order to keep it as simple as possible fuses etc. have been omitted because at this stage, it is the 'idea' that is the focus of attention.

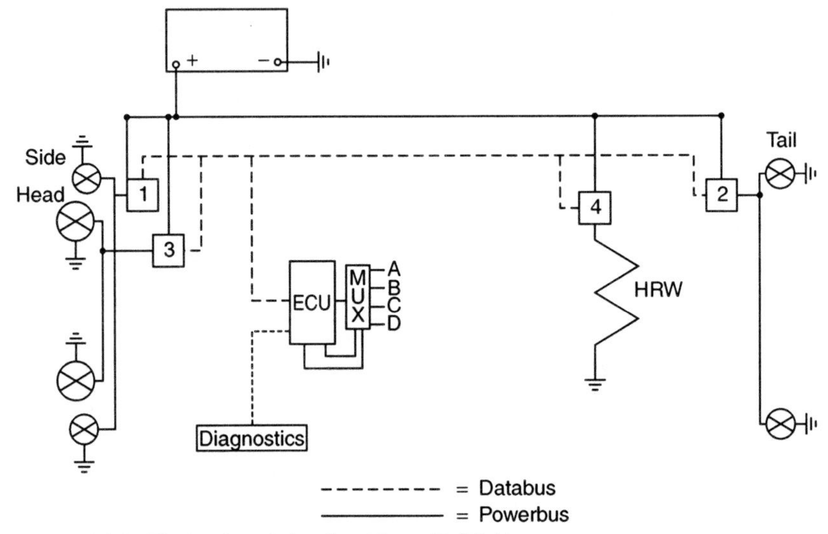

Fig. 2.12 The multiplexed wiring concept

As the legend for the diagram states, the broken line represents the data bus. This is the electrical conductor (wire) which conveys messages along the data bus to the respective remote control units. These messages are composed of digital data (0s and 1s) as described earlier.

The rectangles numbered 1, 2, 3 and 4 represent the electronic interface that permits two-way communication between the ECU and the lamps, or the heated rear window. The dash panel switches are connected to a multiplexer (MUX) which permits binary codes to represent different combinations of switch positions to be transmitted via the ECU onto the data bus. For example, switches for the side and tail lamps on and the other switches off could result in a binary code of 1000, plus the other bits (0s and 1s) required by the protocol, which are placed on the data bus so that the side lamps are energized. Operating other switches, e.g. switching on the heated rear window would result in a different code and this would be transmitted by the ECU to the data bus, in a similar way. As the processor is moving the data bits at a rate of around 10 000 per second it is evident that, to the human eye, any changes appear to occur instantaneously.

2.11.2 DATA BUSES FOR DIFFERENT APPLICATIONS

The traction control system described in Chapter 1 operates in 'real-time' and this means that data must move between elements of the system at high speed. Increasing use is being made of the high data speed network system, CAN, for systems such as traction control. CAN originate from Bosch and falls into SAE Class C. It utilizes a two-wire twisted pair for the transmission of data.

The Rover 75 is a modern vehicle that uses several different data buses, as described below.

1. A two-wire CAN bus that can operate at high data transmission speeds of up to 500 kbaud (500 000 bits/s) (Fig. 2.13).

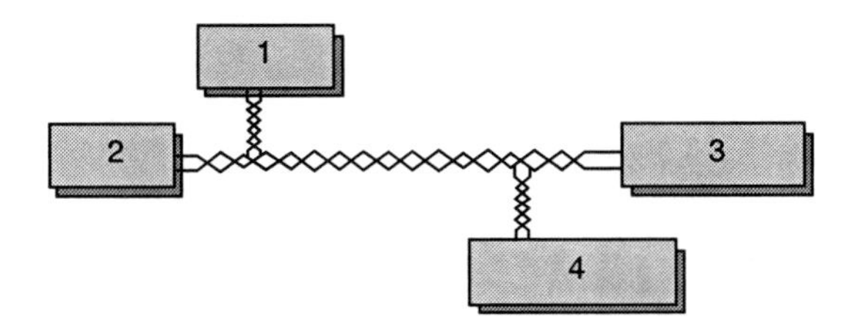

1. Automatic transmission control unit
2. Engine control module
3. ABS/ traction control ECU
4. Instrument pack

Fig. 2.13 The CAN bus system

2. A single-wire bus for doors, lights, sun roof etc. This bus operates at a data speed of 9.6 kbaud (Fig. 2.14).
3. A single-wire bus for diagnostic purposes. This bus operates at 10.4 kbaud (Fig. 2.15).

The twisted pair of the CAN bus system minimizes electrically-initiated interference and virtually eliminates the possibility of messages becoming corrupted.

2.11.3 ENCODING SERIAL DATA

Apart from the details about protocols that are given in section 2.10.4, consideration has to be given to practical methods of transmitting the 'messages' around the data buses. Factors such as speed of transmission (bit rate), electrical interference and preservation of the integrity of messages has to be considered. Two of the methods currently in use are:

- non-return to zero (NRZ) (Fig. 2.16);
- controller area network (CAN) (Fig. 2.17).

The differences between them are largely to do with the ways of representing the logic levels (0 and 1, or high and low) that are used in computing. In the NRZ method a binary code of 0,1,1,1,0,0,1 would be transmitted as shown in Fig. 2.16.

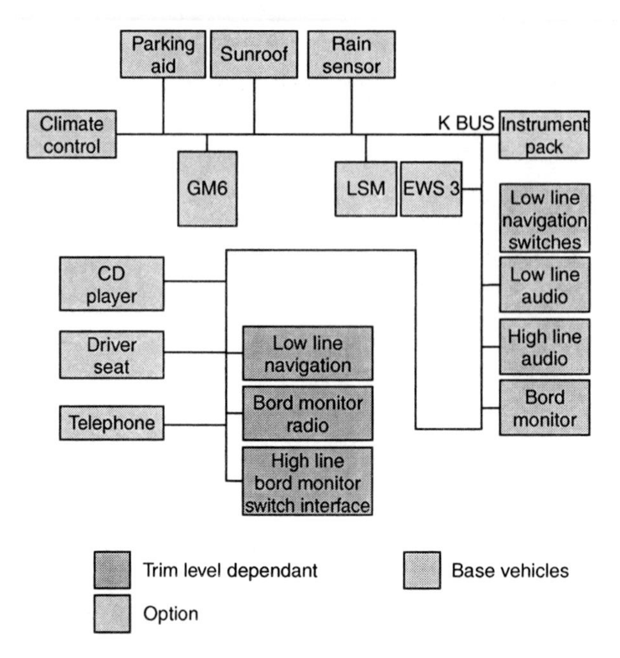

Fig. 2.14 The single wire BMW K bus

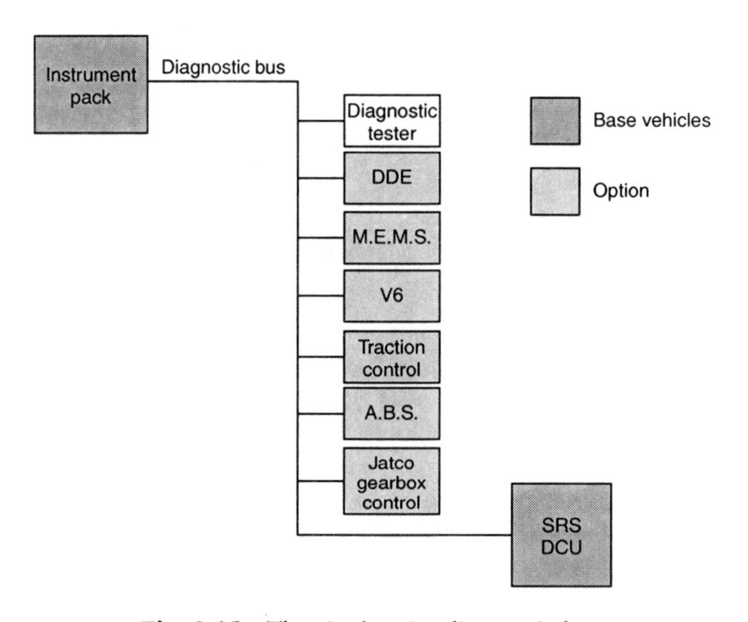

Fig. 2.15 The single wire diagnostic bus

The point to note is that each bit is transmitted for one bit time without any change.

In CAN, two wires are used for data transmission. One wire is known as CAN-high (CAN-H) and the other as CAN-low (CAN-L). A CAN bit sequence is shown in Fig. 2.17.

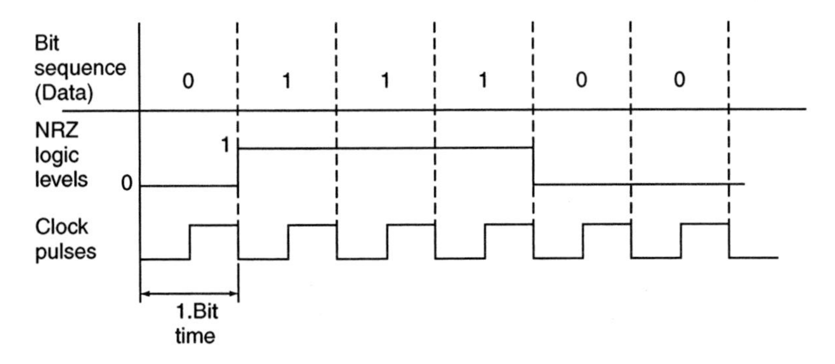

Fig. 2.16 NRZ transmission of a binary bit sequence

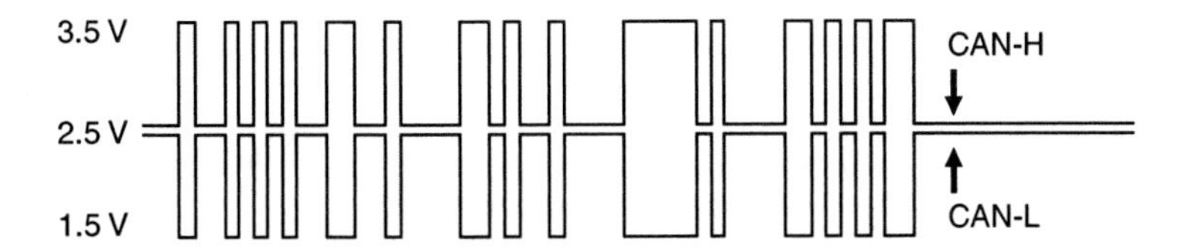

Fig. 2.17 A CAN bit sequence

The CAN-H wire switches between 2.5 and 3.5 V and the CAN-L wire switches between 2.5 and 1.5 V. When CAN-H and CAN-L are at 2.5 V, there is no voltage difference between them and this represents computer logic 0. Computer logic 1 is created when there is a 2 V difference between the two wires as happens when CAN-H is at 3.5 V and CAN-L is at 1.5 V.

2.12 Prototype network systems

In order to provide further insight into the way in which vehicle systems are networked it will be helpful to consider the following details of a concept vehicle that was developed by LucasVarity. Figure 2.18 gives an impression of the system.

Several references to networked systems, e.g. traction control, stability control etc., have already been made. The system shown in Fig. 2.18 is suitable for virtually any vehicle, including trucks and buses. The system comprises four subsystems.

1. The Lucas EPIC electronically-programmed injection control system, which is a computer controlled engine management system for diesel engines, similar to the one described in section 1.11.
2. The Lucas flow valve anti-lock braking system. On this advanced prototype vehicle, the ABS system is provided with a second solenoid valve at each front wheel that permits independent application of the brakes by using the ABS pump to supply pressure.
3. A clutch management system (CMS). This replaces the normal clutch pedal linkage with a computer controlled, hydraulically actuated system. The manual

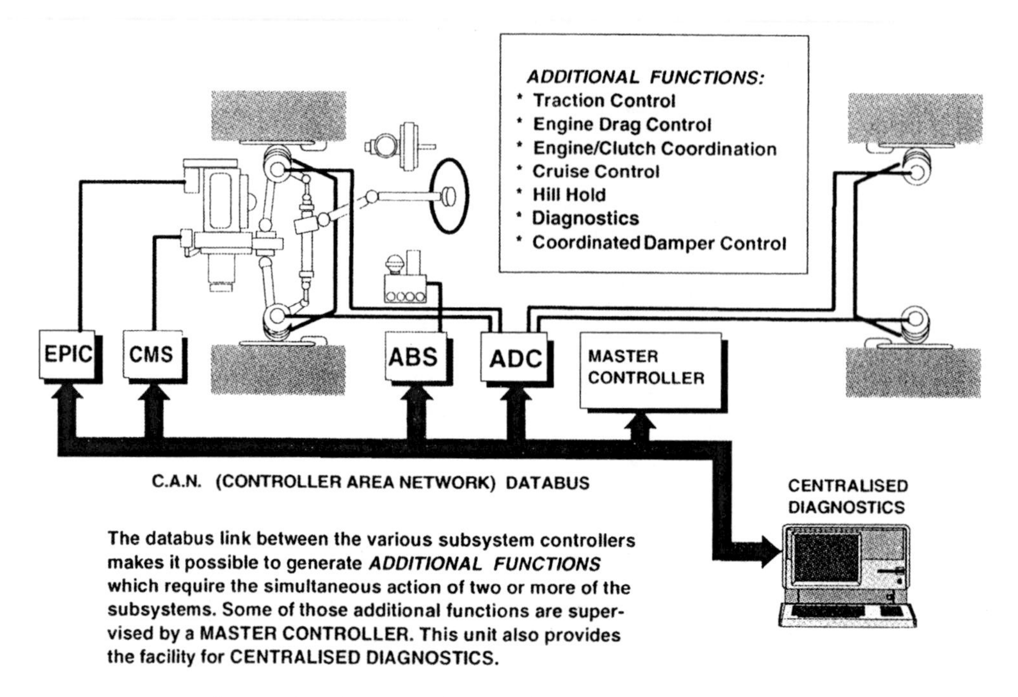

Fig. 2.18 The LucasVarity advanced prototype vehicle

gearshift is retained, but there is no clutch pedal. The driver still lifts the foot from the accelerator pedal when changing gear. The advantage of this is that the driver gains two-pedal control without the fuel consumption penalty that is associated with automatic transmission. The driver also retains full control over gear change operations.

4. Adjustable rate dampers are fitted. The damping rate is adjusted by the computer (ECM) to provide optimum damping during rapid steering input, braking and acceleration.

Master controller

Each of the above systems has a CAN interface which permits them to be connected to the master controller. A network of twisted pair cables connects each of the above subsystems to the master controller and this allows the transfer of sensor information and control signals with reliable safety checking and minimal wiring. The master controller thus receives information from the subsystems via the CAN bus (cables).

The master controller is directly connected to a switch pack (for cruise and damper control), two accelerometers and an inclinometer (for hill detection). This means that the master controller 'knows' the complete status of the vehicle and the driver's requirements. The vehicle status information is processed by the master controller to generate control signals which are sent to the subsystems. These 'master' signals over-ride the normal operation of the subsystems to operate another tier of systems known as the integrated systems. In the event of CAN failure, each subsystem defaults to stand-alone operation.

The integrated systems

The four subsystems, i.e. EPIC, ABS, damper control and clutch management, are integrated (made to work together) to provide seven additional functions of vehicle management. The computer programs that do this controlling are executed by the master controller. These seven integrated systems are:

- traction and stability control
- cruise control
- power shift
- engine drag control
- hill hold
- damper control
- centralized diagnostics.

Traction and stability control

The ABS wheel speed sensor signals inform the ABS computer of wheel to surface conditions. At low speeds, or when only one wheel spins (such as pulling away with one driving wheel on ice and the other on dry tarmac), the brake is automatically applied to the spinning wheel. At higher speeds, or when both wheels spin, engine power is reduced to eliminate wheel spin. These two strategies combine to give improved traction and acceleration, and safer cornering at higher speeds.

Cruise control

The vehicle speed sensor information is used by the engine control (EPIC) to maintain a constant vehicle speed that is selected by the driver. The cruise control switch pack provides commands for setting the desired cruise speed and for switching on and off as required.

Power shift

The power shift function automatically reduces engine power during gear changes. This means that the driver no longer has to lift the foot from the accelerator pedal and need only move the gear lever to effect a gear change. This reduction of engine power, via the controller, overcomes the difficulty of synchronizing the accelerator and gear lever movements. It is also possible to provide for 'blipping' of the throttle to give smooth downward gear changes.

Engine drag control

This eliminates wheel locking due to engine braking on very slippery surfaces, and improves anti-lock braking performance. The reduced engine drag is achieved by increasing engine power slightly to maintain the correct level of wheel slip for maximum retardation and stability. In extreme cases, the clutch can be disengaged to remove the inertia from the engine driveline. This allows the wheels to respond more quickly to the anti-lock brake control and gives improved steering capability and reduced stopping distances.

Hill hold
Hill hold uses brake actuation to apply the rear brakes automatically when coming to a halt on a hill. When re-starting on the hill, information from the inclinometer sensor, the EPIC system, the ABS controller and the clutch management controller is used to determine the point at which the brakes should be released to give a smooth pull away, with no roll back.

Damper control
The damper control system uses data from the CAN data bus to control the damping rate settings. The dampers are switched to 'firm' setting for optimum response to rapid steering input, braking and acceleration. On returning to 'normal' cruising, the 'soft' damper setting is selected for improved ride comfort.

Centralized diagnostics
The centralized diagnostics uses the master controller to monitor all of the networked systems. Data bus information is interpreted in order to detect and respond to faults in the subsystems and the network communications. By this means, fail safe operation is achieved and correct fault recognition is made. The diagnostic section is provided with an interface which permits an interrogation tool, the Lucas Laser 2000, to access diagnostic information. Future developments of this system are anticipated and these will include electronic power assisted steering, electronic braking, and active anti-roll bars.

2.13 Summary

The details of data bus communication that are outlined in this chapter are highly specialized and it is work that is normally of most concern to designers. However, vehicle repair technicians do encounter the terminology and it is helpful to have an insight into some of the details. Fortunately, diagnostic equipment manufacturers need to take care of the aspects that affect the suitability of their equipment for diagnostic tests on the vehicles that it is made for. It is probably in the area of equipment selection that a vehicle repair technician is most likely to feel the need for familiarity with serial data terminology. Equipment specifications sometimes contain references to some of the material contained in this chapter and it should help when arranging for a demonstration of an instrument's capabilities, which should always be a step in the process of making a decision about purchasing an item of equipment.

2.14 Review questions (see Appendix 2 for answers)

1. Serial data transmission:
   (a) requires a separate wire for each bit that is transmitted between the computer and a peripheral such as a fuel injector?
   (b) is faster than parallel transmission of the same data?
   (c) is transmitted one bit after another along the same wire?
   (d) is not used in vehicle systems?
2. In a multiplexed system:
   (a) a data bus is used to carry signals to and from the computer to the remote control units?
   (b) each unit in the system uses a separate computer?
   (c) more wire is used than in a conventional system?
   (d) a high voltage is required on the power supply?
3. In networked systems messages are divided into smaller packages to:
   (a) prevent problems that may arise because some messages are longer that others?
   (b) avoid each computer on the network having to have an interface?
   (c) make the system faster?
   (d) save battery power?
4. When DTCs are stored in an EEPROM:
   (a) the DTCs are removed when the vehicle battery is disconnected?
   (b) an internal circuit must be activated to clear the fault code memory?
   (c) they can only be read out by use of a multimeter?
   (d) they can only be removed by replacing the memory circuit?
5. A traction control system:
   (a) uses a high speed bus?
   (b) is not networked because it does not need to work with other systems on the vehicle?
   (c) is not used on front-wheel drive vehicles?
   (d) can only operate on vehicles equipped with a differential lock?
6. The computer clock is required:
   (a) to permit the time of day to be displayed on the instrument panel?
   (b) to allow the time and date to be stored for future reference?
   (c) to create the electrical pulses that regulate the flow of data?
   (d) to generate the voltage levels required for operation of a data bus?
7. In the CAN system:
   (a) a 'twisted pair' of wires is used so that the correct length of cable can be placed in a small space?
   (b) a 'twisted pair' of wires is used to provide the two different voltage levels and minimize electrical interference?
   (c) the 'twisted pair' of wires carries the current that drives the ABS modulator?
   (d) it is used only for the engine management system?

8. The RAM of the ECM computer is:
   (a) the part of memory where sensor data is held while the system is in operation?
   (b) only used for storing of fault codes?
   (c) held on a floppy disc?
   (d) not a volatile memory?

3
Self-diagnosis and fault codes

Whilst a computer controlled system is operating normally, the processor is constantly monitoring the electrical state of input and output connections at the various interfaces of the ECM. This monitoring (reading) of the inputs and outputs occurs so that the instructions that the computer processor has to perform, such as to compare an input value with a programmed value stored in the ROM, means that the ECM is ideally placed to 'know' what is happening at many parts of the system that is connected to it. If, for example, a throttle position sensor is producing a reading that does not tie in well with engine speed and load signals that the ECM is reading, the software (program) in the ROM can be written so that an alternative section of program is followed (loop), and a predetermined code can be stored in a section of working memory (RAM) that is allocated for this purpose. The piece of coded information that is stored is known as a 'fault code' or 'diagnostic trouble code' (often abbreviated to DTC).

It will be evident that these DTCs are a valuable source of information when trouble occurs and it is necessary for us to consider the methods that are available which provide access to them. It is worth mentioning at this point, that tools are available that permit the readings that the ECM 'sees' to be viewed on an oscilloscope while the system is in operation, or placed in a memory for later viewing and analysis.

3.1 Access to DTCs

Early computer controlled systems often had quite small amounts of memory allocated to the storage of DTCs, but as integrated circuit technology has developed, the amount of memory space allocated to the storage of DTCs has also increased to the stage where hundreds of DTCs can be held in an ECM. In order that the DTCs are not lost when the ignition is switched off, the section of RAM that they are stored may be energized directly from the battery, via a fuse. This type of memory is sometimes referred to as 'keep alive memory' (KAM). In other cases the DTCs may be stored in an EEPROM.

These DTCs are safely stored until a deliberate action is taken that instructs a circuit in the ECM to generate an electrical pulse, probably 25 V, that will clear the DTCs.

As a result of the different technology used in automotive computer controlled systems, it is possible to find a number of different methods of accessing DTCs. Three methods are in general use.

1. Displaying the code as flashes on a dashboard indicator lamp.
2. Connecting an LED or test lamp externally and observing the number of flashes and pauses.
3. Connecting a code reading machine, and/or a scan tool, to the diagnostic port on the ECM.

The actual procedure varies very considerably from one type of vehicle to another. However, the following examples give an insight into the general principles involved.

3.1.1 METHOD 1: THE DASHBOARD LAMP

The system under examination is the Toyota electronic fuel injection system (EFI) shown in Fig. 3.1.

This fuel system is fitted to the 4A-GE, 1600 cc, 16 valve, 122 bhp engine in vehicles such as the Toyota Corolla GT Hatchback (AE82) 1985 to 1987. The ECU has built-in self-diagnosis, which detects any 'problems' and displays a signal on the 'check engine' warning light (Fig. 3.2). This lamp is in a convenient position on the instrument panel as shown in Fig. 3.3.

When the ignition is switched on, the light will come on. If there are no faults the light will go out when the engine starts. If the 'engine check' lamp stays on this is a warning that a fault is present. To find out what the fault is, it is necessary to put the system into diagnostic mode. This requires some preliminary work, as follows.

1. (a) Check that the battery voltage is above 11 V.
   (b) Check that the throttle valve is fully closed (throttle position sensor switch points closed).
   (c) Ensure that transmission is in neutral position.
   (d) Check that all accessory switches are off.
   (e) Ensure that engine is at its normal operating temperature.
2. Turn the ignition on, but do not start the engine.
3. Using a service wire connect together (short) the terminals T and E_1 of the 'check engine' connector (Fig. 3.4).
   [Note that the 'check engine' connector is located near the wiper motor (AE) or battery (AA), these being different vehicle models.]
4. Read the diagnostic code as indicated by the number of flashes of the 'check engine' warning light.

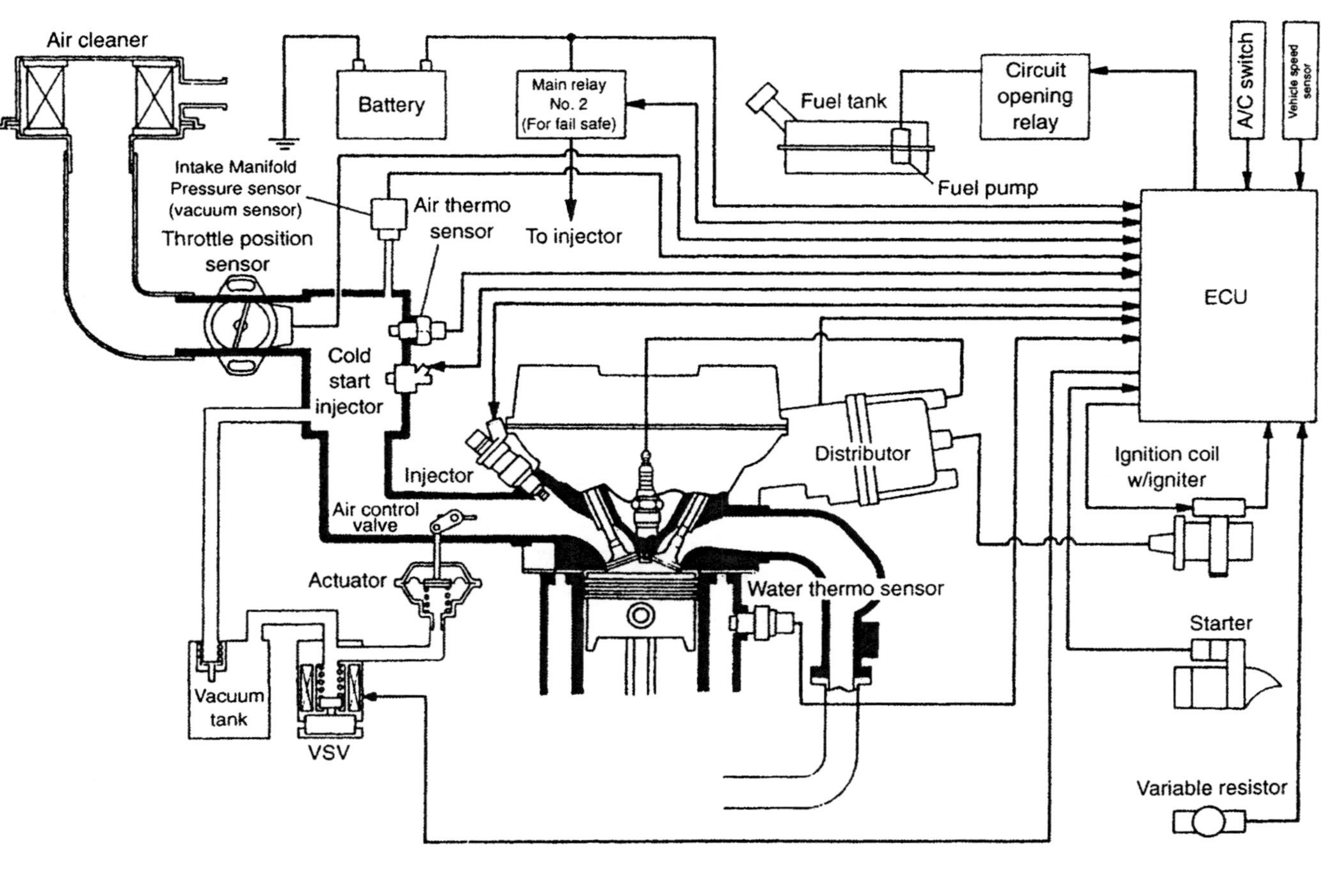

Fig. 3.1 Electronic fuel injection system

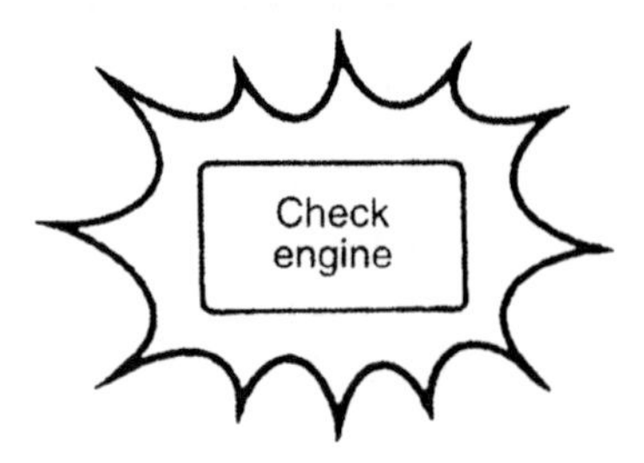

Fig. 3.2 The diagnostic lamp

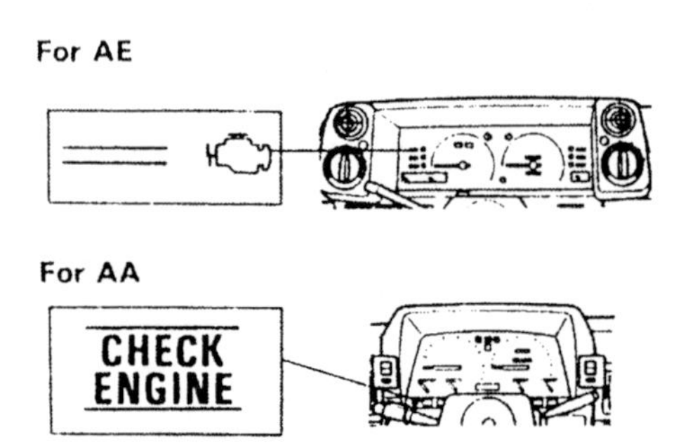

Fig. 3.3 Engine diagnostic lamp on the instrument panel

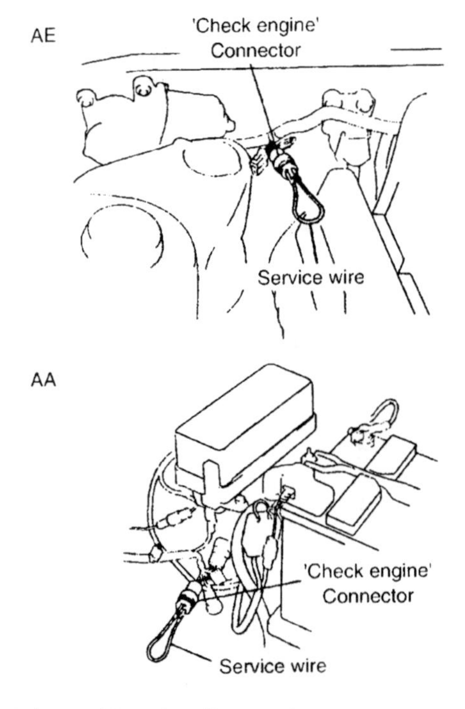

Fig. 3.4 Making the diagnostic output connection

Diagnostic codes

Diagnostic code number 1 (Fig. 3.5) is a single flash every 3 seconds. It shows that the system is functioning correctly and it will only appear if none of the other fault codes are identified. Figure 3.6 shows the fault codes for code 2 and code 4.

The 'check engine' lamp blinks a number of times equal to the fault code being displayed; there are therefore two blinks close together (1 second apart) for code 2, and a pause of 3 seconds and then four blinks to show fault code 4. The fault code will continue to be repeated for as long as the 'check engine' connector terminals (T and E_1) are connected together. In the event of a number of faults occurring simultaneously, the display will begin with the lowest number and continue to the higher numbers in sequence.

Figure 3.7 shows a section of the Toyota workshop manual that gives the diagnostic codes. Reading from left to right it will be seen that each code is related to a section of the system. The column marked 'See page' refers to the section of the workshop manual where further aid to diagnosis will be found.

When the diagnostic check is completed the 'service wire' must be removed from the 'check engine' connector and then the diagnostic code must be cancelled.

After the fault has been rectified, the diagnostic code stored in the ECU memory must be cancelled. This is done, in the case of this Toyota model, by removing the appropriate fuse. The fuse must be removed for a period of 10 seconds or more, depending on the ambient temperature, with the ignition switched off. (The lower the ambient temperature the longer the period for which the fuse is left out.)

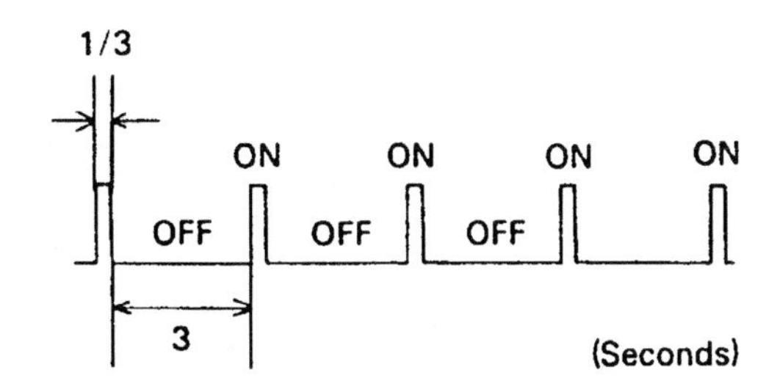

Fig. 3.5 Diagnostic code number 1 (system normal)

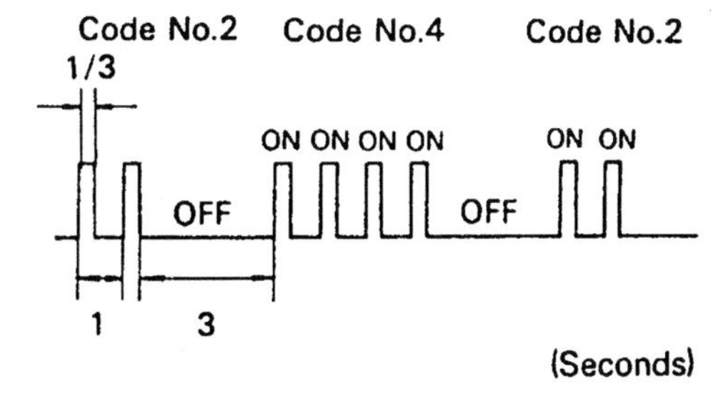

Fig. 3.6 Diagnostic codes for codes 2 and 4

Code No.	Number of blinks "CHECK ENGINE"	System	Diagnosis	Trouble area	See page
1	ON 1/3 ON ON ON OFF OFF OFF OFF 3 (Seconds)	Normal	This appears when none of the other codes (2 thru 11) are identified.	—	—
2	1/3 1 (Seconds)	Pressure sensor signal	Open or short circuit in pressure sensor.	1. Pressure sensor circuit 2. Pressure sensor 3. ECU	FI-31
3		Ignition signal	No signal from ignitor four times in succession.	1. Ignition circuit (+B, IGt, IGf) 2. Igniter 3. ECU	FI-37
4		Water thermo sensor signal	Open or short circuit in coolant temperature sensor signal.	1. Coolant Temp. sensor circuit 2. Coolant Temp. sensor 3. ECU	FI-35

Fig. 3.7 Diagnostic codes as given in a Toyota workshop manual

3.1.2 METHOD 2: FAULT CODES DISPLAYED THROUGH A LOGIC PROBE OR TEST LAMP

Figure 3.8 shows the circuit, and details of the method that can be used to obtain fault codes, from a Wabco braking system.

The procedure, as taken from the Wabco publication 'Blinkcode for Goods Vehicles and Buses ABS/ASR "C"-Generation' gives the procedure as follows.

1. In the case of a vehicle not having an ASR lamp installed: connect a filament bulb (2W...5W) to pin 3 of the ECU (see top circuit diagram). This can be achieved using Wabco inter-adaptor installed between ECU and ECU connector (Ignition:OFF!).
2. By connecting pin 14 to vehicle ground (earth) for longer than 5 seconds this can be achieved via the switch on the inter-adaptor (Ignition:ON!).
3. The blinkcode can be read and noted until the user is in no doubt as to the transmitted fault code! The fault code can be erased by disconnecting pin 14 from vehicle ground during the blinkcode transmission.

To prevent inadvertent erasure of fault codes, the ignition should be switched off during blinkcode transmission. Faults actually present must first be repaired before further fault codes can be read. Following each repair the blinkcode should be reactivated to ensure that no further faults exist and the fault memory of the ECU is cleared. When all faults have been read and erased, the code for 'System OK' is transmitted (i.e., X-0-0).

After each repair the system operation should be further verified by a test drive during which the ABS and ASR lamps should extinguish once the vehicle has reached 7 km/h.

The blinkcode frame is made up as shown in Fig. 3.9.

3.1.3 METHOD 3: FAULT CODE READERS AND SCAN TOOLS

Fault code readers vary in complexity from inexpensive devices that read out flash codes, such as the Gunson 'Fault Finder' shown in Fig. 3.10, to

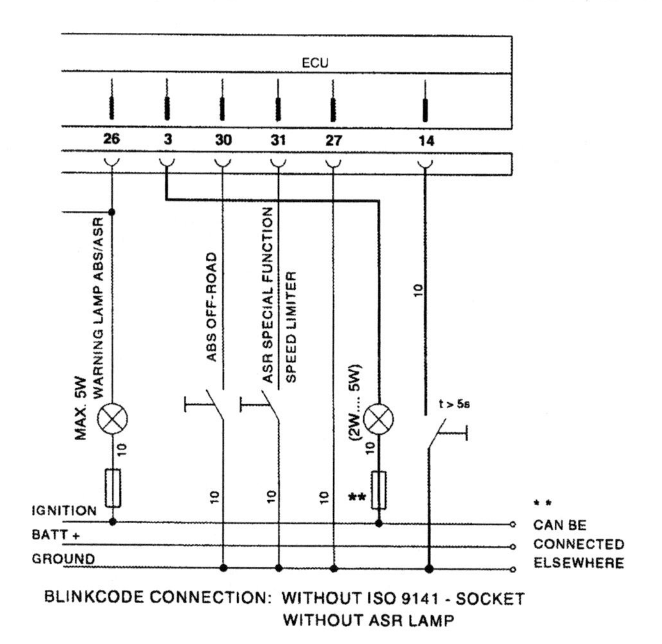

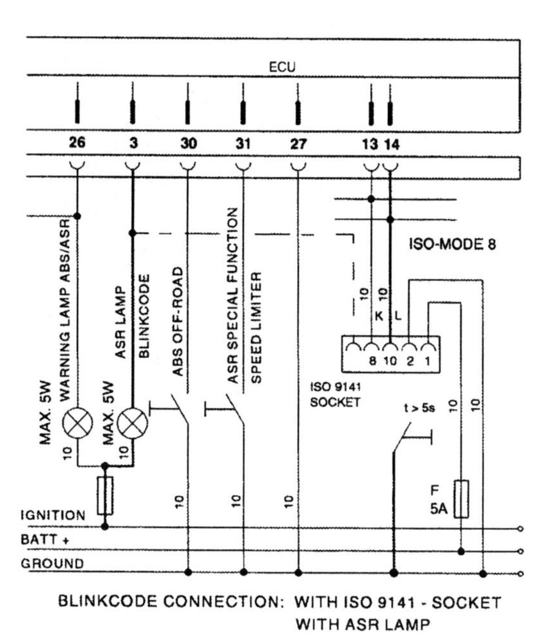

Fig. 3.8 The external lamp circuit for reading blinkcodes

microprocessor-based machines, such as the one described below. Gunson also market similar machines, as do many other suppliers and this is a factor that is covered in the next chapter.

The aim here is to give a reasonable description of the work involved in obtaining diagnostic information through the serial port. In effect, one connects

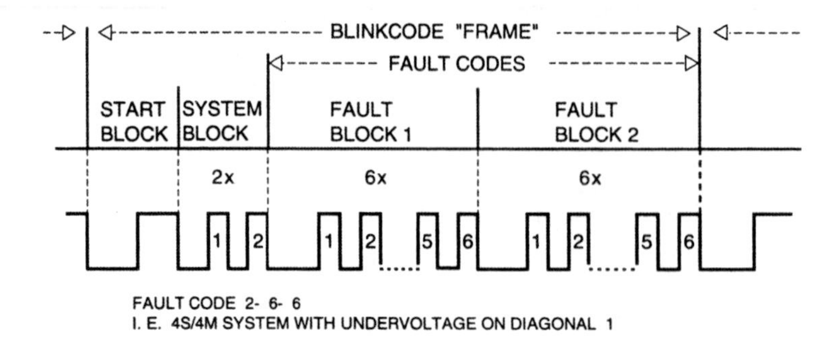

Fig. 3.9 The fault code frame for the Wabco system

Code	Where to look	Remedy
17	Manifold Absolute Pressure (MAP) sensor	Refer to TPN (6) *
18	Battery voltage (V BATT) low	Check charging system and battery
19	Keep Alive Memory (KAM) failure	Check whether battery was disconnected. Check KAM/ROM fuse. If OK, module is faulty

Fig. 3.10 A fault code reader (the Gunson 'fault finder')

up the tool and follows the instructions in the handbook and on the instrument's display panel. However, in order to give an insight into the work involved I am including further details.

Figure 3.11 shows the principal parts of the diagnostic kit. There is a handheld tester, a lead to connect the tester to the vehicle's diagnostic connector, a 'smart card' that matches the tester to the vehicle system under test, a printer to provide a permanent record of the test results, and leads for making connections to the battery and from the tester to the printer. This is accompanied by an instruction manual (Fig. 3.12), although it needs to be stated that once the test program has started, the display screen on the tester provides a step by step menu to guide the operator through the test sequence.

The 'smart card' is the equivalent of computer software and it enables the tester to use the ECM processor power to interrogate circuits. The test instrument

Fig. 3.11 The diagnostic kit (scan tool)

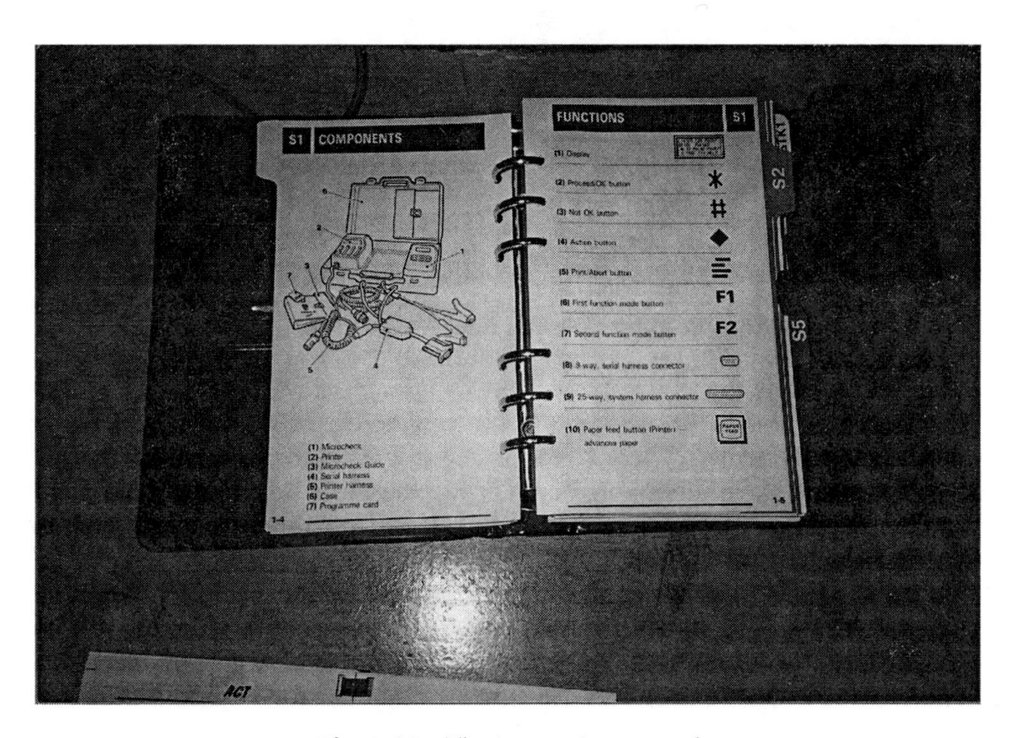

Fig. 3.12 The instruction manual

is thus able to test all circuits that are served by the ECM. The test connection plug is also known as the serial port because test information is fed out serially (one bit after the other, e.g. 10110011). A considerable advantage of the serial port is that it permits testing without the need to disconnect wiring.

Unless the operator is familiar with the vehicle, it will be necessary to refer to a location chart in order to locate the diagnostic connector. Figure 3.13 shows the Rover 200 series connection point.

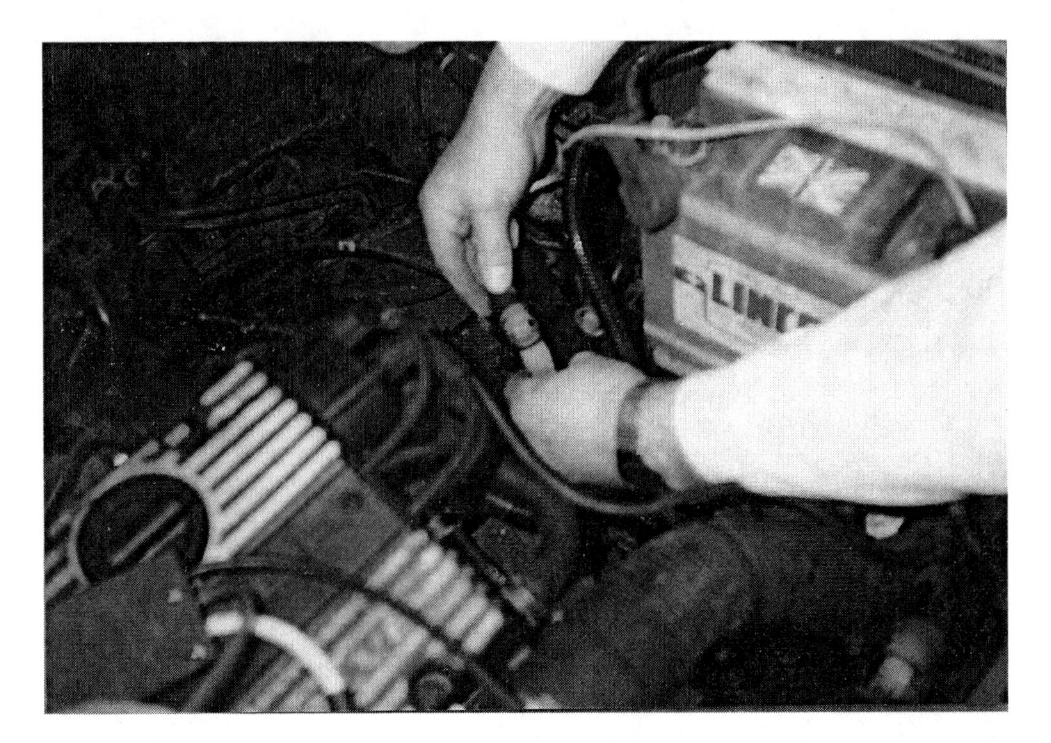

Fig. 3.13 The diagnostic connector

The source of power for the tester is the vehicle battery and Fig. 3.14 shows the leads being connected. The tester is placed in the position shown for the sole purpose of taking the photograph. For test purposes it is held in the hand and it should be noted that care must be taken to place the instrument in a safe position when not being held in the hand.

The next step is to connect the diagnostic lead to the vehicle's diagnostic connector and Fig. 3.13 shows this being done. Prior to commencing the test the diagnostic lead that relates to the specific vehicle model will have been selected, as will the smart card that customizes the test instrument to the vehicle. Figure 3.15 shows the smart card being inserted.

It will be understood that different makes of vehicle require different types of diagnostic leads and smart cards. This is necessary because the diagnostic

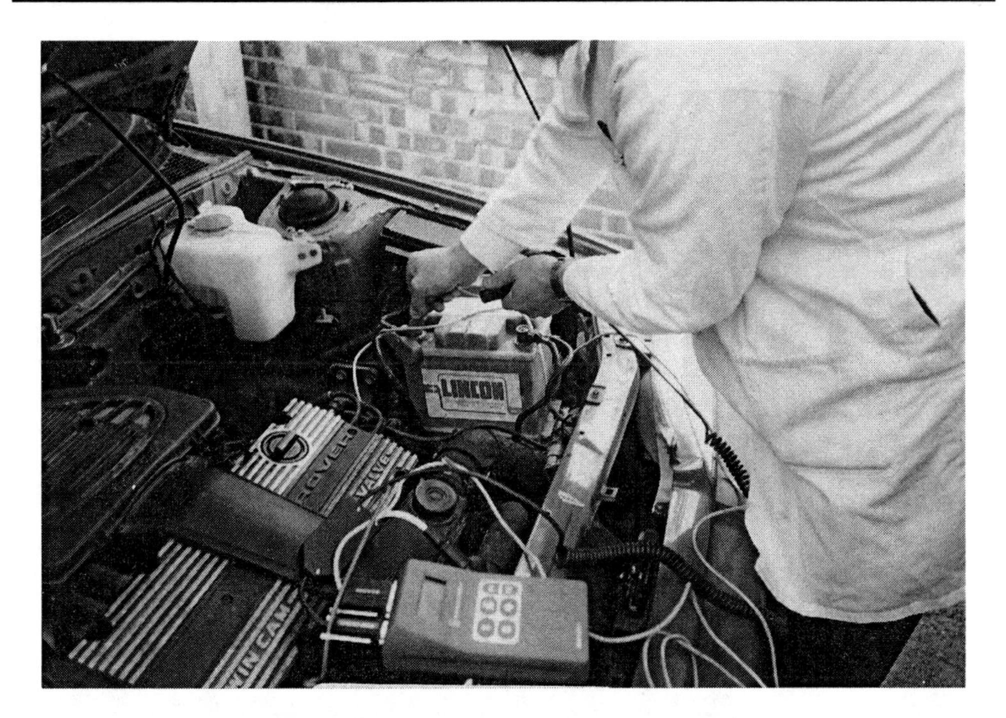

Fig. 3.14 Connecting to the power supply

Fig. 3.15 Inserting the 'smart' card to adapt the test instrument to the specific vehicle application

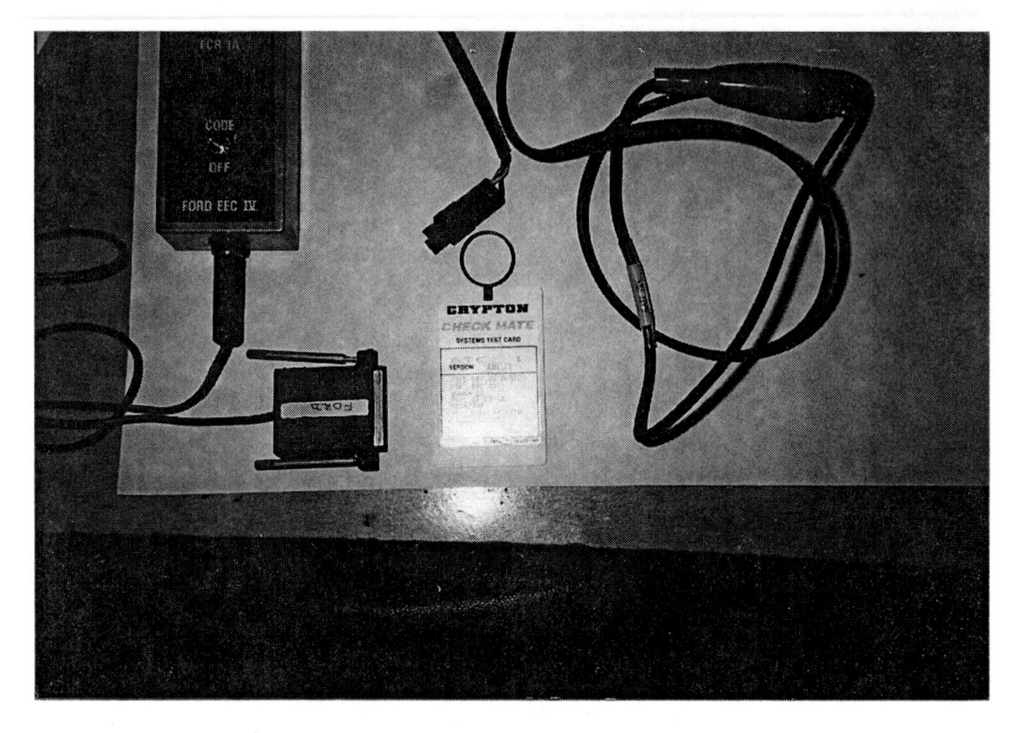

Fig. 3.16 The diagnostic lead and smart card for a Ford vehicle

connections vary from vehicle to vehicle, as does the test program. Figure 3.16 shows the diagnostic lead and smart card that customizes the test instrument to a Ford vehicle. A reasonable range of diagnostic leads and smart cards is available to make this type of equipment suitable for use on a range of vehicle makes. It will be appreciated that this is an important consideration for the independent garage which is not linked to a particular vehicle manufacturer.

An important part of any systematic approach to fault finding, is the gathering of evidence. Figure 3.17 shows the printer being connected. It is from the printout that a permanent record of the test results will be obtained, as shown in Fig. 3.18.

When all leads have been correctly connected and steps taken to ensure that leads are clear of drive belts, hot engine parts etc., the test may commence. The manual (Fig. 3.12) gives a description of the instrument controls and, when all preparations are made, the test instrument screen displays a message which guides the operator through the test sequence.

The test procedure may require the operator to operate certain vehicle controls. Figure 3.19 shows the accelerator being depressed. Here it will be seen that a certain amount of movement around the vehicle is required during a test sequence. It is therefore important that care is exercised to ensure that leads do not become tangled, and that the vehicle is placed so that freedom of movement around it is ensured.

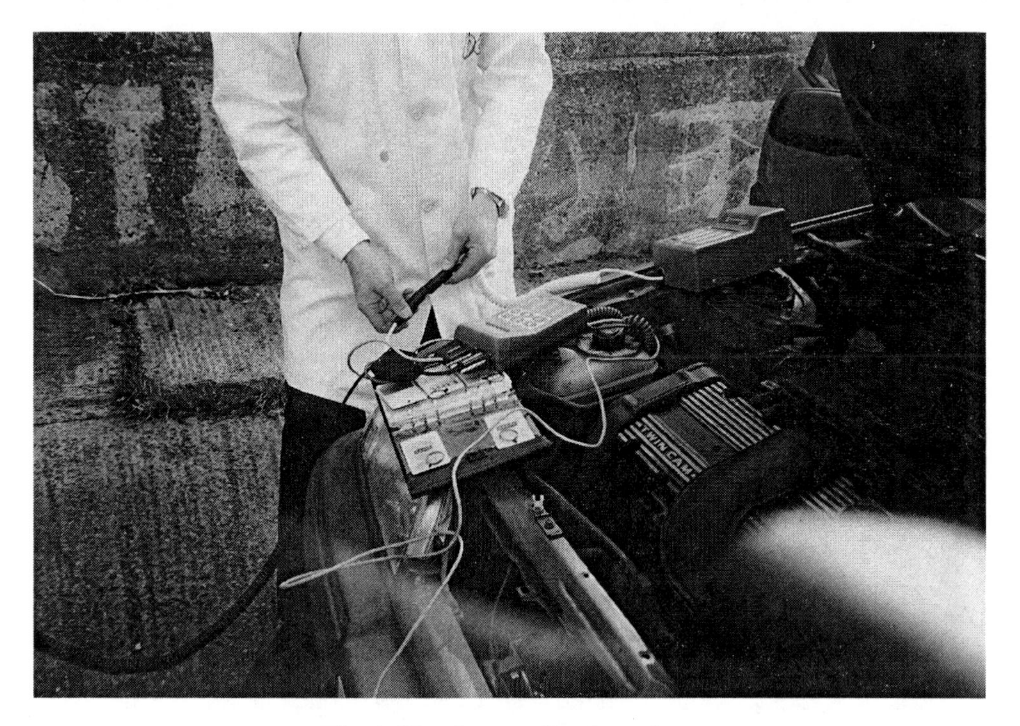

Fig. 3.17 Connecting the printer

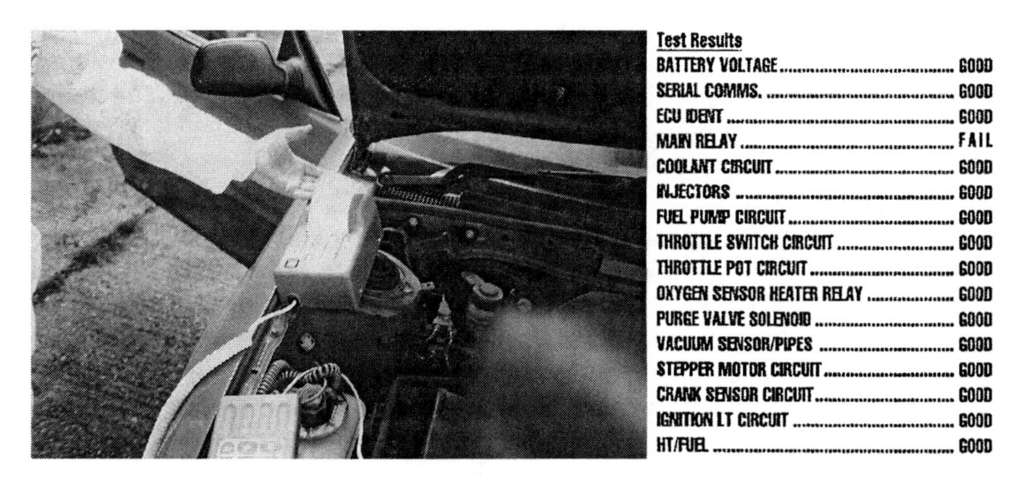

Test Results

BATTERY VOLTAGE	GOOD
SERIAL COMMS.	GOOD
ECU IDENT	GOOD
MAIN RELAY	FAIL
COOLANT CIRCUIT	GOOD
INJECTORS	GOOD
FUEL PUMP CIRCUIT	GOOD
THROTTLE SWITCH CIRCUIT	GOOD
THROTTLE POT CIRCUIT	GOOD
OXYGEN SENSOR HEATER RELAY	GOOD
PURGE VALVE SOLENOID	GOOD
VACUUM SENSOR/PIPES	GOOD
STEPPER MOTOR CIRCUIT	GOOD
CRANK SENSOR CIRCUIT	GOOD
IGNITION LT CIRCUIT	GOOD
HT/FUEL	GOOD

Fig. 3.18 The permanent copy of the test results

On completion of the test the printout is produced and analysis of the results can proceed. When the diagnostic and repair work is completed, the fault code is cleared, normally by following 'on screen' instructions. The instrument is then removed and the vehicle prepared for the road test to establish that the repair work has been effective.

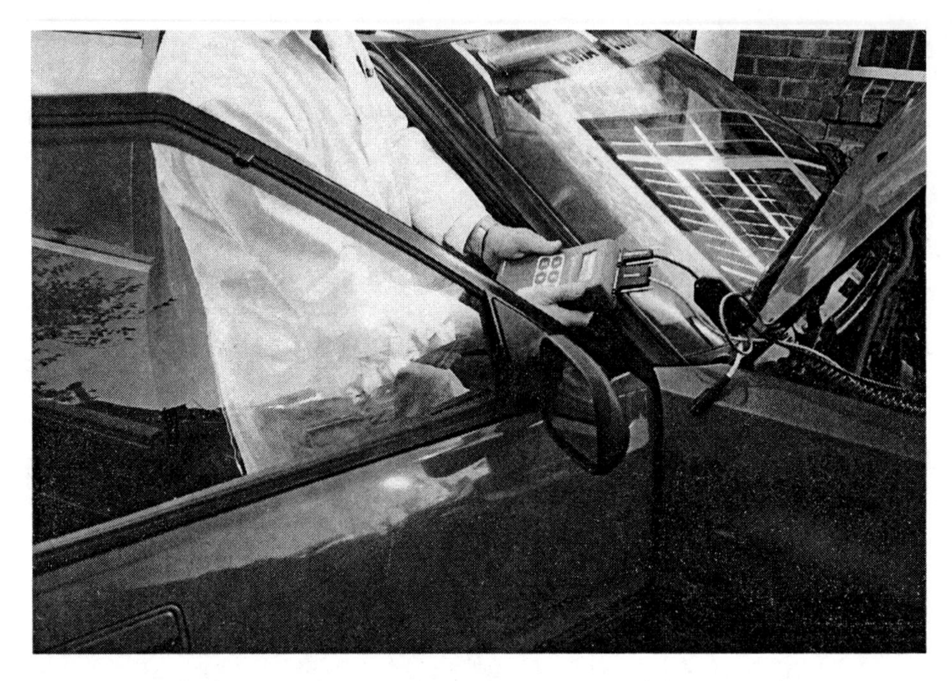

Fig. 3.19 Operating the vehicle controls during the tests

3.2 Developments in self-diagnosis

The above descriptions give a reasonable overview of the methods of 'reading' diagnostic trouble codes. I now propose to look at more recent developments that have arisen from legislation and advances in technology.

From the foregoing review of methods for accessing fault codes it will be evident that there are many different methods in use. In many cases, vehicle manufacturers have developed a version of diagnostic equipment that is unique to their range of vehicles. When an equipment manufacturer makes a piece of diagnostic equipment which has capabilities to perform diagnostic work on a range of vehicles, it is normally accompanied by an extensive range of leads and adapters so that it can be configured to suit a particular vehicle.

For several years now there have been arguments for and against bringing a degree of standardization into automotive computer applications, particularly in the area of access to diagnostic information. Generally speaking, there are two driving forces that cause changes to occur; one is legislation and the other is changes in technology. In the automotive field, as in other areas where computers are used, the technology has changed rapidly.

In the area of legislation, the world-wide concern with the effects of atmospheric pollution has had a major effect on vehicle design and, quite naturally, enforcement authorities are anxious to ensure that vehicles comply with the current emissions regulation. The result is that there have been major developments that cause vehicles in the USA to be equipped with a standard means of access to emissions related data and it is anticipated that Europe will follow a similar path in the near future. As the USA has a major influence on events in technology, it is to be expected that their developments will have an effect on vehicle technology in Europe and elsewhere. There is plenty of evidence to show that this is the case and readers should benefit from a review of some recent developments, such as on board diagnostics (OBD).

The term on board diagnostics refers to the self-diagnosing capabilities that are carried in the computers on the vehicle, and the aids that are provided to make the diagnostic data available to authorized users. Off-board diagnostics is equipment such as scan tools, oscilloscopes and other test equipment. In most cases both types of equipment are required for vehicle repair work. To date (January 2000) there have been two versions of OBD, i.e., OBD I and OBD II. Both apply to the USA, but introduction of similar legislation for Europe is imminent.

3.2.1 OBD I

This required vehicles produced from 1988 onwards to be equipped with electronically (computer) controlled systems that were capable of monitoring themselves. Any malfunction (defect) that affected exhaust emissions must be displayed on a warning lamp, known as the malfunction indicator lamp (MIL), on the dashboard. The malfunction must be stored in the ECM's memory and it must be readable with the aid of 'on board' facilities, e.g. a flash code on a lamp.

3.2.2 OBD II

OBD II strengthens the requirements of OBD I on vehicles of model year 1994 and afterwards. OBD II applies to spark-ignition cars and light vans, and from 1996 onwards to diesel-engined vehicles. The main features are that the following emissions related systems must be continuously monitored:

- combustion
- catalytic convertor
- oxygen (lambda) sensors
- secondary air system
- fuel evaporative control system
- exhaust gas recirculation system.

The requirements for diesel-engined vehicles vary and glow plug equipment may be monitored instead of the catalytic converter.

Features of OBD II are as follows.

- The malfunction indicator lamp (MIL) is provided with an additional 'flashing' function.
- The DTCs can be read out by a standard form of scan tool, via a standardized interface that uses a 16-pin diagnostic connector of the type shown in Fig. 3.20.
- The emissions-related components must be monitored for adherence to emissions limits in addition to monitoring them for defects.
- Operating conditions (performance data) can be logged and stored in a 'freeze frame'.

CARB plug:
Pin 7 and 15: Data transfer in accordance with DIN ISO 9141-2
Pin 2 and 10: Data transfer in accordance with SAE J 1962
Pin 1, 3, 6, 8, 9, 11-14 are not assigned to CARB. (OBD II data administration guideline "OBD II-DV")
Pin 4: Vehicle ground (body)
Pin 5: Signal ground
Pin 16: Battery positive

Fig. 3.20 The SAE J 1962 standardized diagnostic link connector (DLC)

The fault codes consist of five digits.

Example: P 0 2 8 3

Digit 1 identifies the vehicle system.
Digit 2 identifies the sub-group
Digit 3 identifies the sub-assembly
Digits 4 and 5 identify the localised system components.

Digit	Possibility	Meaning
1	B C P U	Body Chassis Drive/OBD II (Powertrain) Future systems (Undefined)
2	0 1 2 3	Fault code under SAE testing Fault code under manufacturer's testing Fault code under manufacturer's testing Reserved fault code
3	1 2 3 4 5 6 7	Fuel and air-addition measurement Fuel and air-addition measurement Ignition system Additional exhaust control Speed and idling regulation Computer and output signals Transmission
4 and 5	01 to 99	Designation of system components

Fig. 3.21 The structure of standard fault codes for OBD II

The malfunction indicator lamp (MIL) should light up when the ignition is first switched on and then go out after about 3 seconds, during which time the ECM is performing a series of 'self checks'. After this, when the engine is running, the MIL should only light up when a malfunction occurs. If the MIL does not light up when the ignition is first switched on, it is an indication that there is a fault in the MIL, or in the ECM itself, assuming that the battery is not flat.

From the repair shop point of view, OBD II provides some features which should produce benefits. Examples of such benefits are: (1) a standardized diagnostic interface and connector (see Fig. 3.20); and (2) standardized fault codes. The fault codes, as presented at the scan tool, comprise five digits, e.g. P0125. Digit 1, at the left-hand end, identifies the vehicle system. Digit 2 identifies the subgroup. Digit 3 identifies the subassembly. Digits 4 and 5 identify the localized system components.

Figure 3.21 shows how a range of fault codes can be constructed by using the recommended standard approach.

The example quoted, i.e., P0125, has the following meaning under this coding system: 'insufficient coolant temperature for closed loop fuel control'. There are many hundreds of codes and full details are given in the SAE J 2012 publication.

3.3 Diagnostic equipment and limitations of DTCs

The preceding sections of this chapter may give the impression that a completely new set of tools and equipment is needed to deal with OBD II and possibly EOBD (European on-board diagnostics). Fortunately this is not necessarily the case because diagnostic equipment, such as the Bosch KTS300 machine that is shown in Fig. 3.22 is capable of dealing with fault code retrieval, analysis and diagnostic work on non-OBD II systems that are equipped with an ISO 9141 serial link and, with the aid of suitable adaptors, OBD II systems also. This and similar equipment is described in greater detail in Chapter 4.

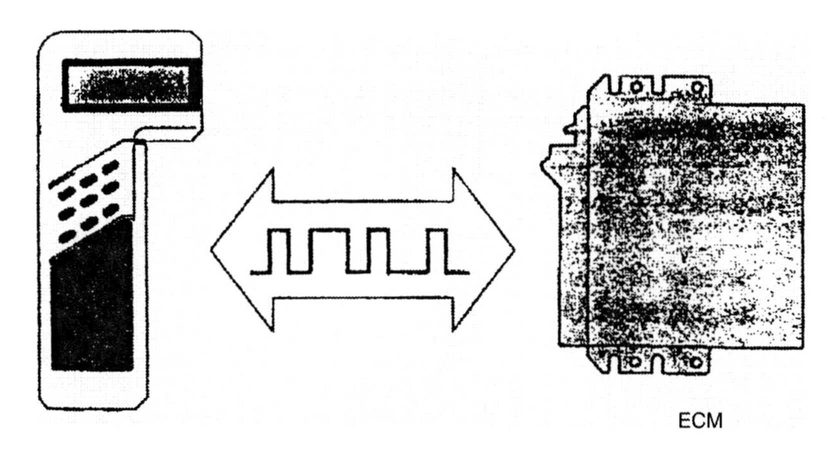

Fig. 3.22 The Bosch KTS300 portable diagnostic tool

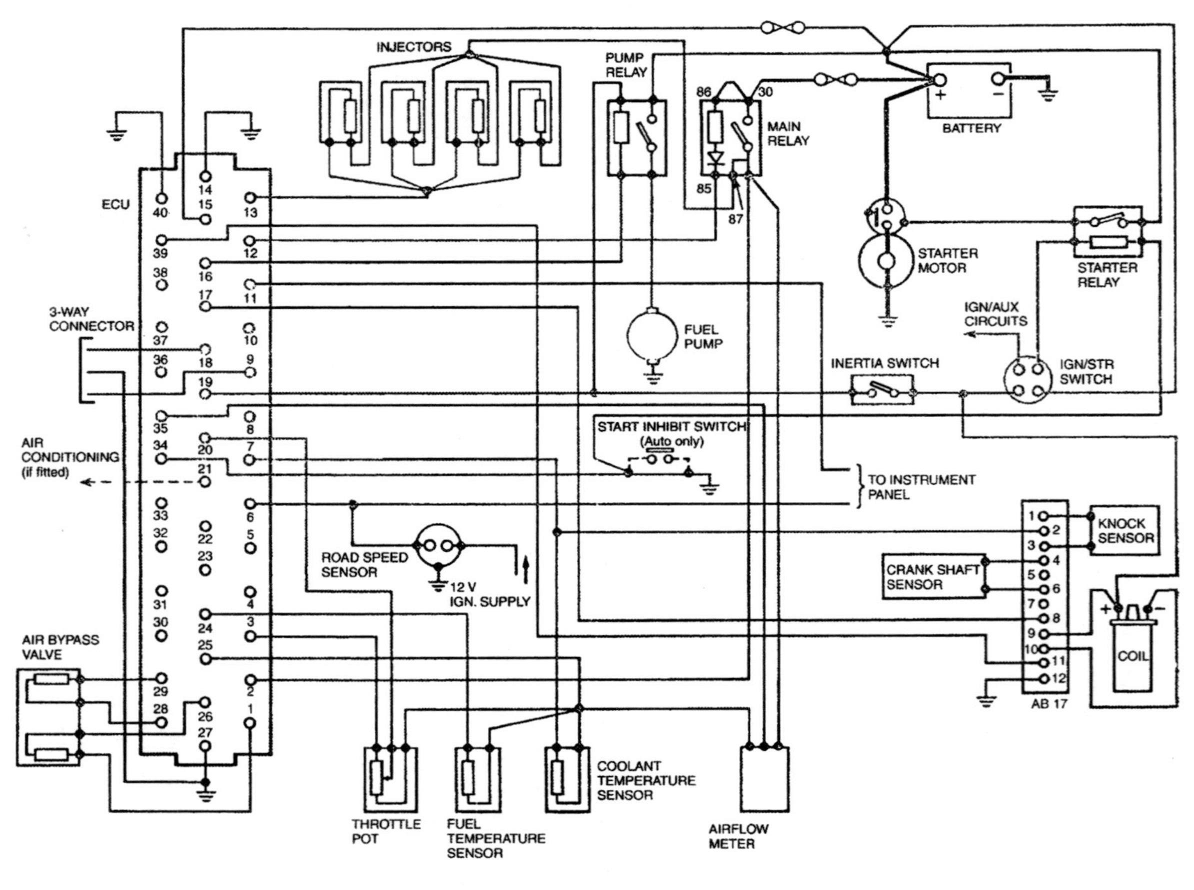

Fig. 3.23 Coolant temperature sensor circuit

If we take the above example of the coolant sensor and consider the implications of the code that tells us that there is a low coolant temperature and consider what might be involved, we should see that reading the DTCs is often one of several steps on the path to diagnosis and repair of defects. The output signal from the sensor is conveyed to the ECM via a cable which probably has at least two connectors on it as shown in Fig. 3.23. What the ECM reads is what is presented to it through the sensor circuit. If there is a defect in the sensor circuit, such as high resistance, this could cause the ECM to receive a voltage reading that represents the coolant temperature, which is lower than the actual coolant temperature.

This is true of practically all sensors, actuators and related circuits that form part of automotive computer controlled systems which the control computer (ECM) relies on for its operation. The DTCs are thus an aid to fault diagnosis; they rarely state exactly what the trouble is, and normally additional tools and equipment are needed in order to trace the causes of faults and effect a repair. This topic is covered in detail in Chapter 4.

3.4 Review questions (see Appendix 2 for answers)

1. Diagnostic trouble codes are:
   (a) computer codes that can be displayed at a fault code reader?
   (b) only readable through the serial connector of the ECM?
   (c) generated at random whenever there is a fault on any part of the vehicle?
   (d) information that tells the user exactly what the fault is?
2. A limited operating strategy:
   (a) permits the vehicle to be driven until a repair can be made?
   (b) cuts out fuel injection above a certain engine speed?
   (c) retards ignition timing to stop combustion knock?
   (d) refers to the limited nature of fault codes for diagnostic purposes?
3. Microprocessor based diagnostic testers can:
   (a) only read fault codes?
   (b) read fault codes and perform actuator tests via the serial port of the vehicle?
   (c) reset the values stored in a mask programmable ROM?
   (d) only read out diagnostic data from CAN systems?
4. The standardized serial port for diagnostics that is used with OBD II has:
   (a) a 3-pin connector?
   (b) no specified number of pins but its position on the vehicle is specified?
   (c) a 16-pin connector?
   (d) no pins specifically allocated to the OBD II emissions systems?
5. In order to read out diagnostic trouble codes (fault codes) it is necessary to:
   (a) earth the K line and read the flashing light?
   (b) carry out the manufacturer's recommended procedure?
   (c) have the engine running?
   (d) take the vehicle for a road test first?

6. An ABS ECM should have good self-diagnostics because:
   (a) the sensor output signals cannot be measured independently?
   (b) it is difficult to simulate actual anti-lock conditions with a stationary vehicle?
   (c) if the ABS warning light comes on it will stop the vehicle?
   (d) the fault codes are always stored in an EEPROM?
7. A 'freeze frame' is:
   (a) a set of ROM data that is used in very cold weather?
   (b) data that is used by the ECM when there is an emergency?
   (c) a set of data about operating conditions that is placed in the fault code memory when the self-diagnostics detects a fault?
   (d) a diagnostic feature of very early types of electronic control only?
8. In standardized fault codes:
   (a) digit 1, at the left-hand end, identifies the vehicle system?
   (b) the digit at the far-right hand end identifies the system?
   (c) all computer controlled vehicle systems must use them?
   (d) the identifying digits can appear in any order?

4
Diagnostic tools and equipment

This chapter covers the types of tools and equipment and other aids that are available to assist technicians to perform accurate and efficient diagnosis and repair of automotive systems. It is not intended to be a catalogue, because there is a large number of companies that make and supply equipment. Addresses of several of these companies are given in the Appendix. Where a particular make of tool has been selected as a case study it is because the makers have supplied the information that enables me to provide the type of description that is suitable for this book, and not because I am expressing a preference for any particular make of tool.

4.1 Diagnostic tools that connect to the ECM

As stated in Chapter 3, the control computer ECM has considerable self-diagnostic power and the DTCs that may be stored in its memory are a valuable source of diagnostic information. The fact that the ECM processor is constantly monitoring inputs and outputs, means that the data that it uses can also be used by any other computer that is using the same protocols (language). This is a facility that microprocessor-based diagnostic tools possess and it enables them to be used to read system behaviour while the system is in operation and also to capture and save data for detailed analysis. In order for this to happen the ECM must be equipped with a suitable serial diagnostic interface that will permit two-way communication between the test equipment and the ECM (Fig. 2.5 shows the general principle of a serial data connector).

As pointed out in Chapter 3, there is movement towards widespread adoption of the USA OBD II and California Air Resources Board (CARB) standard, but this is not universal and the established standard of ISO 9141 is still to be found in UK and European systems. The ISO 9141 standard permits a scan tool to be connected to the diagnostic plug and with the aid of suitable software, usually on a smart card, enables a technician to access fault codes and other data. The Lucas Laser 2000 machine, as shown in Fig. 4.1, is an example of diagnostic equipment that can be used for a range of diagnostic work.

Fig. 4.1 Laser 2000 uses the on-board diagnostic facilities incorporated in the ECU's controlling vehicle systems. The sophistication of these facilities varies considerably across vehicle manufacturers

Various types of work can be performed with the aid of this machine as follows.

- Reading out fault codes and explanatory text.
- Monitoring (reading) live data as the system is in operation, and displaying the data as bar charts. Several different variables and parameters may be selected and displayed simultaneously for comparison and to aid analysis.
- Data storage. Up to four sets of data can be stored together with time and date of storage.
- Snap shot mode. When in use, during a road test to locate intermittent faults, the machine can be triggered so that data for a period of time before the fault occurred and a period of time after is recorded. This data can then be reviewed on screen so that the traces can be examined for any abnormalities. This aspect of use is covered in Chapter 7.
- Actuator operation. Microprocessor-based machines, such as Lucas Laser 2000, can be used to activate injectors and other devices so that they can be checked independently.

4.1.1 ACCESSING DIAGNOSTIC DATA WHEN THERE IS NO SERIAL PORT

In the UK, the average life of a vehicle is approximately 10 years and this means that vehicle technicians are likely to encounter a range of computer controlled automotive systems, some of which may not have a serial diagnostic port. To provide for this contingency, the Lucas Laser machine can be adapted for use on computer controlled systems that do not have a serial port by the use of an additional piece of equipment, known as the Lucas Laser 1500. This piece of equipment is shown in Fig. 4.2 and can perform the following functions.

- The Laser 1500 is a data acquisition unit which may be connected between the ECM and the main harness connector to the ECM.
- Once the interface is secure, the Laser 2000 machine is connected to the Laser 1500 and the computer controlled system can be subjected to a similar range of tests to those that have a serial port. System specific adapter cables are required together with overlays that make the Laser 1500 adaptable to a range of systems. Figure 4.2 shows the general principle.

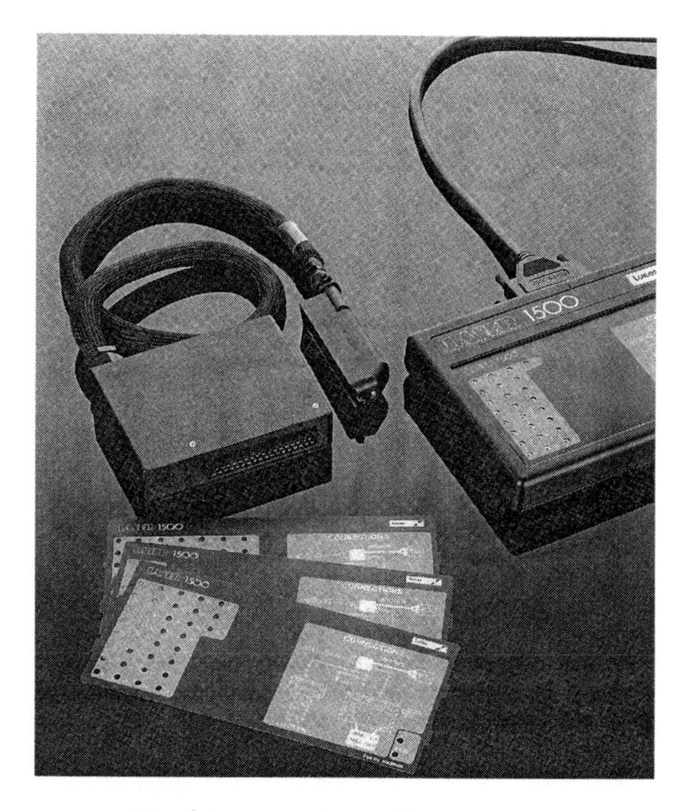

Fig. 4.2 Lucas Laser 1500 equipment

4.1.2 OBD II TYPE DIAGNOSTIC EQUIPMENT

Much of the diagnostic capability that is known as OBD II has been in use for some time. The main effect of the OBD II legislation seems to have been to focus attention on the desirability of standardization, particularly in the areas of means of access to diagnostic data on vehicle systems and the availability of supporting documentation. This means that some of the techniques described under this heading will, in some cases, be available on non-OBD II systems.

Firstly the principal features that relate to OBD II are outlined. This is followed by a description of the equipment and its diagnostic functions. The principal features are as follows.

1. A standard 16-pin SAE J 1962 diagnostics plug-in point, described in Chapter 3.
2. The diagnostic plug should be accessible from the driving seat and the preferred location is as shown in Fig. 4.3.

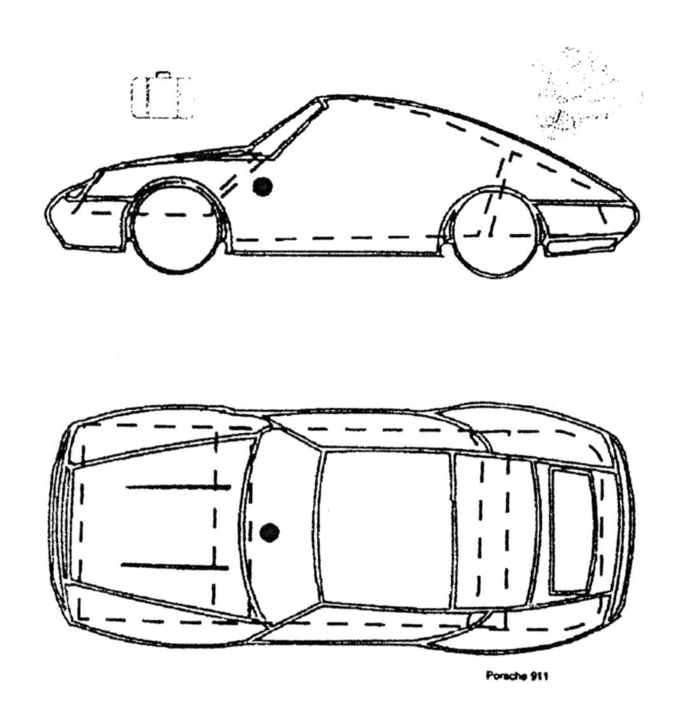

Fig. 4.3 The preferred location of the SAE J 1962 diagnostic interface

3. The communication between the scan tool and the ECM takes place according to one of a small number of protocols.
4. Initializing (starting) the diagnostic communication takes place via the diagnostics equipment as follows: hexadecimal 33 is signalled to the ECM, by the scan tool at a transmission speed of 5 binary bits per second, i.e., 5 baud.

5. The ECM then sends a 'header label' to the scan tool in response to the intialization prompt. This header label consists of information about the baud rate and two keywords.
6. To check that the communication is correctly set up, the scan tool inverts the second keyword and sends it back to the ECM. (Inversion in binary language means turning 0s into 1s and vice versa.)
7. The ECM sends the inverted memory address (hexadecimal 33) back to the scan tool.

As far as actual practice is concerned, the user merely prepares for the test. The above procedure takes place automatically via the control buttons on the scan tool.

Some of the main points about the OBD II test tool are that it must:

- automatically recognize the type of data transfer used by the engine control system that is being checked;
- display any fault codes relating to emissions;
- display current (live) values relevant to exhaust emissions;
- be able to erase the fault codes;
- contain an 'on-line' help facility that can be called up from the instrument panel.

Readers who are familiar with scan tools will recognize that many of these features have been available on scan tools for some years. As modern vehicles are frequently equipped with several computer (ECM) controlled systems a technician will expect a scan tool to be able to perform diagnostic functions on a range of systems. The additional pins on the 16-pin diagnostic interface permit access to other systems on the vehicle.

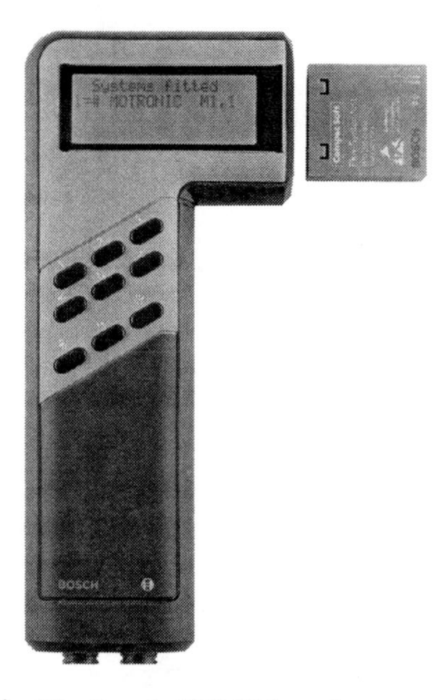

Fig. 4.4 The Bosch KTS 300 pocket system tester

Figure 4.4 shows the Bosch KTS 300. For OBD II purposes, the KTS 300 pocket system tester is supplied with two leads, as shown in Fig. 4.5. One of these leads (Bosch reference 1 684 463 361) provides connection to the control units related to exhaust emissions. The other lead, which includes the adapter box shown in Fig. 4.6, provides the diagnostic link to other control units such as ABS and transmission control etc.

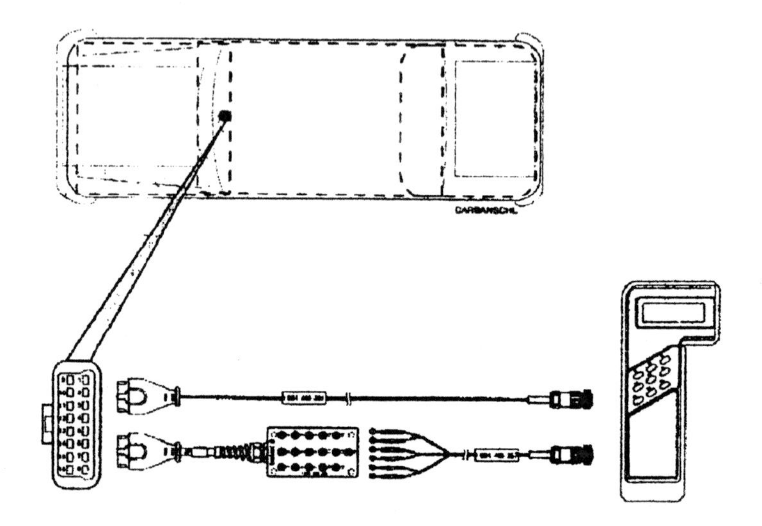

Fig. 4.5 The leads for the Bosch KTS 300 pocket system tester

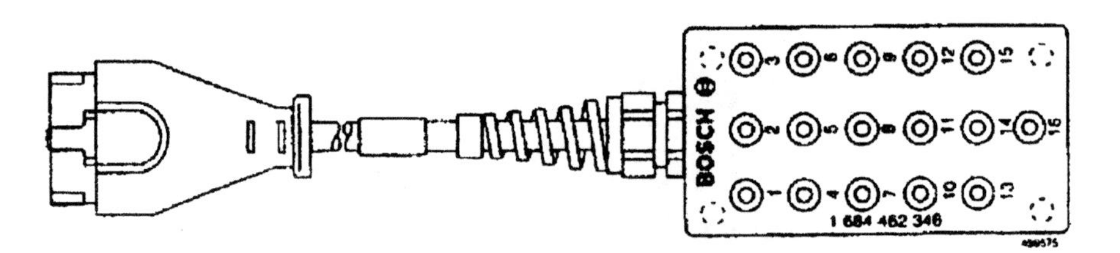

The ends of the adapter box are attached to sleeves into which, depending on the pin assignment of the CARB plug, the banana plugs of the universal line are inserted.

Fig. 4.6 The CARB adaptor box

When the test instrument is connected and diagnostic communication is established, the instrument guides the user through the test procedure. In addition to the OBD II functions, the KTS 300 tester can also be used on vehicle systems that are provided with an ISO 9141 diagnostic connector. The extent of the diagnostics that can be performed on these systems is dependent on the vehicle that is being tested and the availability of diagnostic data. Vehicle specific connector leads are also required.

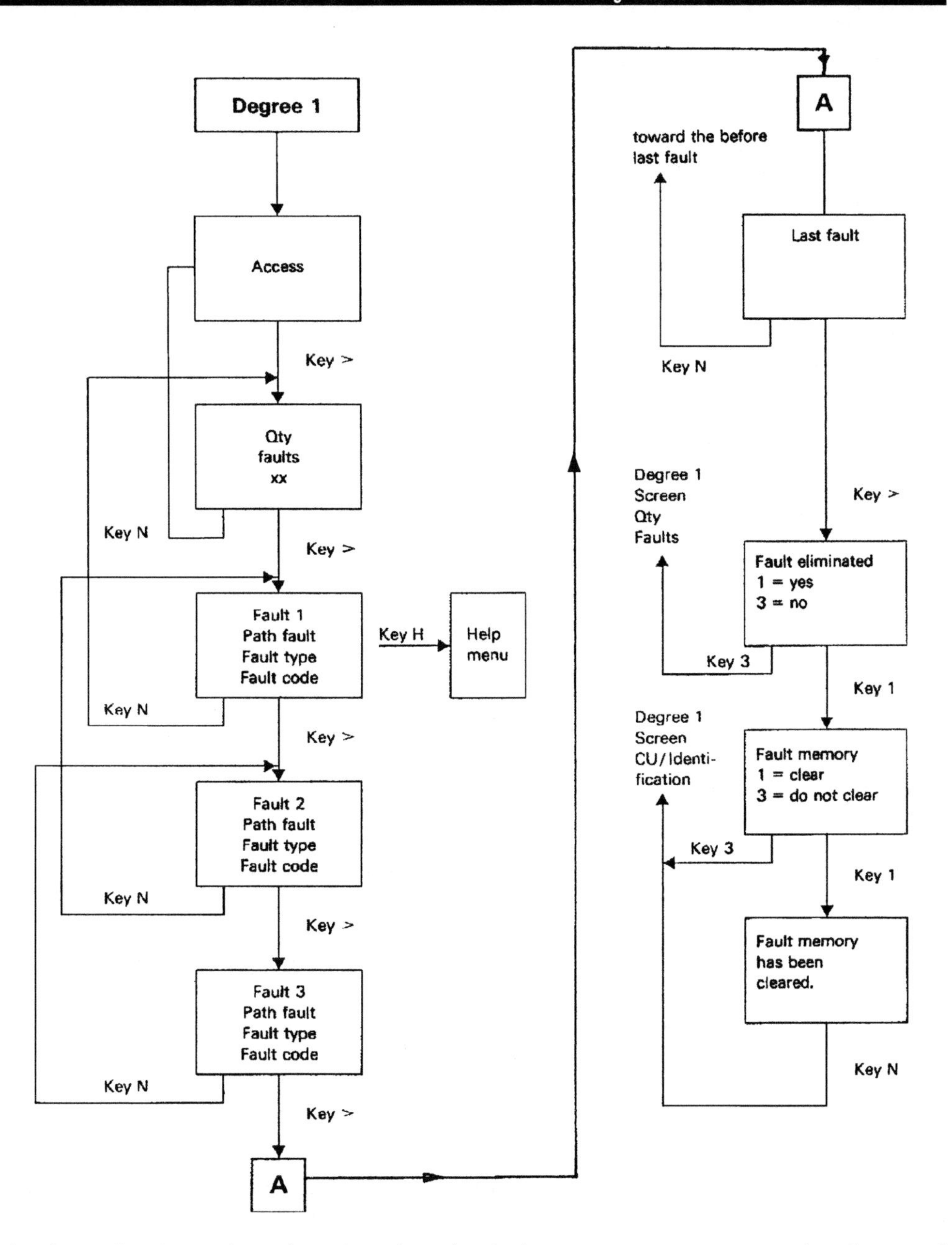

Fig. 4.7 The flow chart that describes the fault memory processing under degree of extension 1

The software is described as having two degrees of extension. With degree of extension 1, the KTS 300 provides for the selection of the fault code memory and the 'help' menu, processing the communications between the tester and the ECM. The flow chart shown in Fig. 4.7 shows what is available under degree of extension 1. (Where 'key' is stated it means that this is the key to press to move to the next step.) From the flow chart it may be seen that fault codes (DTCs) are 'read out' and cleared, as required. Pressing the '>' key takes the user forward to the next screen menu and pressing the 'N' key takes the user back to the beginning of the previous step.

With degree of extension 2, the user is presented with three options as shown in Fig. 4.8. Option 1 gives the fault memory processing capacity of degree of extension 1, plus extra items. Options 2 and 3 give access to screen menus that provide considerable diagnostic capacity.

If you examine the flow chart shown in Fig. 4.9, you will gain an impression of the diagnostic work that can be performed under option 1. For example, if the

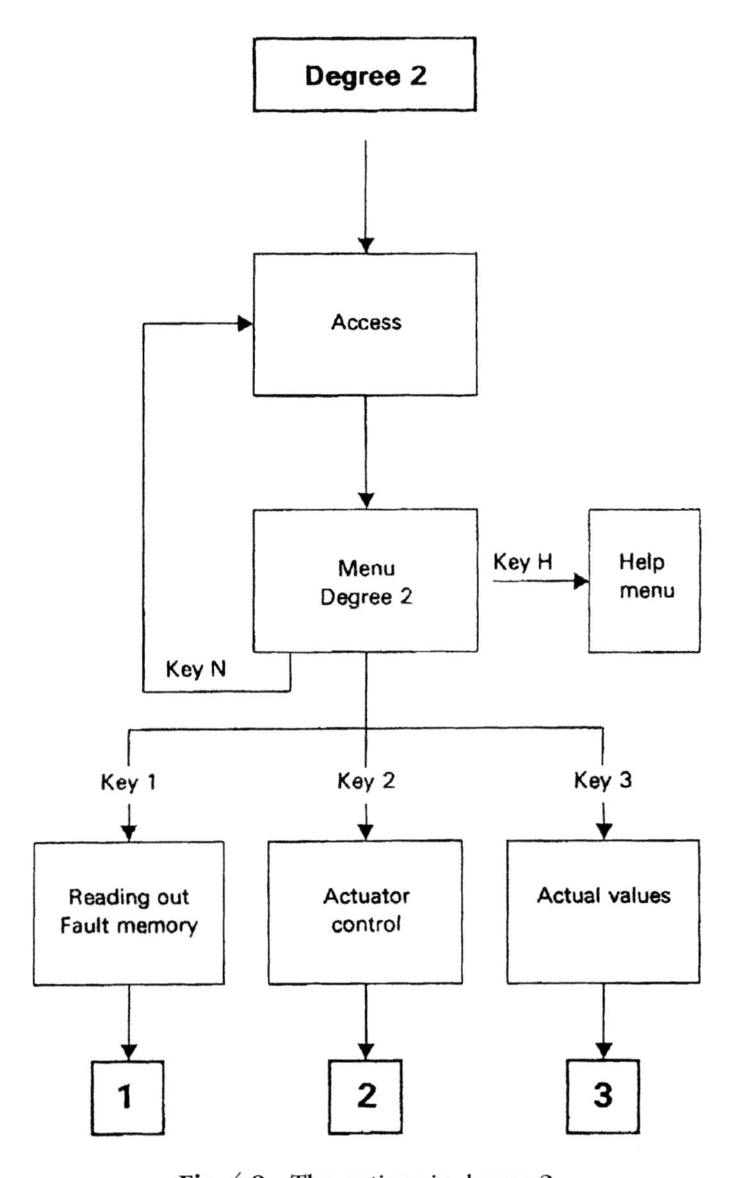

Fig. 4.8 The options in degree 2

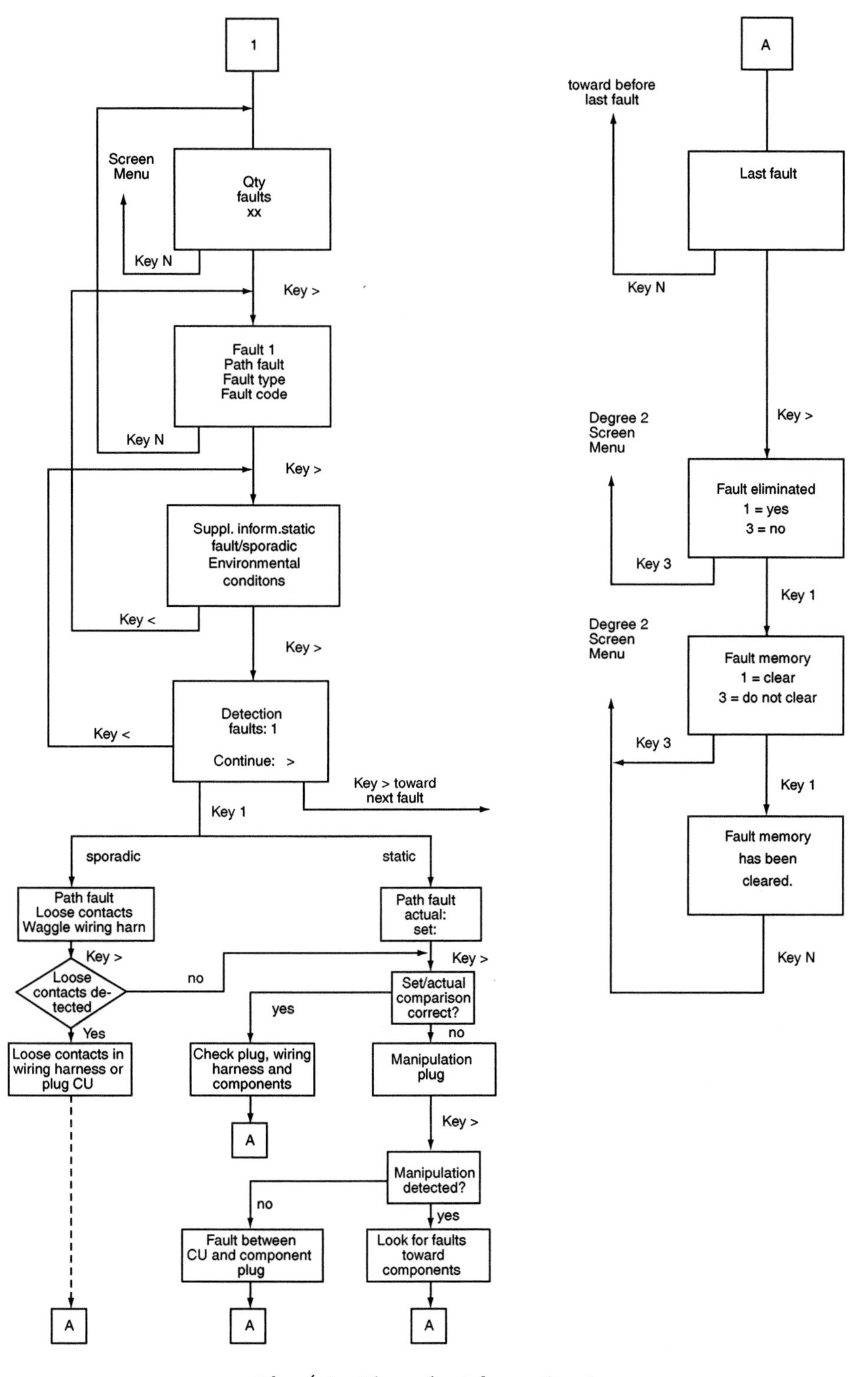

Fig. 4.9 Flow chart for option 1

fault is sporadic (intermittent) it is possible to 'wiggle' the connectors. If loose connections are detected, the '>' key will direct the user down the dotted line route to A and then to A at the top right-hand side and so on, through the steps on the flow chart.

Option 2 permits the KTS tester to perform a range of actuator tests.

Option 3 permits actual component values to be read from the ECM, e.g. duty cycle, injection time etc., and compare them with values that are stored in the KTS 300 software.

4.2 Breakout boxes

The breakout box is a diagnostic aid that is normally connected to the main ECM harness connector as shown in Fig. 4.10.

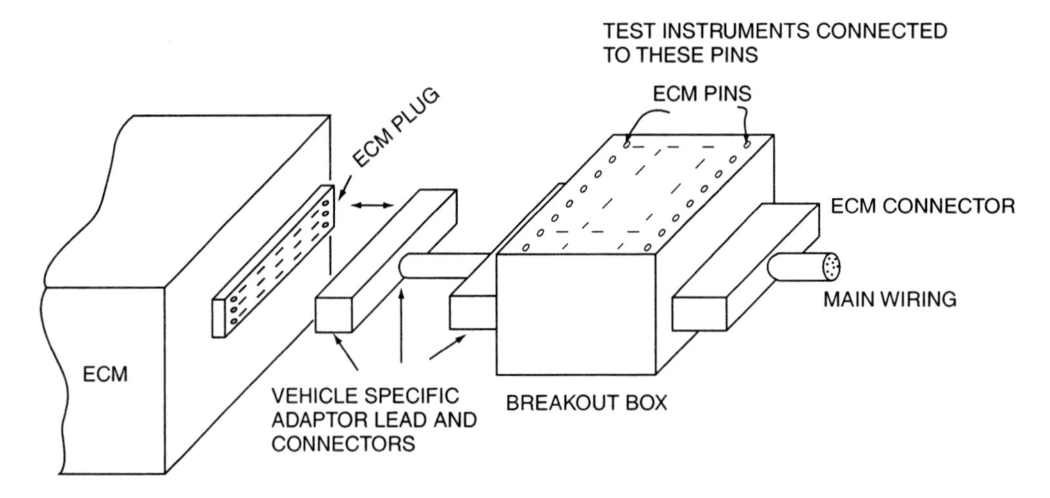

Fig. 4.10 A breakout box

When the box is inserted it permits electrical access to most elements, such as sensors actuators and circuits, that make up the system that is being controlled by the ECM in question. A major advantage of the breakout box is that it permits resistance, continuity and dynamic tests of sensor and actuator circuits to be performed more easily because it overcomes the need to access the contacts inside the harness to ECM connector.

The breakout box's electrical connections are numbered to match those in the ECM connector, and instructions for performing circuit tests frequently refer to ECM pin numbers. For example, on a certain system the resistance reading between pins 12 and 22 should be 750 ohms. This would require the ohmmeter leads to be connected between connectors 12 and 22 on the breakout box - with the ignition switched off, of course, and the breakout box disconnected from the ECM. The reading of 750 ohms should then be obtained. The action to take if it is not, is considered in Chapter 7.

Some breakout boxes are designed for resistance checks only. In such cases the the main harness connector is removed from the ECM and is then connected to the breakout box with no additional connection to the ECM.

4.3 The digital multimeter

In section 3.3, the limitations of fault codes are described and an indication is given of the extra work that may be required to locate the cause of a defect. A digital multimeter is a valuable asset in this additional diagnostic work. Digital meters are preferred for work on electronic systems because they have a high impedance (internal resistance) and this prevents them from placing a load on electronic devices.

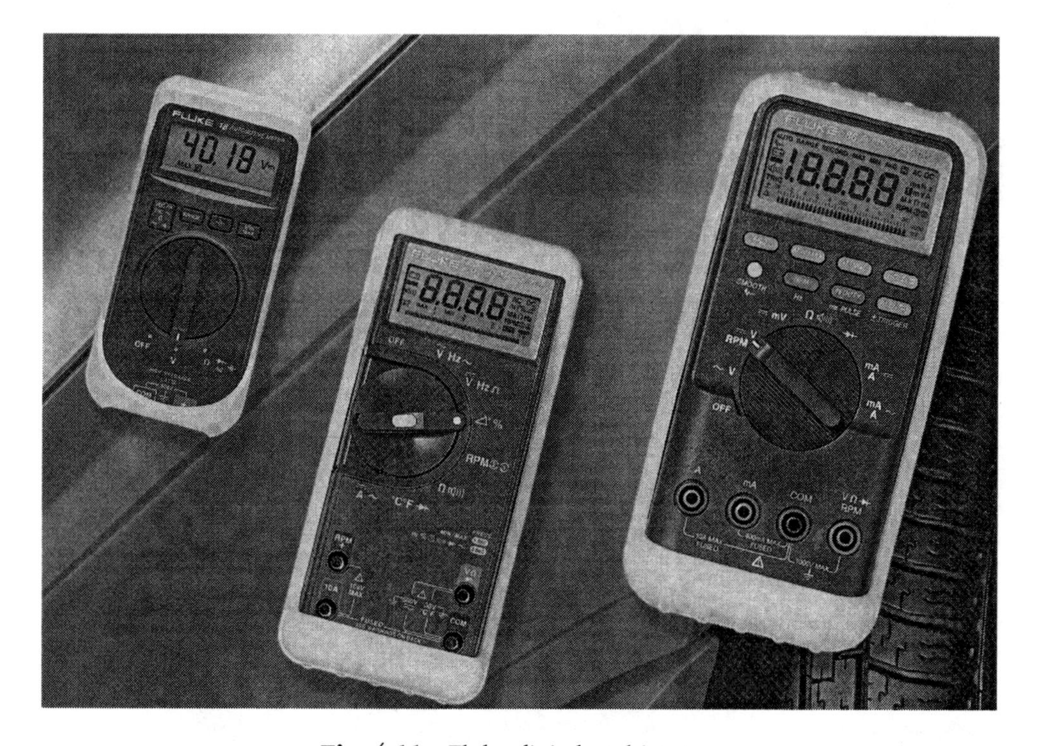

Fig. 4.11 Fluke digital multimeters

The 'Fluke 78' shown in Fig. 4.11 is an example of a digital multimeter that has been specially developed for work on automotive systems. Some uses of this type of meter are covered in later chapters.

These particular multimeters are supported by a range of test leads and adaptors that are shown in Fig. 4.12. It should be noted that test connections need to be securely made to ensure that good electrical contact is made. Test leads should also be supported so that they do not drop off during testing, or get tangled up in moving and/or hot parts.

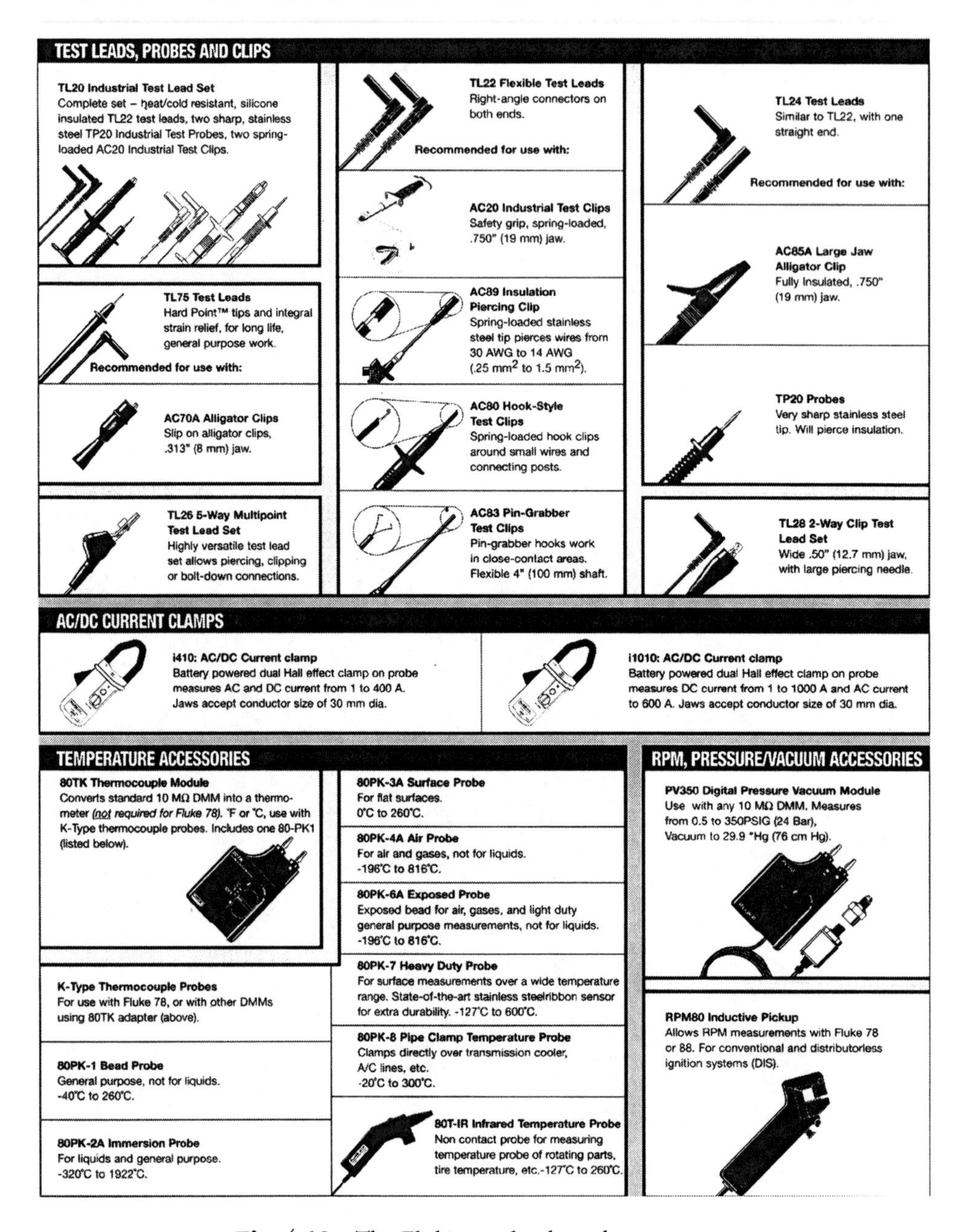

Fig. 4.12 The Fluke test leads and connectors

4.4 Portable flat screen oscilloscopes

Oscilloscopes of the cathode ray type have long been part of the equipment for workshop-based diagnostic work. Recent developments in liquid crystal displays (LCDs) and thin film transistor (TFT) technology have made small portable oscilloscopes possible and they are available from several suppliers at reasonable cost. For electronic system diagnosis they have considerable value because they are versatile and can measure sensor and other components' performance very accurately.

Fig. 4.13(a) The Bosch PMS 100 portable oscilloscope

The majority of the scope patterns that appear in Chapters 4 and 5 are derived from the Bosch PMS 100 oscilloscope, shown in Fig. 4.13(a). Figure 4.13(b) shows some screen menus for this scope and these give a good indication of the types of test that the instrument can be used for. Other oscilloscopes, such as the Lucas YWB 220 and the Crypton CPT 50, can be expected to perform a similar range of tests.

4.5 Diagnostic tool and oscilloscope combined

The Bosch KTS 500 is an example of a recent development that puts even greater diagnostic power at the disposal of the technician. This instrument is plugged into the diagnostic connector as shown in Fig. 4.14.

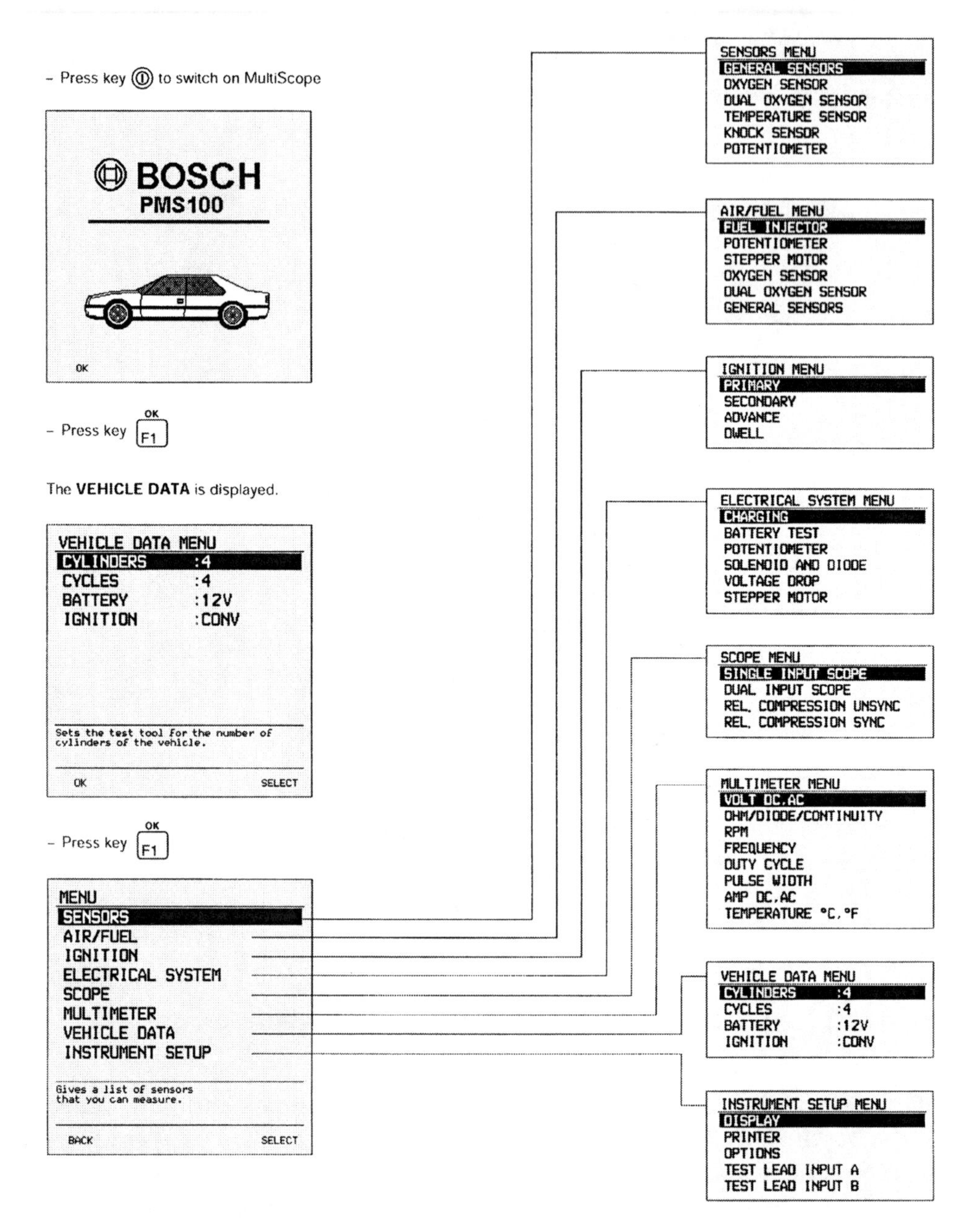

Fig. 4.13(b) An overview of available test functions from the MENU key of the Bosch PMS 100 portable oscilloscope

This instrument is backed up by fault finding aids that can be viewed on screen, e.g. trouble shooting and repair guides, circuit diagrams etc. The instrument has a large memory capacity and it can display live data and also store live data 'captured' during a road test and display it on screen, or print it out for analysis in the workshop. The chart in Fig. 4.15 shows how vehicle identification followed

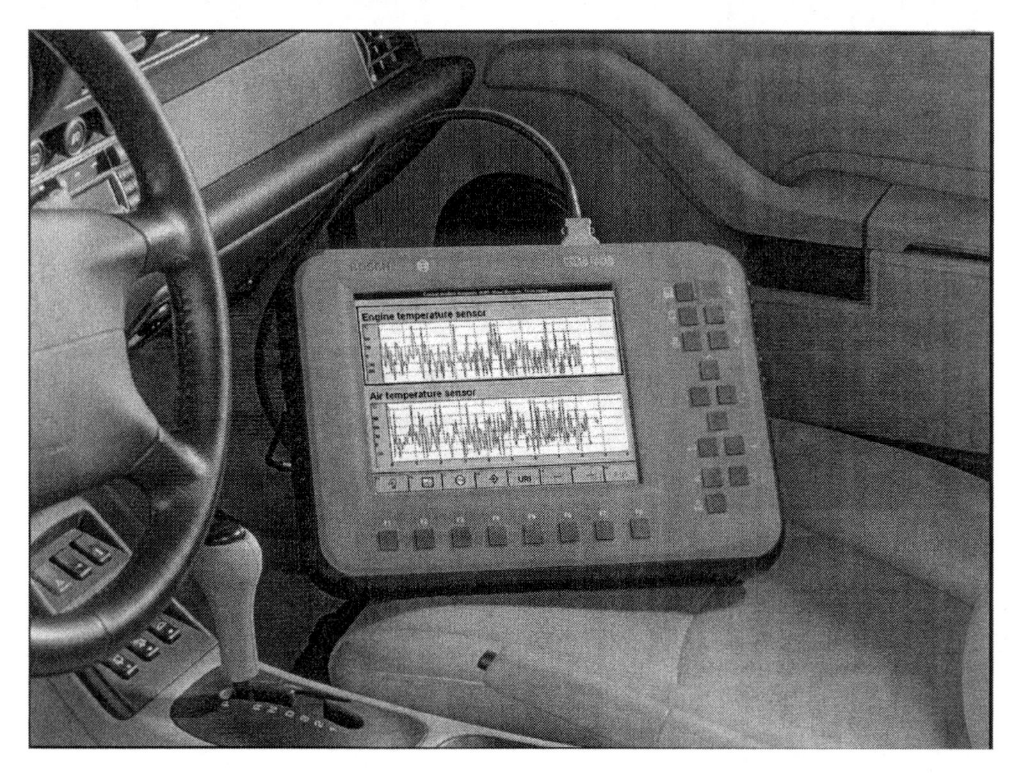

Fig. 4.14 The Bosch KTS 500 control unit diagnosis tester

by fault code read-out can be followed-up by taking readings of actual values. The use of data from this machine is discussed further in Chapter 7.

4.6 Pressure gauges

A good quality, accurate pressure gauge is an aid that is frequently used in diagnostic and routine maintenance work. An example of such use is the checking of fuel gallery pressure as shown in Fig. 4.16. The fuel pressure is a critical factor in determining the amount of fuel that passes through an injector nozzle, in a given time. Fuel gallery pressure is quite high, often several bar, and the procedure for connecting the gauge safely varies considerably from one make of vehicle to another. For this reason I do not propose to describe details of the test, except to emphasize that it is important to follow the manufacturers' prescribed procedure.

4.6.1 VACUUM PUMPS AND GAUGES

A vacuum pump and gauge of the type shown in Fig. 4.17 can be used for a range of purposes, such as checking a MAP sensor performance, EGR valves and other devices that are vacuum operated.

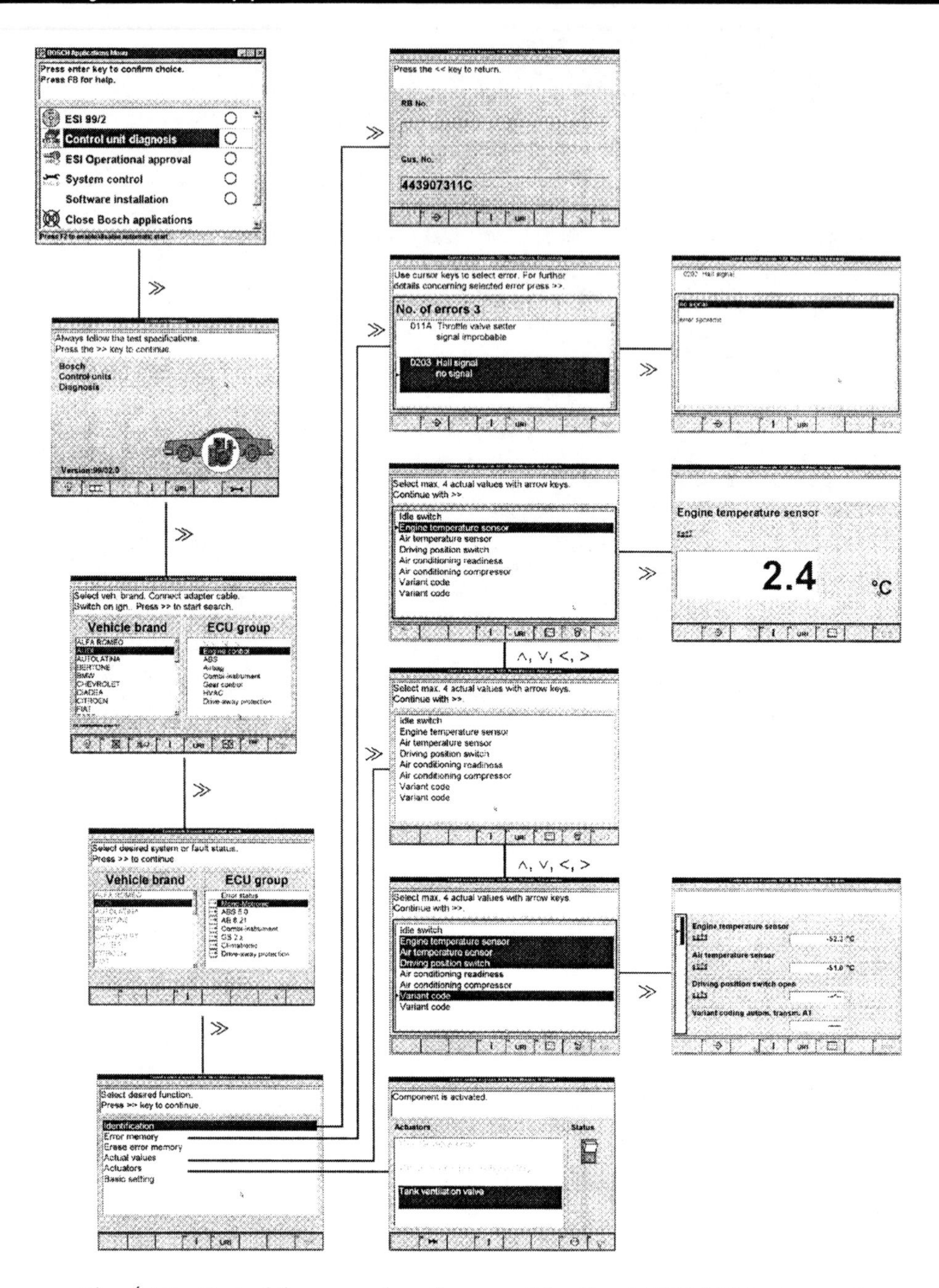

Fig. 4.15 Part of the procedure for use of the Bosch KTS 500 instrument

When used to check the performance of a manifold absolute pressure sensor, or an exhaust gas recirculation valve, the vacuum pump is used to replace the manifold vacuum, as shown in Fig. 4.18.

When the vacuum gauge and pump are used to test a MAP sensor, the vacuum pipe from the inlet manifold is removed and the pump and gauge are connected to

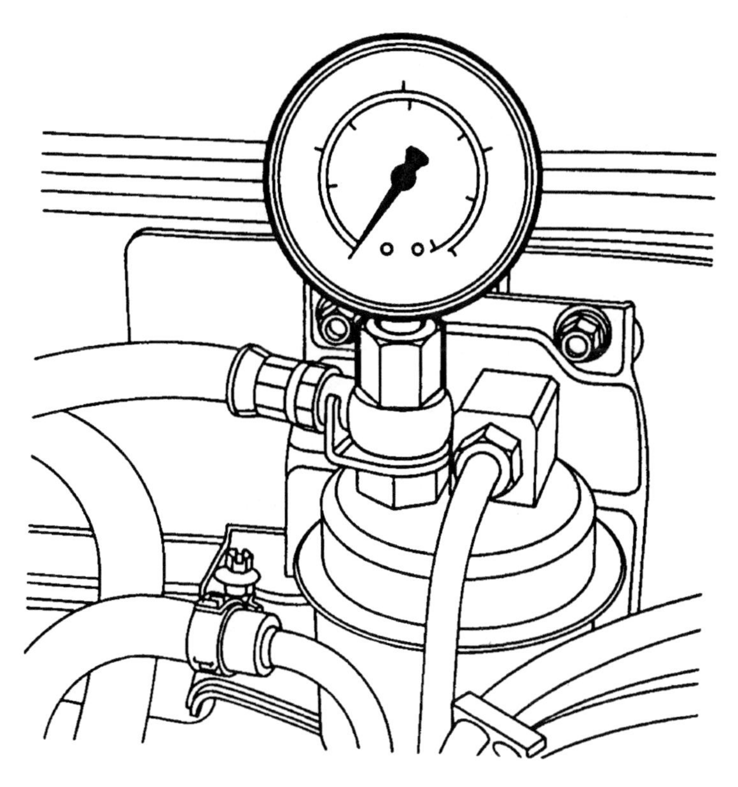

Fig. 4.16 Pressure gauge and fuel gallery pressure check

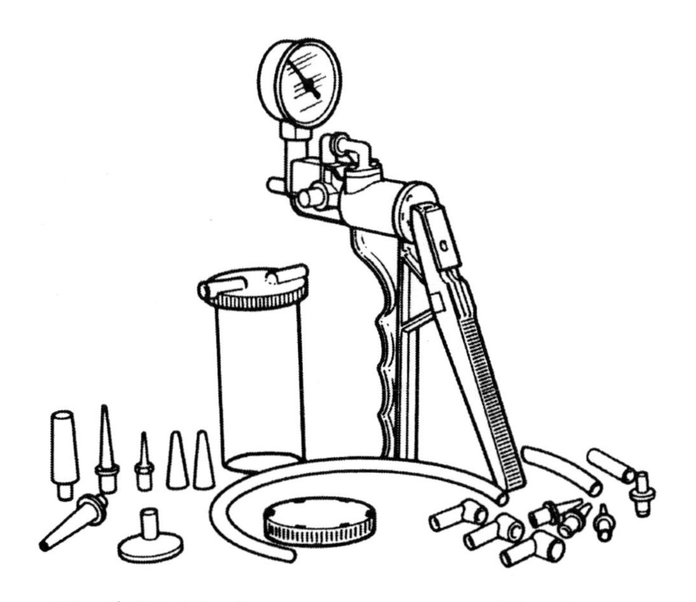

Fig. 4.17 The Lucas vacuum gauge and its adaptors

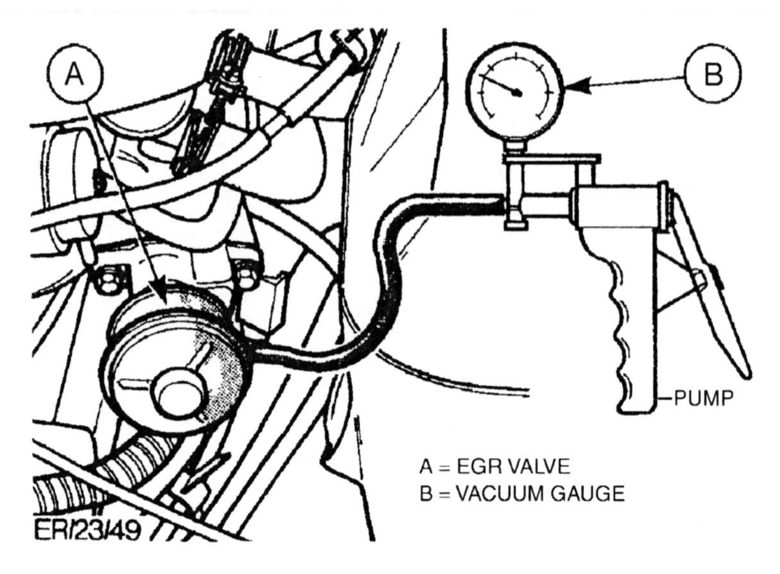

Fig. 4.18 Using a vacuum pump to replace manifold vacuum in a test on an EGR. valve

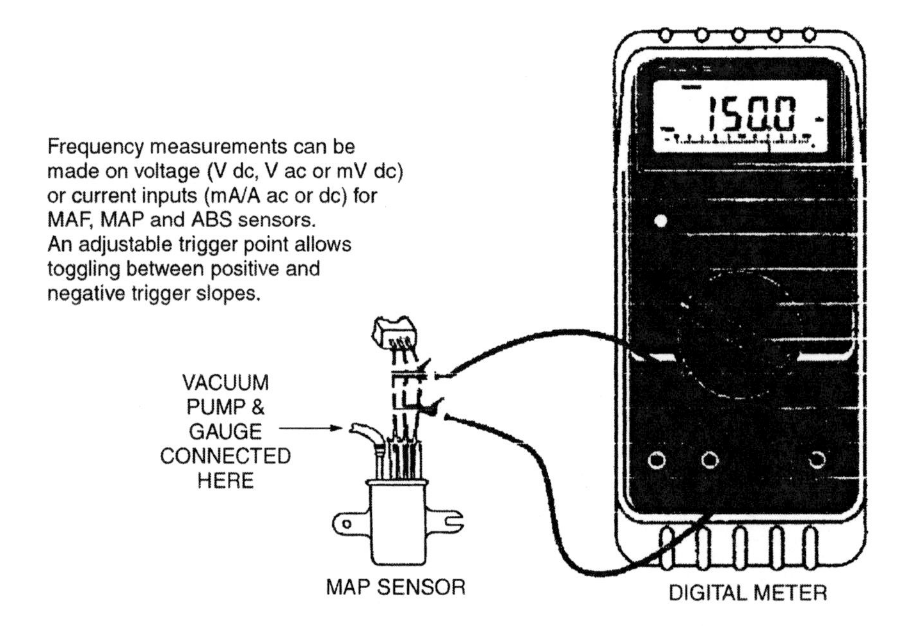

Fig. 4.19 A manifold absolute pressure sensor test

the vacuum connection of the sensor. The meter is then connected to the signal wire of the sensor with electrical energy supplied but the engine not running. Figure 4.19 gives an impression of the set-up.

Starting with zero vacuum, i.e., at atmospheric pressure, the frequency reading should be written down. The vacuum should then be increased in steps of 25 mm of mercury (vacuum gauge reading) and the frequency output at each step noted until the maximum vacuum of approximately 750 mm of mercury is reached. The figures can then be compared with those that relate to the sensor under test. The frequency normally lies in a range of approximately 70 - 150 Hz, where the lower reading applies to maximum vacuum.

4.7 Calibrating test instruments

One should be reasonably certain that a new instrument will be accurate. However, during the course of normal use, it is not uncommon for instruments to become less accurate and it is good practice to check instrument readings periodically against 'known good' ones. This process which is known as 'calibration' should be performed on all measuring instruments at regular intervals. Figure 4.20 shows the procedure for calibrating the test leads of the Bosch PMS 100 oscilloscope.

The supplier of any test instrument will normally provide a calibration service and often this service is included in the equipment service engineer's routine.

4.8 Location charts and wiring diagrams

A location chart of the type shown in Fig. 4.21 shows the position on a vehicle where the individual sensors and actuators etc. are situated. Such charts are useful when making visual inspections and also when locating the device to check connections or to back probe for test purposes.

Circuit diagrams, or wiring diagrams as they are commonly known, are an essential aid to checking circuits. Figure 4.22 shows a typical wiring diagram. Many modern items of diagnostic equipment now include location charts and circuit diagrams in the software, so that 'on screen' displays are readily to hand.

4.9 Sources of diagnostic data

Manufacturers and suppliers of the type of equipment described in this chapter normally provide a service that ensures that equipment users have a ready supply of diagnostic data relating to a range of vehicles. In some cases a subscription ensures that updates are provided as and when required. In addition, most companies provide a 'hotline' service to help with diagnostic problems. A number of publishing companies also produce books of data and fault codes and these are readily available from companies such as those listed in the Appendix.

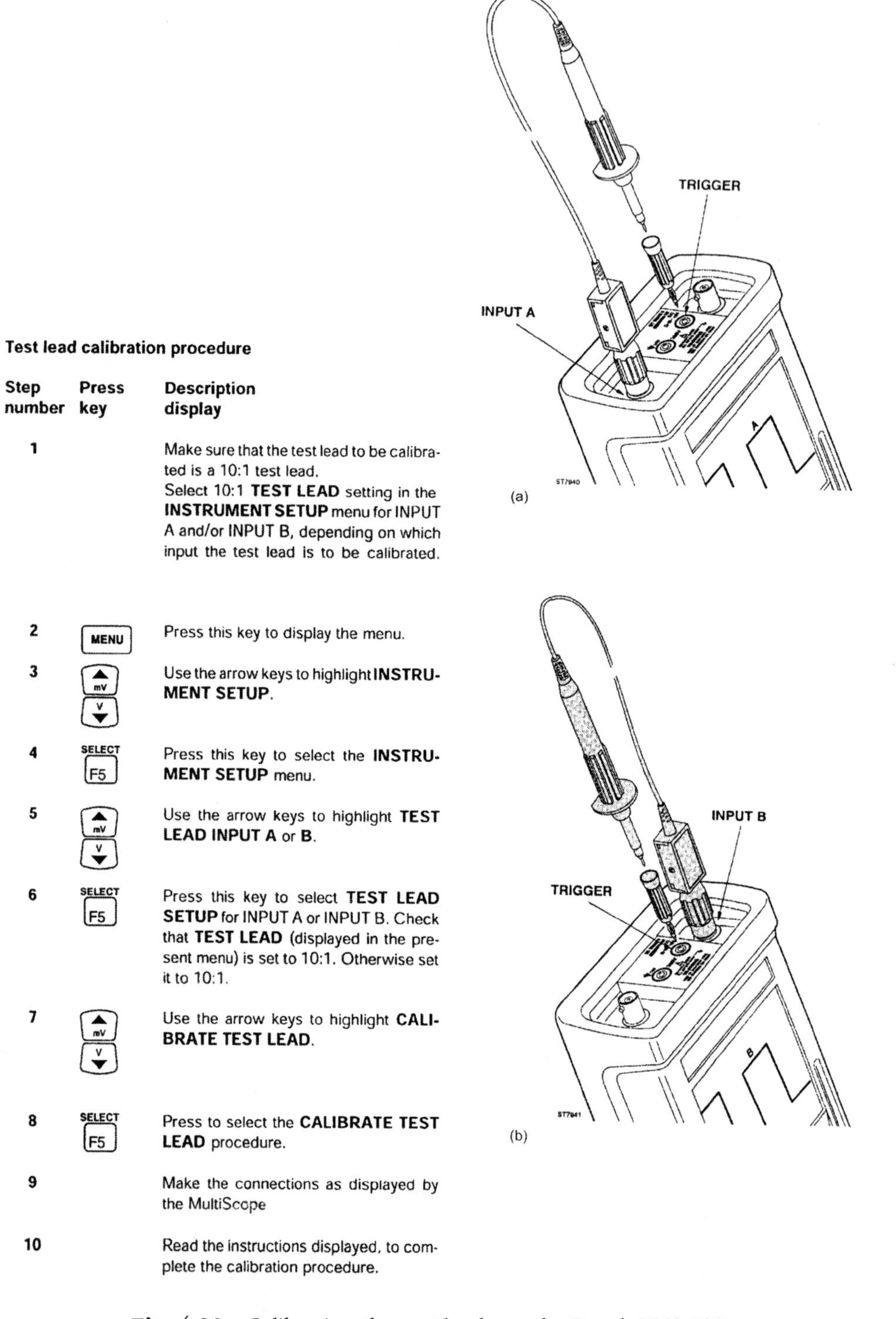

Test lead calibration procedure

Step number	Press key	Description display
1		Make sure that the test lead to be calibrated is a 10:1 test lead. Select 10:1 **TEST LEAD** setting in the **INSTRUMENT SETUP** menu for INPUT A and/or INPUT B, depending on which input the test lead is to be calibrated.
2	MENU	Press this key to display the menu.
3	▲ mV / V ▼	Use the arrow keys to highlight **INSTRUMENT SETUP**.
4	SELECT F5	Press this key to select the **INSTRUMENT SETUP** menu.
5	▲ mV / V ▼	Use the arrow keys to highlight **TEST LEAD INPUT A** or **B**.
6	SELECT F5	Press this key to select **TEST LEAD SETUP** for INPUT A or INPUT B. Check that **TEST LEAD** (displayed in the present menu) is set to 10:1. Otherwise set it to 10:1.
7	▲ mV / V ▼	Use the arrow keys to highlight **CALIBRATE TEST LEAD**.
8	SELECT F5	Press to select the **CALIBRATE TEST LEAD** procedure.
9		Make the connections as displayed by the MultiScope
10		Read the instructions displayed, to complete the calibration procedure.

Fig. 4.20 Calibrating the test leads on the Bosch PMS 100

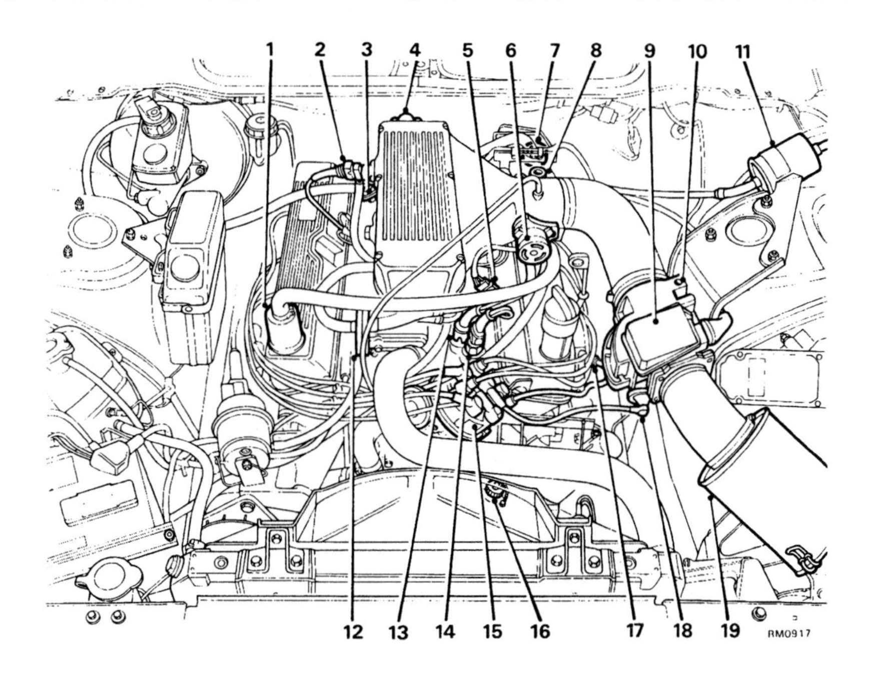

1. Breather flame trap
2. Cold start injector
3. Fuel pressure regulator
4. Overrun valve
5. Throttle potentiometer connection
6. Throttle potentiometer
7. Engine air breather
8. Idle speed adjustment screw
9. Air-flow meter
10. Idle mixture adjustment screw
11. Fuel filter
12. Extra air valve
13. Coolant temperature switch
14. Thermo time switch
15. Distributor
16. Diagnostic plug
17. Spark plug—No. 1 cylinder
18. Ignition coil
19. Air cleaner

Fig. 4.21 A typical location chart

4.10 Exhaust gas emissions and emission system testing

4.10.1 PETROL ENGINE EMISSIONS

Petrol is a hydrocarbon fuel that is composed of approximately 85% carbon and 15% hydrogen, by mass. In order for the petrol to burn efficiently and release the energy that drives the engine, it must be supplied with the chemically correct amount of oxygen. The oxygen that is used for combustion comes from atmospheric air and this contains approximately 23% oxygen and 77% nitrogen, by mass. The chemical equations that govern the combustion of fuel areas follows.

$$\text{For carbon } C + O_2 = CO_2$$

For correct (stoichiometric) combustion this can be interpreted as: 1 kg of carbon requires 2.67 kg of oxygen and produces 3.67 kg of carbon dioxide.

$$\text{For hydrogen } 2H_2 + O_2 = 2H_2O$$

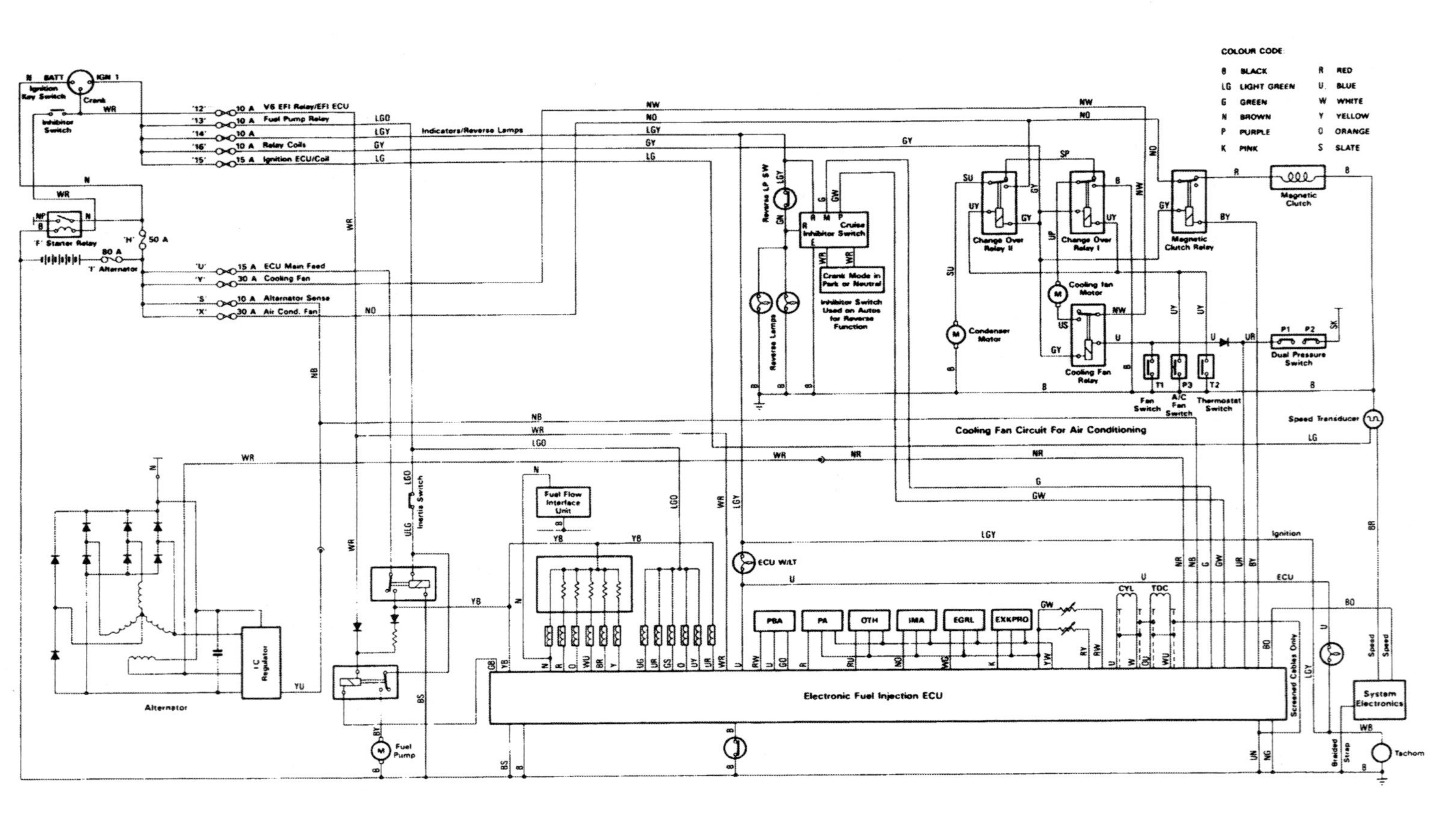

Fig. 4.22 A typical wiring diagram

In this case stoichiometric combustion is achieved when 1 kg of hydrogen is supplied with 8 kg of oxygen to produce 9 kg of H_2O (steam).

In this simplest case, there are two main products of combustion that appear in the exhaust gas, CO_2 and steam. The steam is eventually converted back to water in the atmosphere and it does not figure in the analysis of exhaust gas. This leaves the CO_2, but this is not the complete picture because the air contains nitrogen and this also appears in the exhaust gas. This leaves us with two main constituents of exhaust gas, carbon dioxide and nitrogen (CO_2 and N_2).

Arising from the composition of the fuel and the atmospheric air that provides the oxygen, correct combustion requires approximately 14.7 kg of air for each 1 kg of fuel. The air to fuel ratio is calculated from the mass of air divided by the mass of fuel. When this ratio is 14.7:1, it is given the value of lambda = 1. When the mixture is rich, i.e., extra fuel, lambda is less than 1 and when the mixture is weak, i.e., less fuel, lambda is greater than 1. The value of lambda is much quoted in literature that deals with catalytic converters and engine management.

During its operation, a petrol engine meets a range of different conditions and this causes the air–fuel ratio to vary either side of 14.7:1 (lambda less than 1 to lambda greater than 1). These variations from the chemically correct air–fuel ratio result in other gases such as carbon monoxide (CO) and unburnt hydrocarbons (HC) appearing in the exhaust gas. Thus, under operating conditions, the exhaust gas will contain small amounts of CO and HC in addition to the CO_2 and N_2.

Unfortunately, under high temperature conditions that occur in the engine cylinder, nitrogen reacts with oxygen to form oxides of nitrogen (NOx) and this also appears in the exhaust gas, mainly under high engine load. NOx is reduced by a three-way catalyst and other measures, such as exhaust gas recirculation and possibly secondary air injection into the exhaust system.

In the UK, a vehicle is tested annually for exhaust emissions and enforcement authority officials can stop a vehicle at any time to carry out spot checks. An exhaust gas analyser is used for these tests and most garage equipment manufacturers make, or market this type of equipment. They are called four-gas anlysers because they are able to detect quantities of CO_2, HC, CO and HC that are in a sample of exhaust gas. In most cases, the exhaust gas analyser will also give a value of the air–fuel ratio, or lambda, and this can be a very useful aid when attempting to trace emissions-related faults, such as defective exhaust gas oxygen sensor signals, catalyst problems, blocked air filter, and faulty fuel injectors.

For Department of Transport tests, the test stations use manufacturers' figures to confirm compliance with the standards and these vary slightly. In this case exact data is a prerequisite for testing. However, if the system is out of limits at idle speed, most systems include a means of adjusting the mixture to bring the CO% reading back inside the limits. Figure 4.23 gives and indication of the type of adjustment that is provided on an early model Rover car.

In addition to the diagnostic work that can be performed with the aid of the exhaust gas analyser, the self-diagnostic capability of the ECM also provides

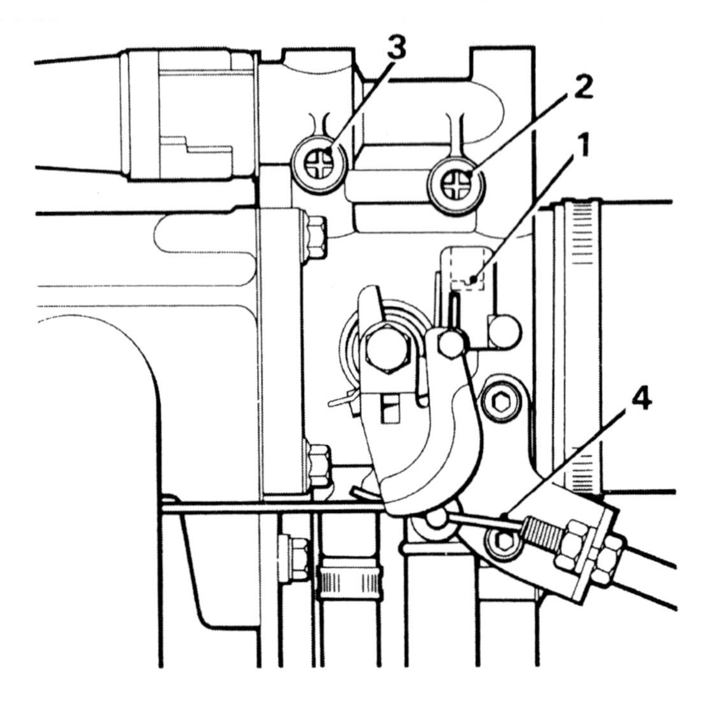

1. Throttle stop screw
2. Idle speed adjuster screw
3. Idle mixture adjuster screw
4. Kickdown cable (automatic only)

Fig. 4.23 The idle air adjustment to correct CO%

valuable information through the fault codes as it constantly monitors the performance of the engine management system.

4.10.2 DIESEL ENGINE EMISSIONS

Whilst diesel fuel is derived from the distillation of the same crude oil as petrol, it has quite different properties in terms of volatility and viscosity. It has a similar composition to petrol, in terms of carbon and hydrogen content, but the method of mixing air and fuel in the combustion chamber and the high pressures required to produce spontaneous combustion cause there to be quite different conditions inside the combustion chamber. These different conditions cause smoke (soot) and NO_X to be the main areas of concern with regard to diesel engine emissions.

NO_X is reduced by exhaust gas recirculation and accurate control of fuelling. The test for compliance with the regulations requires the use of an analyser that measures the opacity (denseness) of the smoke and for test purposes the opacity is measured in units that are given the symbol K. When the tests were first introduced in the UK, two maximum readings were used: for naturally aspirated engines (no turbocharger) the level was $K = 3.2$; and for turbocharged engines the maximum permitted level was $K = 3.7$.

No.	Cause	Difficult starting	Irregular Idle and Fast Idle	Insufficient Maximum Speed	Erratic Running/Surging	Excessive Smoke	Excessive Noise	Lack of Power	Excessive Fuel Consumption	Stalling	Slow Engine Die-down	Engine cannot be Shut Off	No.	Check
1	Lack of fuel	●			●								1	Fuel level
2	Manual stop control* faulty	●											2	In run position and linkage free
3	Stop solenoid* faulty	●			●							●	3	Audible operation when switched. Check electrical supply
4	Stop solenoid* valve leaking											●	4	Engine stops when supply lead removed
5	Wrong starting procedure	●											5	Starting procedure correct. (Start advance and starting aid operation)
6	Air in fuel system	●	●	●	●			●		●			6	Fuel system is vented, all joints and unions airtight, no leakage at diaphragms
7	Fuel inlet restriction	●	●	●	●			●		●			7	Filter not choked and feed pipes clear
8	Fuel contamination	●	●	●		●		●		●			8	Diesel fuel being used, not petrol; free of water, dirt, ice and wax
9	Low cranking speed	●											9	Battery, starter, cable connections. Correct engine lubricating oil
10	Starting aid* ineffective (Plugs or Thermostart)	●				●			●				10	Correct functioning. Fuel supply* and electrical connections
11	Injection timing incorrect	●	●		●	●	●	●	●	●			11	Pump to engine timing
12	Timing belt* slipped several teeth	●											12	Belt condition and tension
13	Feed pump* faulty	●	●	●				●					13	Feed pump pressure
14	Fuel return restricted	●	●	●				●		●			14	Rotary pump backleak, tank returns, filter vents are clear
15	Fuel circuit incorrect	●	●	●	●			●					15	Inlet and backleak pipes correct way round. Banjo bolts of correct CAV type
16	Engine condition	●	●	●	●	●	●	●	●	●			16	Cylinder compression. Valve timing and clearances. Air filter not choked. Injector seating
17	Exhaust system defective							●	●				17	System unrestricted
18	Fuel atomisation	●	●		●	●	●	●	●				18	Injectors: correct type, opening pressure, spray condition, evenly tightened
19	Fuel tank blockage	●	●		●			●					19	Tank vent and outlet filter unrestricted
20	HP pipe type/firing order incorrect	●	●	●	●	●		●	●				20	HP pipes type and fitted in correct firing order
21	HP pipe restriction		●		●			●					21	HP pipes not kinked or bore reduced at nipples
22	HP vent* leaking	●	●		●			●	●	●			22	Vent screw tight
23	HP pipe leaking	●	●		●			●	●				23	HP pipe joint tightness
24	LP leakage							●	●				24	Feed and return pipes, filter, feed pump and tank for leakage
25	Idling speed incorrect		●						●	●			25	Idling speed
26	Anti-stall* setting incorrect		●							●	●		26	Recovery from acceleration, engine warm (if necessary, reset idling and anti-stall)
27	Maximum speed setting incorrect			●				●					27	Engine flight speed
28	Accelerator linkage faulty		●	●	●			●			●		28	Lever tight on pump and reaches stop screws. Linkage wear, adjustment, freedom
29	Engine vibration		●		●		●						29	Engine mountings tight and effective
30	Vibration		●		●								30	Vibrations transmitted to engine
31	Overloading							●					31	Vehicle payload
32	Vehicle brakes binding			●				●	●				32	Brake freedom wheel by wheel, handbrake off
33	Injection pump mountings loose		●		●								33	Tightness of pump drive mounting bolts
34	Injection pump defective	●	●	●	●	●	●	●	●	●	●	●	34	If all other relevant checks are satisfactory, remove pump for specialist check

* if fitted

Fig. 4.24 A diesel fault tracing chart (Lucas CAV)

From early to mid 1990s, the Institute of Road Transport Engineers conducted a large sample survey and found that 99% of naturally aspirated engines and 96% of turbocharged engines were within the limits. More recently these figures have been lowered and the figures now (March 2000) are 2.5 for naturally aspirated engines and 3.0 for turbocharged engines. As with the exhaust gas analyser for petrol engines, the smoke test meter can be a useful aid to diagnosis. The chart shown in Fig. 4.24 shows the items of engine equipment that are likely to be at fault in the event of excessive smoke.

Smoke meters are items of equipment that most garage equipment manufacturers produce or market and, like the exhaust gas analyser, they are an essential item of equipment in a garage that is approved for vehicle testing.

The Garage Equipment Association of Daventry, Northamptonshire is the trade organization for garage equipment manufacturers and suppliers in the UK.

4.11 Review questions (see Appendix 2 for answers)

1. For diesel engine emissions testing in UK MOT test stations it is necessary:
   (a) to use a chassis dynamometer?
   (b) to use an approved smoke meter?
   (c) to take a sample of exhaust products and do a chemical analysis?
   (d) to hold a piece of white paper over the exhaust outlet?
2. The 'flight recorder' function on some diagnostic equipment:
   (a) permits the capture of live data from just before an intermittent fault happens and for a period afterwards?
   (b) can only be used on very expensive vehicles?
   (c) keeps a permanent record of the vehicle's service history?
   (d) can only be used in the workshop?
3. A digital multimeter is preferred for work on computer controlled systems because:
   (a) it has a low impedance (internal resistance)?
   (b) it has a high impedance?
   (c) they are cheaper than moving coil instruments?
   (d) the test leads make it easier to backprobe at a sensor?
4. An oscilloscope is useful for petrol injector tests because:
   (a) the amount of fuel injected is shown on the screen?
   (b) the shape of the injector pulse can be seen and the duty cycle is calculated?
   (c) the screen display shows up misfiring of the spark plug?
   (d) it saves time?
5. A vacuum pump and gauge permit:
   (a) devices such as a manifold absolute pressure sensor (MAP) to be tested without the engine running?
   (b) idle control valves to be reset to the CO% emissions level?
   (c) the evaporative purge control solenoid to be tested?
   (d) the wastegate valve on turbocharged engines to be tested?

6. All test instruments should be calibrated at regular intervals:
   (a) to stop them wearing out?
   (b) to keep the battery charged?
   (c) to ensure that they are making accurate measurements?
   (d) because regulations are constantly changing?
7. Fuel gallery pressure in a petrol injection system:
   (a) does not need checking because it is controlled by a pressure regulator?
   (b) can be checked by means of a pressure gauge?
   (c) does not vary?
   (d) is set at approximately 200 bar?
8. Backprobing of sensor connections should only be done with great care because:
   (a) it may upset the readings?
   (b) damage to the insulation may allow moisture to enter and cause corrosion?
   (c) closed-loop oxygen sensors need to be disconnected in order to get a correct reading?
   (d) silicone grease is expensive?

5
Sensors

Sensors are the components of the system that provide the inputs that enable the computer (ECM) to carry out the operations that make the system function correctly. In the case of vehicle sensors it is usually a voltage that is represented by a code at the computer's processor. If this voltage is incorrect the processor will probably take it as an invalid input and record a fault.

The fact that the controller itself receives an incorrect sensor signal normally means that the sensor and/or the circuit that connects it to the controller is not working properly and, as with many other parts of electronic systems, it may not be the sensor itself that is at fault. However, it is probable that a fault code has been produced that says 'sensor fault'. This just means that the sensor signal that reached the controller was defective. It is quite possible that the sensor is functioning correctly, but the circuit connecting it to the ECM is defective. There is, therefore, good reason for knowing how sensors work, what type of performance they give when they are working properly, and how to check their performance so that efficient diagnosis and repair can take place.

5.1 Electromagnetic sensors

Electromagnetic sensors are often used to sense the speed and/or angular position of a rotating object. Two common uses are: (1) crankshaft position for ignition and fuel injection control; and (2) road wheel rotational speed relative to vehicle frame for anti-lock braking (ABS) and traction control (TCS). The interactions between electricity and magnetism are used in various ways to produce the desired sensing effect. However, there are two types of sensor that are widely used in vehicle systems: variable reluctance and Hall type sensors.

5.1.1 THE VARIABLE RELUCTANCE TYPE SENSOR

This type of sensor is used in many vehicle applications, such as ignition systems, engine speed sensors for fuelling, and wheel speed sensors for anti-lock braking etc. Air has a greater reluctance (resistance to magnetism) than iron and this fact is made use of in many sensors. The basic principle of operation of a variable reluctance type sensor (Fig. 5.1) may be understood from the following description.

The principal elements of the sensor are:

- an iron rotor with lobes on it;
- a permanent magnet;
- a metallic path (the pole piece) for carrying the magnetic flux;
- a coil, wound around the metallic path, in which a voltage is induced.

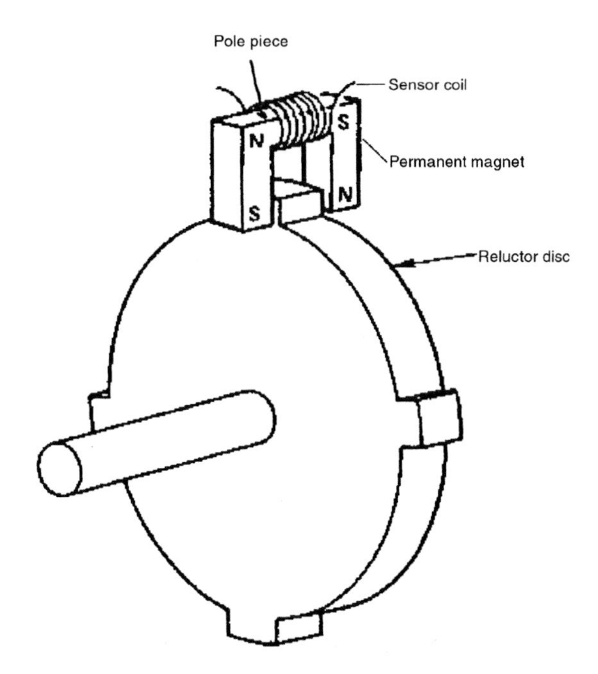

Fig. 5.1 The basic principle of the variable reluctance sensor

The reluctor disc has a number of tabs on it and these tabs are made to move through the air gap in the magnetic circuit. The movement of the reluctor tabs, through the air gap is achieved by rotation of the reluctor shaft. The voltage induced in the sensor coil is related to the rate of change of magnetic flux in the magnetic circuit. The faster the rate of change of magnetic flux the larger will be the voltage that is generated in the sensor coil. When the metal tab on the reluctor rotor is outside the air gap, the sensor voltage is zero. As the tab moves into the air gap the flow of magnetism (flux) increases rapidly. This causes the sensor voltage to increase, quite quickly, to a maximum positive value. Figure 5.2 shows the approximate behaviour of the voltage output as the reluctor is rotated.

Figure 5.2(a) shows the reluctor tab moving into the air gap. As the metal tab moves further into the gap the voltage begins to fall and, when the metal tab is exactly aligned with the pole piece, the sensor voltage falls back to zero.

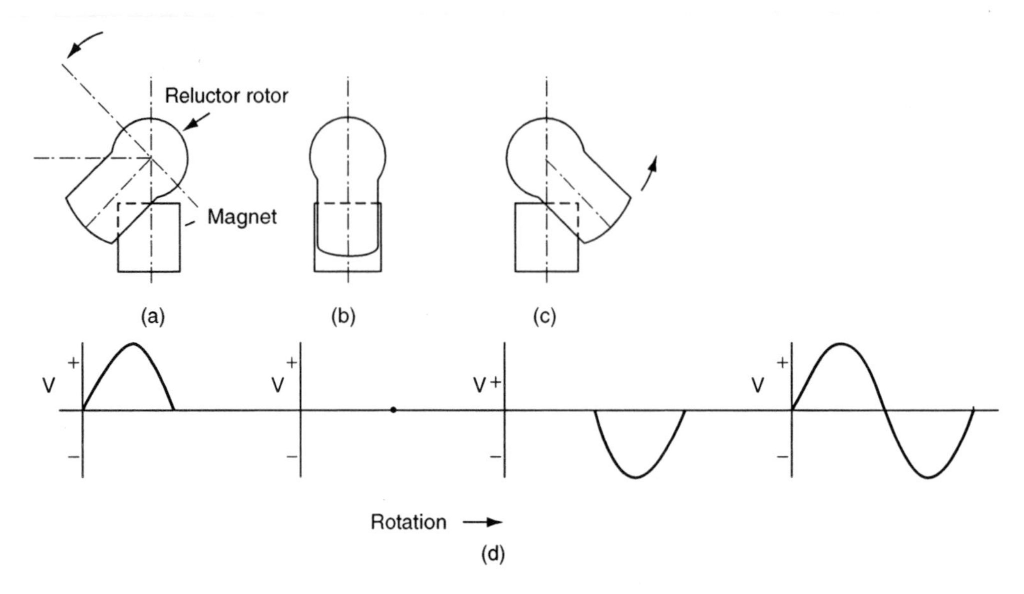

Fig. 5.2 The voltage pattern from a variable reluctance sensor

(Although the magnetic flux is strongest at this point, it is not changing and this means that the voltage is zero.) Figure 5.2(b) shows that there is zero voltage when the reluctor tab is in alignment with the pole piece. As the metal tab continues to rotate out of the air gap and away from the pole piece, the rate of change of the magnetic flux is rapid, but opposite in direction to when the tab was moving into the air gap. This results in the negative half of the voltage waveform as shown in Fig. 5.2(c). When the tab has moved out of the air gap the sensor voltage returns to zero. While the rotor shaft continues to turn another tab will enter the air gap and the above process will be repeated. If the sensor coil is connected to an oscilloscope the pattern observed will be similar to that shown in Fig. 5.2(d).

Crankshaft position sensor

Figure 5.3 shows a crankshaft sensor. Here the reluctor disc is attached to the engine flywheel. The permanent magnet, the pole piece and the sensor coil are attached to the cylinder block. As each metal tab on the reluctor disc passes the sensor pole piece a voltage is induced in the sensor winding.

The size of this voltage, induced in the sensor winding, depends on engine speed; the faster the engine speed the higher the sensor voltage. Each time a reluctor passes the pole piece an alternating current waveform is produced and at high engine speed the voltage produced by the sensor can be of the order of 100 V and some sensor circuits are designed to restrict the maximum voltage. In order to provide a top dead center (TDC) reference, there is a missing tab on the reluctor disc which means that the TDC position is marked by the absence of a voltage and this 'gap' is used to indicate to the ECM that the TDC position has been reached. The voltage waveform to be expected from this type of sensor is

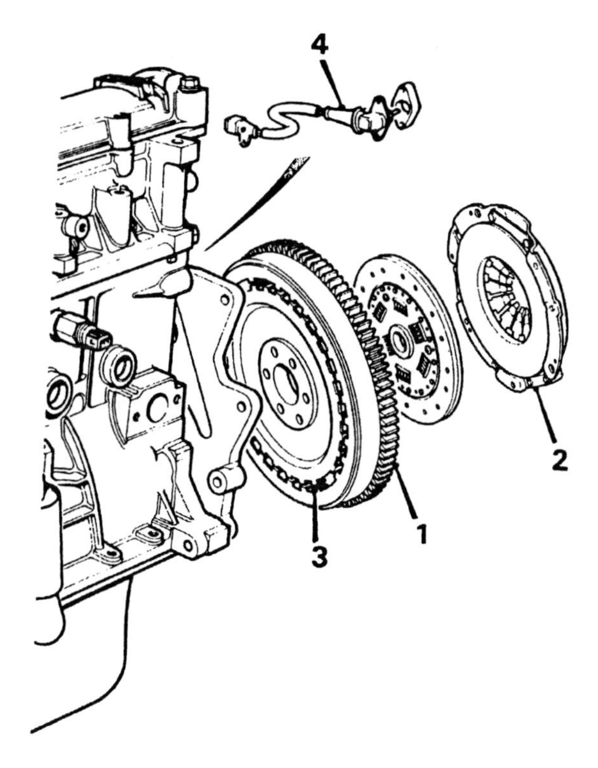

Fig. 5.3 Variable reluctance crank speed and position sensor

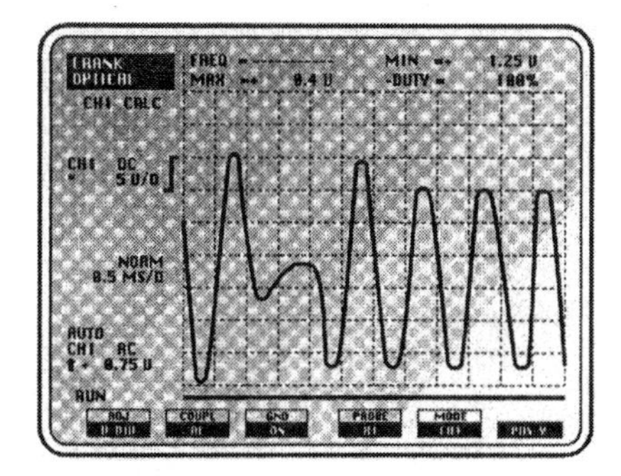

Fig. 5.4 A crank sensor voltage pattern

shown in Fig. 5.4. The missing wave at the TDC position is evident at the left-hand side of the pattern.

ABS wheel sensor

The principle of operation of many ABS wheel sensors is the same as for the crank sensor. However, the purpose for which it is used is somewhat different. To obtain the most effective braking, and to allow the driver to retain control of the vehicle, the wheels should not lock up under braking. The ABS sensor is used to assess slip between the tyre and the surface on which the tyre is working. The purpose of the ABS sensors is to detect when wheel lock-up is about to occur. This condition is indicated when the rotational speed of the reluctor ring (sensor rotor) is slow in relation to the sensor pick-up, which is fixed to the brake back plate, or equivalent. The layout of the sensor and its voltage waveform is shown in Fig. 5.5.

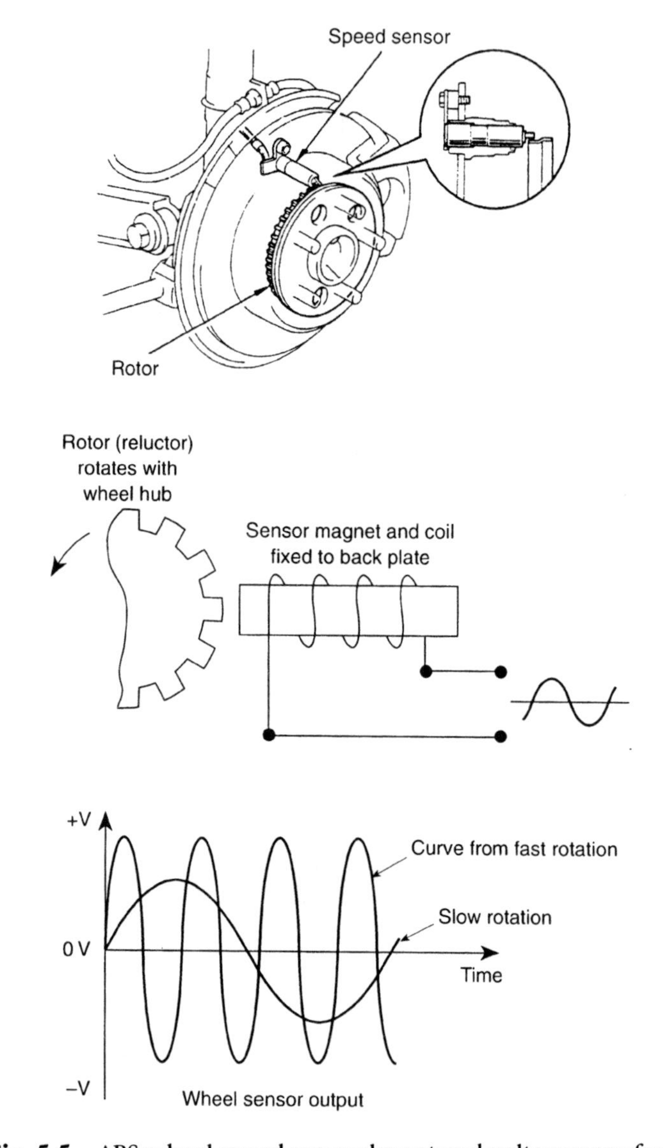

Fig. 5.5 ABS wheel speed sensor layout and voltage waveform

5.1.2 HALL EFFECT SENSORS

Figure 5.6 shows the principle of a Hall type sensor. The Hall element is a small section of semiconductor material such as silicon. When connected as shown in

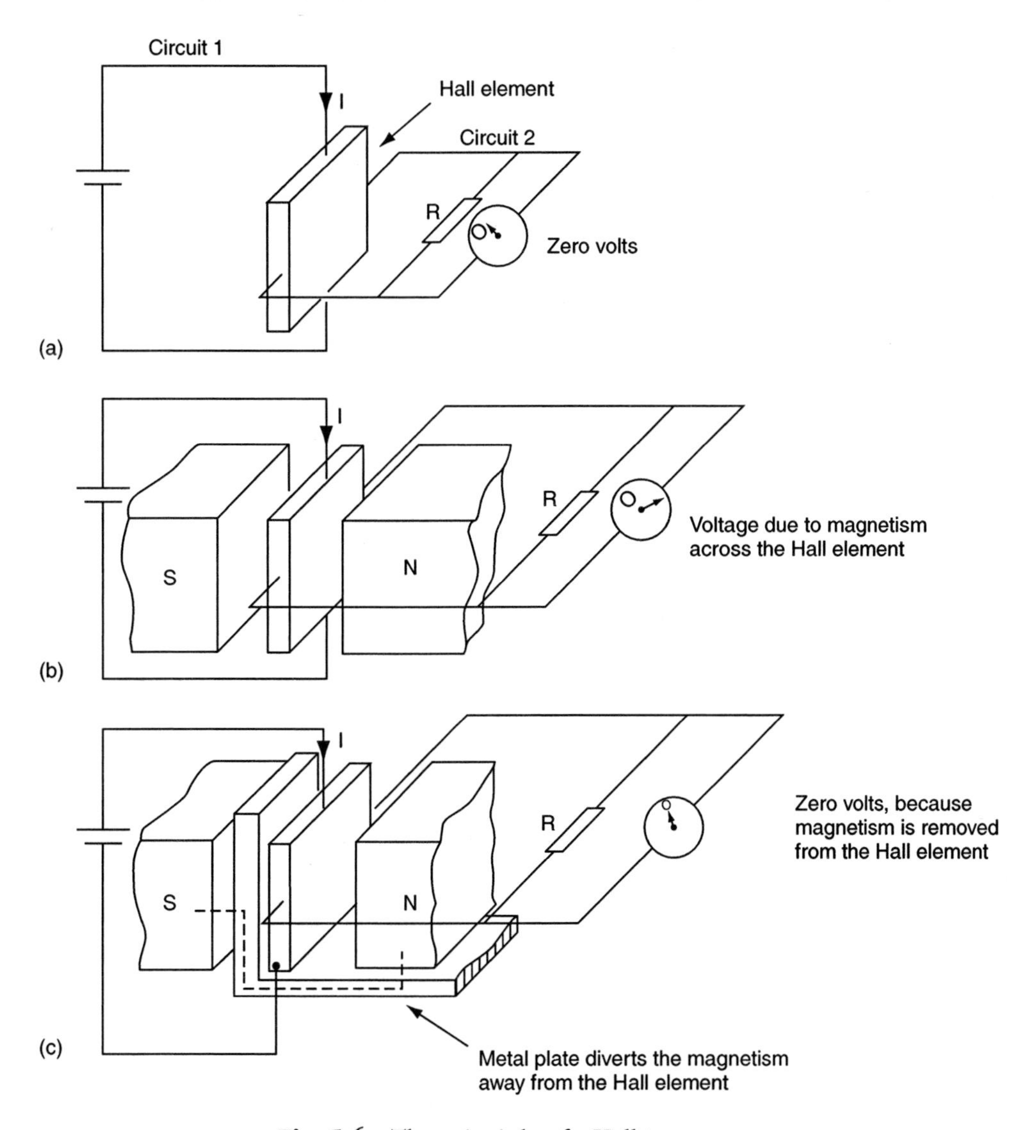

Fig. 5.6 The principle of a Hall type sensor

Fig. 5.6(a), the battery will cause current to flow through the semiconductor Hall element and battery circuit, but there will be no current in the circuit which is at right angles to the battery circuit, as shown by a zero reading on the voltmeter.

When a magnetic field is imposed on the Hall element, as shown in Fig. 5.6(b), a current will flow in circuit 2. When the magnetic effect is prevented from reaching the Hall element, as in Fig. 5.6(c), the current will cease to flow in circuit 2. The result is that the current in circuit 2 can be switched on and off by shielding the Hall element from the magnetic field. When the metal plate that is inserted between the magnet and the Hall element is mounted on a rotating shaft, the Hall current can be switched on and off at any desired frequency. The Hall type sensor produces an output power that is virtually constant at all speeds. Hall effect sensors are used wherever other electromagnetic sensors are used, e.g. engine speed and crank position, ABS wheel sensors, camshaft (cylinder) identification (for ignition and fuelling) etc.

The voltage from a Hall element is quite small and it is common practice for Hall type sensors to incorporate an amplifying and pulse-shaping circuit. The result is that the sensor produces a digital signal, i.e. it is a rectangular waveform as shown in Fig. 5.7.

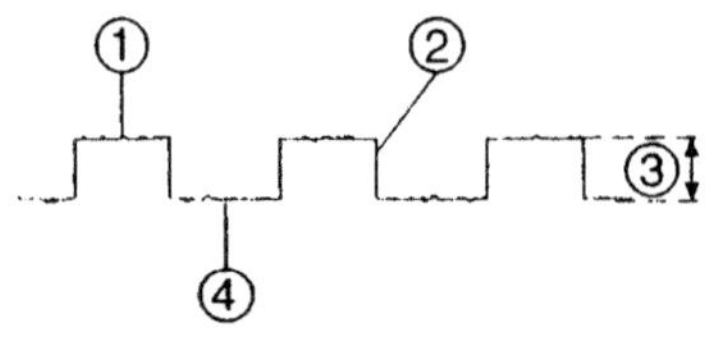

1 The upper horizontal lines should reach reference voltage.
2 Voltage transitions should be straight and vertical.
3 **Peak-Peak** voltages should equal reference voltage.
4 The lower horizontal lines should almost reach ground.

The duty cycle of the signal remains fixed, determined by the spacing between shutter blades.

Frequency of the signal increases as the speed of the engine increases.

Fig. 5.7 A Hall sensor output signal

5.2 Optical sensors

When light is directed onto semiconductor materials, energy is transferred to the semiconductor and this produces changes in the electrical behaviour of the semiconductor. This effect is used in optoelectronic devices, either as a photodiode, or as a phototransistor.

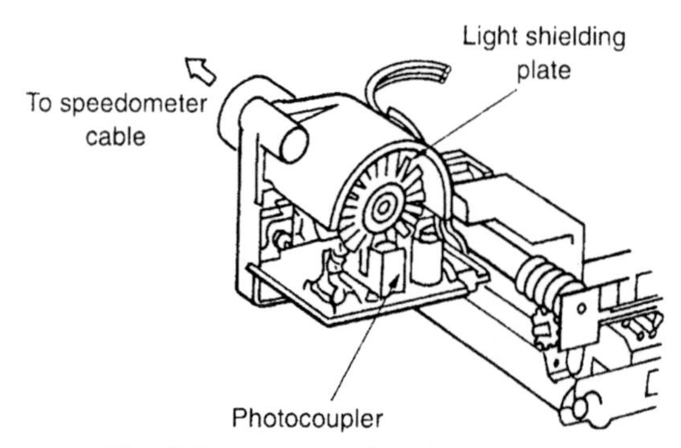

Fig. 5.8 An optoelectronic sensor

Figure 5.8 shows a vehicle speed sensor. The photocoupler consists of an infrared beam that is directed onto a photodiode. The infrared beam is interrupted by the light-shielding rotor (chopper) that is driven by the speedometer drive. In this way the light-sensing element is switched on and off at a frequency that is related to speed.

Optical sensors may be used in any application where electromagnetic sensors are used. They are found in vehicle speed sensing, ignition systems, steering systems etc. These sensors require a power source, and the voltage pattern that is typical of the signal from this type of sensor is shown in Fig. 5.9.

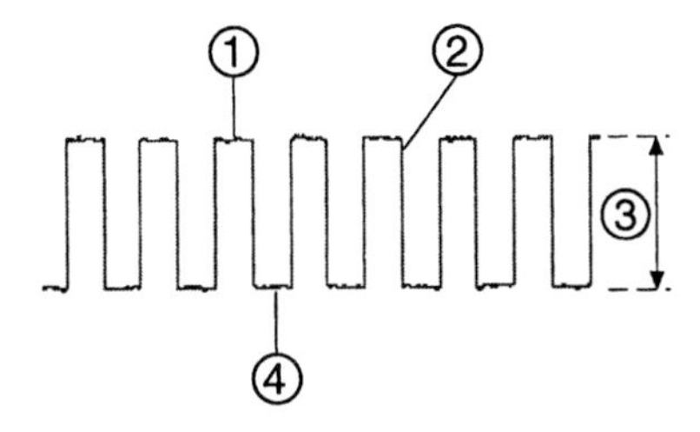

1 The upper horizontal lines should reach reference voltage.
2 Voltage transitions should be straight and vertical.
3 **Peak-Peak** voltages should equal reference voltage.
4 The lower horizontal lines should almost reach ground.

Voltage drop to ground should not exceed 400 mV.
If the voltage drop is greater than 400 mV, look for a bad ground to the sensor or ECU.
Signal frequency increases as the speed of the vehicle increases.

Fig. 5.9 Signal from an optoelectronic (light sensitive) sensor

5.3 Combustion knock sensors

A knock sensor that is commonly used in engine control systems utilizes the piezoelectric generator effect, i.e. the sensing element produces a small electric charge when it is compressed and then relaxed. Materials such as quartz and some ceramics like PZT (a mixture of platinum, zirconium and titanium) are effective in piezoelectric applications. In the application shown, the knock sensor is located on the engine block adjacent to cylinder number 3 (Fig. 5.10). This is the best position to detect vibrations arising from combustion knock in any of the four cylinders.

Because combustion knock is most likely to occur close to TDC in any cylinder, the control program held in the ECM memory enables the processor to use any knock signal generated to alter the ignition timing by an amount that is sufficient to eliminate the knock. When knock has ceased the ECM will advance the ignition, in steps, back to its normal setting. The mechanism by which vibrations arising from knock are converted to electricity is illustrated in Fig. 5.11.

The sensor is accurately designed and the center bolt that pre-tensions the piezoelectric crystal is accurately torqued. The steel washer that makes up the seismic mass has very precise dimensions. When combustion knock occurs,

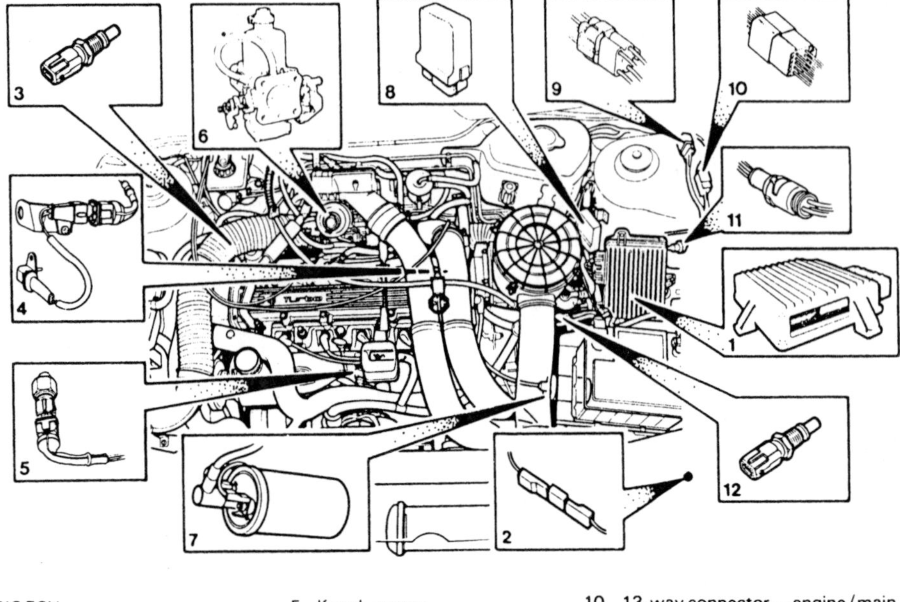

1. ERIC ECU
2. Ambient air temperature sensor (located behind horns)
3. Coolant temperature sensor
4. Crankshaft sensor
5. Knock sensor
6. Carburetter
7. Ignition coil
8. Engine MFU
9. 4-way connector—engine/main
10. 13-way connector—engine/main
11. Serial Diagnostic Link connector
12. Inlet air sensor

Fig. 5.10 The knock sensor on the engine

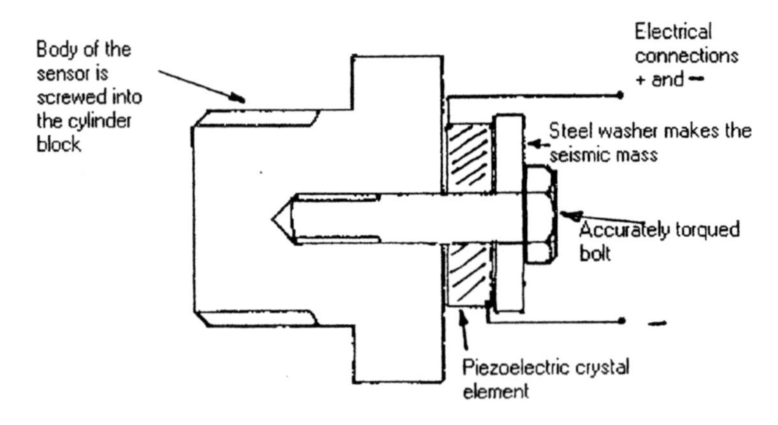

Fig. 5.11 The principle of the piezoelectric combustion knock sensor

the resulting mechanical vibrations are transmitted by the seismic mass, to the piezoelectric crystal. The 'squeezing up' and relaxing of the crystal in response to this action, produces a small electrical signal that oscillates at the same frequency as the knock sensor element. The electrical signal is conducted away from the crystal by wires that are secured to suitable points on the crystal.

The tuning of the sensor is critical because it must be able to distinguish between knock from combustion and other knocks that may arise from the engine mechanism. This is achieved because combustion knock produces vibrations that fall within a known range of frequencies.

5.4 Variable resistance type sensors

When an engine is idling the exhaust gas scavenging of the cylinders is poor. This has the effect of diluting the incoming mixture. The ECU must detect when the throttle is in the idling position, so that alteration of the air-fuel ratio can occur to ensure that the engine continues to run smoothly. At full engine load and full throttle, the mixture (air-fuel ratio) needs enriching, so the ECU also needs a signal to show that the throttle is fully open. These duties are performed by the throttle position switch. Figure 5.12 shows how the action of a throttle position sensor is based on the principle of the potential divider.

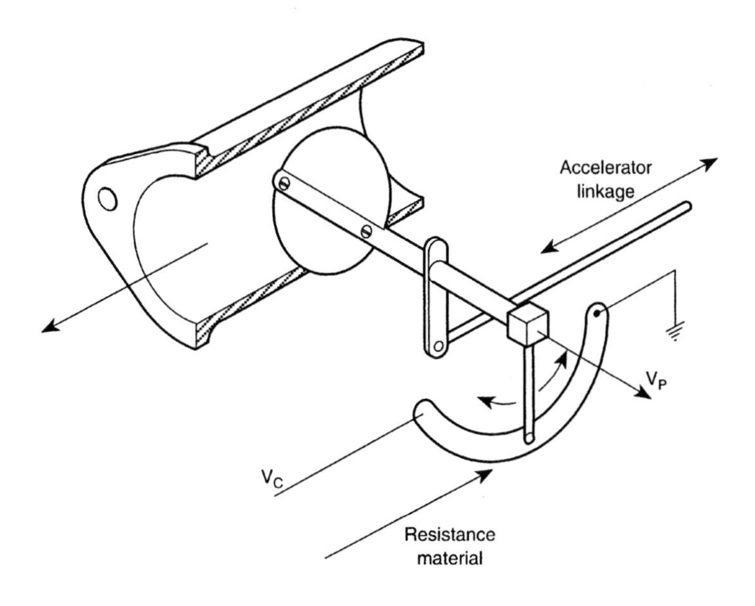

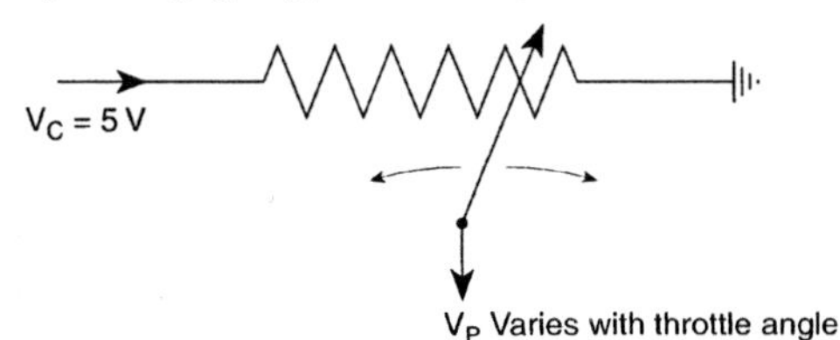

Fig. 5.12 The principle of the throttle position sensor

The sensor produces a voltage which is related to throttle position. The voltage signal is conducted to the ECU where it is used, in conjunction with other inputs, to determine the correct fuelling for a given condition.

There are two types of throttle position sensors in common use, and they are quite different in certain respects. Test procedures that will work on one type will not apply to the other type. Here again it is important to be able to recognize which type is being used in a particular application. Figures 5.13 and 5.14 show details of the throttle position switch.

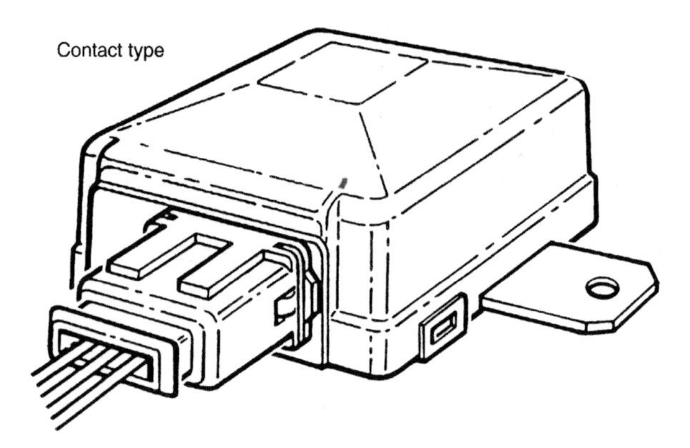

Fig. 5.13 Lucas type throttle position switch

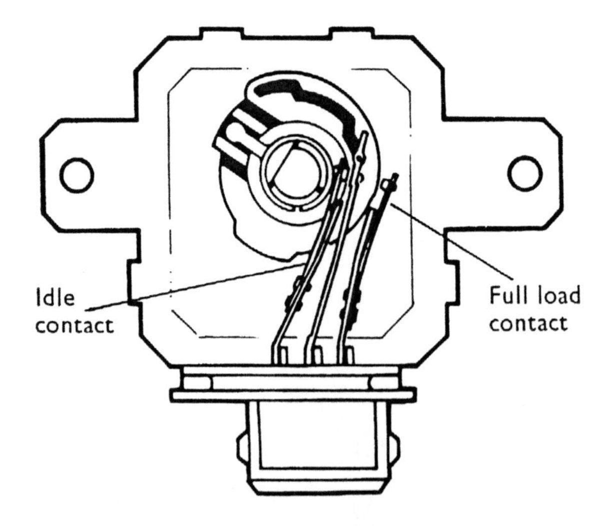

Fig. 5.14 Inside the throttle switch (Lucas)

The points to note about these two types of throttle position sensors is that they are primarily electrical. They do not require any great electrical or electronic knowledge in order to test them. However, the electrical signals which they produce, and which are used by the electronic control module (ECM), must be correct for given conditions. The throttle switch produces 'step' voltage changes at the idling and full throttle position since the ECM program requires precise identification of the throttle open and throttle closed positions. The potentiometer-type throttle position sensor produces a steadily increasing voltage, from idling up to full throttle. It is, therefore, very important that any measurements taken during tests are accurate, and relate correctly to the angular position of the throttle butterfly, as well as the specific type of sensor under test. Figure 5.15 shows the throttle position sensor used on the Toyota 3S-GTE engine.

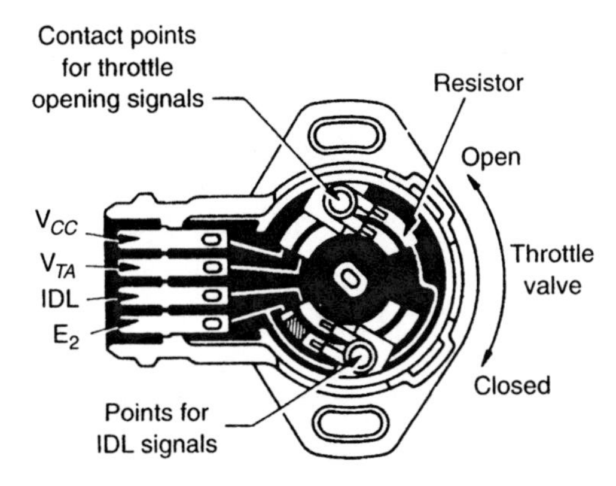

Fig. 5.15 Throttle sensor (Toyota)

V_{cc} is a constant voltage of 5 V supplied by the computer. Terminal E_2 is earthed via the computer. The other two voltages, IDL and V_{TA}, relate to idling and throttle operating angle. Figure 5.16 shows how the voltages at the terminals IDL and V_{TA} relate to the position of the throttle butterfly.

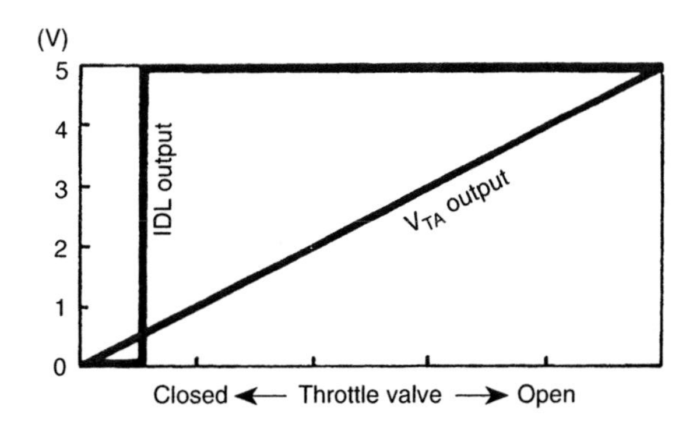

Fig. 5.16 Indication of throttle sensor voltages (Toyota)

The figures shown in the graph are approximate and it should be understood that precise details, of any vehicle being worked on, will be required before meaningful checks can be performed. However, it will be appreciated that this sensor, as with many other sensors, can be checked with quite basic electrical knowledge, and good quality, widely available tools, such as a voltmeter or an oscilloscope. Such tests are described in Chapter 7.

5.5 Temperature sensors

A commonly used device used for sensing temperature is the thermistor. A thermistor utilizes the concept of negative temperature coefficient. Most electrical conductors have a positive temperature coefficient. This means that the hotter the conductor gets the higher is its electrical resistance. This thermistor operates differently; its resistance gets lower as its temperature increases and this is a characteristic of semiconductor materials. There is a well-defined relationship between temperature and resistance. This means that current flow through the thermistor can be used to give an accurate representation of temperature. A typical coolant temperature sensor is shown in Fig. 5.17.

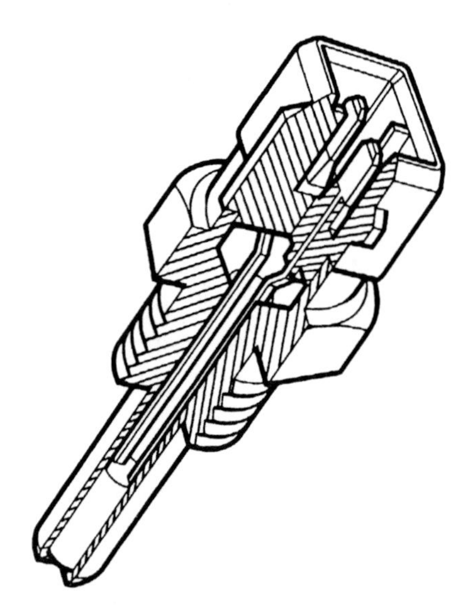

Fig. 5.17 An engine coolant temperature sensor

Figure 5.18 shows the approximate relationship between temperature and resistance. The coolant temperature sensor provides the ECU with information about engine temperature and thus allows the ECU to make alterations to fuelling for cold starts and warm-up enrichment.

The information shown in Fig. 5.18 may be given in tabular form as shown in Table 5.1 (these are approximate figures).

This shows the approximate resistance to be expected between the sensor terminals, for a given temperature. From this it will be seen that it is possible to

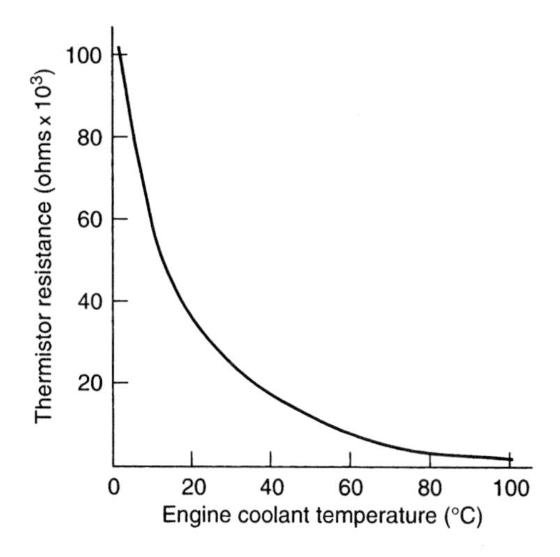

Fig. 5.18 Temperature vs resistance characteristics (thermistor)

Table 5.1 Temperature and corresponding resistance for a coolant sensor

Temperature (°C)	Resistance (ohms)	Voltage
0	6000	4.5
20	2500	3.2
30	1400	3.1
60	800	2.4
80	280	1.2

test such a sensor, for correctness of operation, with the aid of a thermometer and a resistance meter (ohm-meter), provided the exact reference values are known.

5.6 Ride height control sensor

This type of sensor is used in the Toyota height control system that is reviewed in Chapter 1. Figure 5.19 shows the details of the sensor and it may be seen that it uses the principle of the potential divider that is used in a variety of other vehicle sensor applications.

Vertical movement of the vehicle body, relative to the wheels, is converted to rotary movement of the sensor brush by means of the sensor arm. The voltage detected at the sensor signal terminal is dependent on the position of the brush relative to the two ends of the resistive track. Figure 5.20 shows an oscilloscope trace of the signal that the height sensor sends to the control computer. At point (1), the suspension dampers are fully extended. As the vehicle load is increased, the height decreases and this causes the sensor voltage to fall, as shown at (2) on

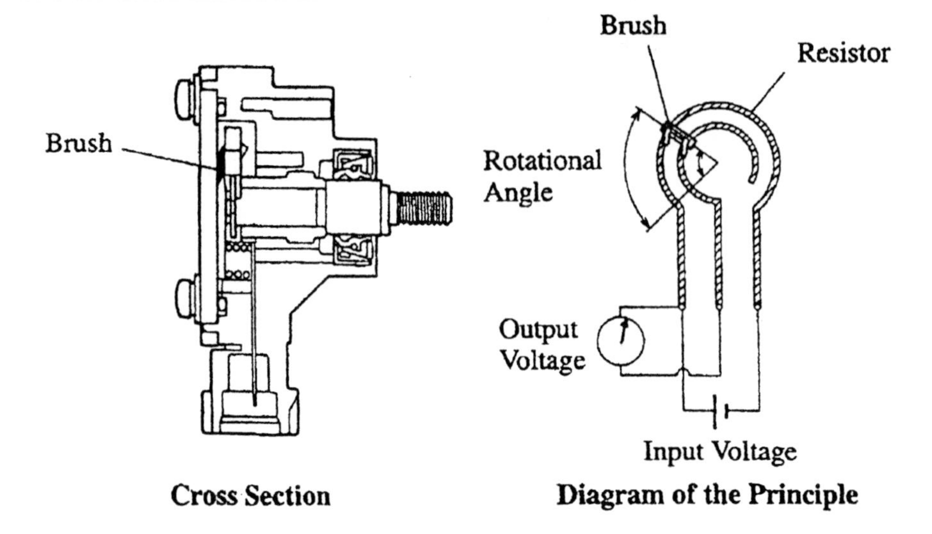

Fig. 5.19 A ride height sensor

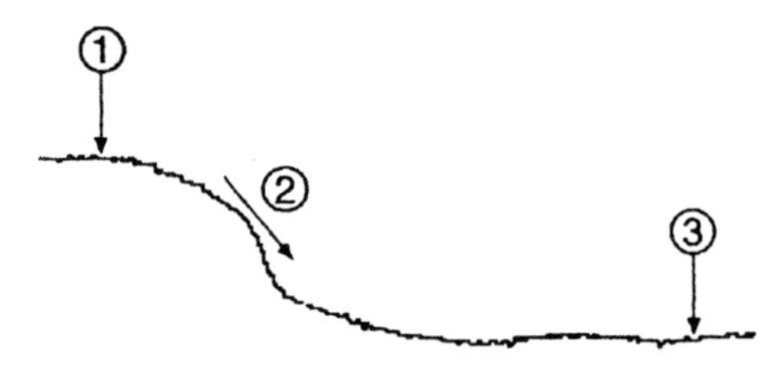

1	Vehicle shocks fully extended, vehicle hight at maximum.
2	Weight added to the vehicle causing vehicle height to lower. Potentiometer on height sensor reducing output voltage.
3	Lower voltage is interrupted by the electronic control module causing the air suspension to extend.

Fig. 5.20 Typical voltage trace for a ride height sensor

the trace. When the suspension is compressed to the lowest height, the voltage level as indicated at (3), causes the control computer to output a signal to the height control actuator and the vehicle height will be restored to the required level.

5.7 Manifold absolute pressure (MAP)

The pressure of the air, or air-fuel mixture, in the engine intake manifold varies with the load on the engine. For example, with the throttle valve fully closed and the engine being used for downhill braking, the manifold pressure will be very low (near perfect vacuum). With the throttle fully open and the vehicle accelerating up an incline the manifold pressure will be higher, i.e. very little vacuum. The pressure inside the intake manifold is known as absolute pressure (i.e. it is measured from absolute zero pressure, or complete vacuum, upwards). Manifold absolute pressure gives a very good guide to (analogy of) engine load and its measurement plays an important part in the operation of digital ignition and engine fuelling systems. In addition, an accurate guide to the amount of air entering an engine can be computed from volume, manifold absolute pressure and temperature. Systems which make use of this method are known as speed density systems, and the MAP sensor effectively replaces the air flow sensor.

The accurate measurement of manifold pressure plays an important part in engine control strategies and the MAP sensor is the device that does this. MAP sensors are available in several different forms. However, there are two types that are commonly used. One of these gives a variable voltage output to represent manifold absolute pressure, and the other uses the frequency of a digital signal to represent it.

5.7.1 THE VARIABLE VOLTAGE MAP SENSOR

The MAP sensor shown in Fig. 5.21 receives a 5 V supply from the ECU. Variations in manifold pressure (vacuum) cause the small silicon diaphragm to deflect. This deflection alters the resistance of the resistors in the sensor's bridge circuit and the resulting electrical output from the bridge circuit is proportional to manifold pressure.

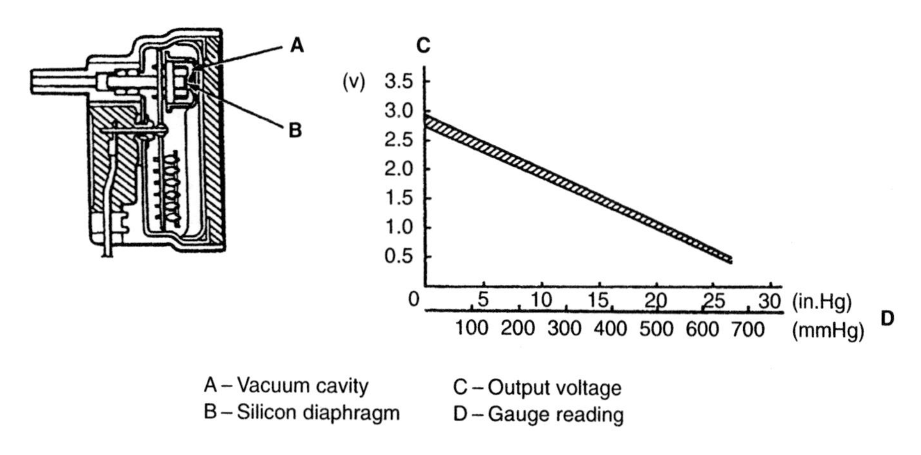

Fig. 5.21 A manifold absolute pressure sensor

Details of the pressure sensing element are shown in Fig. 5.22. Figure 5.22(a) shows four resistors that are diffused into the silicon. The size and chemical composition of the resistors is accurately controlled during the manufacturing process. Figure 5.22(b) shows how the vacuum cavity permits the silicon sensing diaphragm to flex under the influence of pressure in the induction manifold.

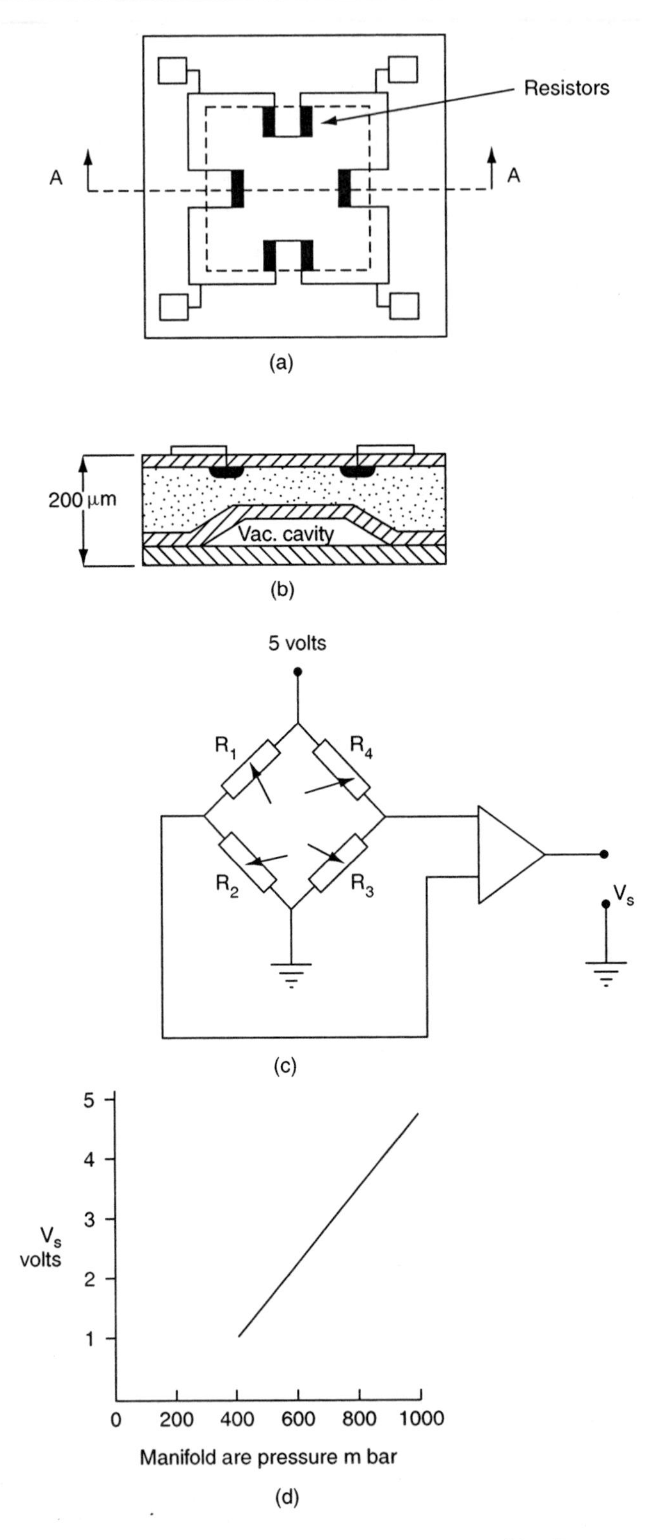

Fig. 5.22 Details of the strain gauge type of MAP sensor

When the sensing diaphragm flexes, the physical sizes of the resistors R1 to R4 change and this affects their electrical resistance. The bridge circuit shown in Fig. 5.22(c) represents the electrical circuit in the sensor. Under the influence of manifold pressure the electrical resistance of R1 and R3 increases and that of R2 and R4 decreases by a similar amount. This alters the voltage levels at the two inputs to the differential transformer and leads to a sensor output voltage that is proportional to manifold absolute pressure. Figure 5.22(d) shows an idealized representation of sensor voltage against manifold absolute pressure and it shows that a MAP sensor can be checked for accuracy of performance by conventional electrical test methods.

5.7.2 OTHER MAP SENSORS

Variable-capacitance type

Figure 5.23 gives an indication of the principle of operation of the variable-capacitance type of MAP sensor.

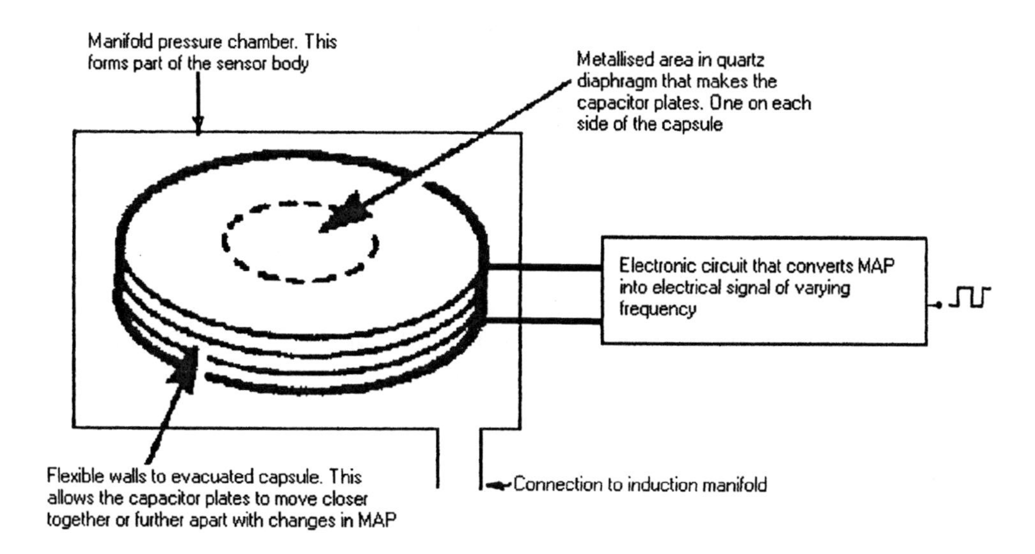

Fig. 5.23 Principle of a capacitive-type MAP sensor

Capacitance $C = e_o\ A/d$, where e_o = permittivity in a vacuum, A = area of the metallized plates and d = the distance between the plates. The metallized plates of the capacitor are placed on each side of an evacuated capsule. This capsule is placed in a chamber which is connected to manifold pressure and, as the manifold pressure changes, the distance d between the capacitor plates changes. This change in distance between the capacitor plates causes the value of the capacitance C to change. The capacitor is connected into an electronic circuit that converts changes in capacitance into an electrical signal.

Variable-inductance type

The variable-inductance type of MAP sensor relies on the principle that the inductance of a coil is altered by varying the position of an iron cylinder placed in the center of the coil. Figure 5.24 illustrates the principle involved.

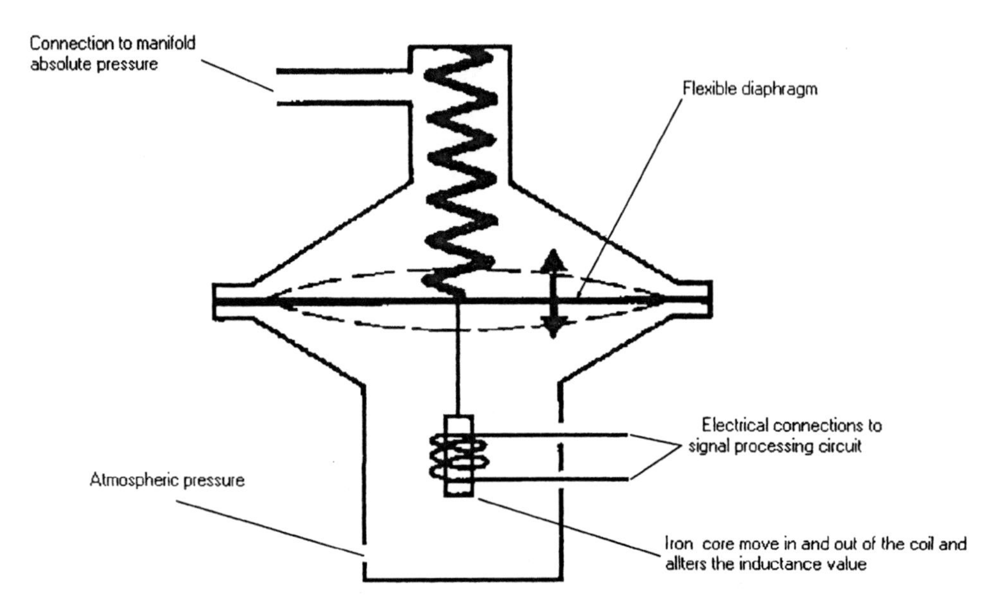

Fig. 5.24 Variable inductance-type MAP Sensor

In this simplified version, the iron cylinder moves in or out of the coil under the influence of the diaphragm and spring. Variations in manifold absolute pressure increase or decrease the 'suction' force acting on the diaphragm and the resultant changes in inductance are related to the manifold absolute pressure. The coil (inductance) forms part of an electronic circuit and this circuit is designed so that the changes in frequency of the square-wave output are accurate representations of manifold absolute pressure.

Figure 5.25 shows the approximate form of the variable frequency output of sensors of this type.

As with all vehicle systems, it is vitally important to have access to the precise figures and data that relate to the system and vehicle being worked on. The information given here is intended as a guide to general principles. In many cases the suppliers and makers of diagnostic equipment will provide a back-up service that will provide test data for a wide range of vehicles and systems. This is a factor that must be considered when making decisions about the purchase of equipment.

5.8 Exhaust gas oxygen sensors

In order for the exhaust emissions catalyst to operate correctly, the air–fuel ratio must be kept close to 15:1 (by mass), and it is the exhaust gas oxygen (EGO)

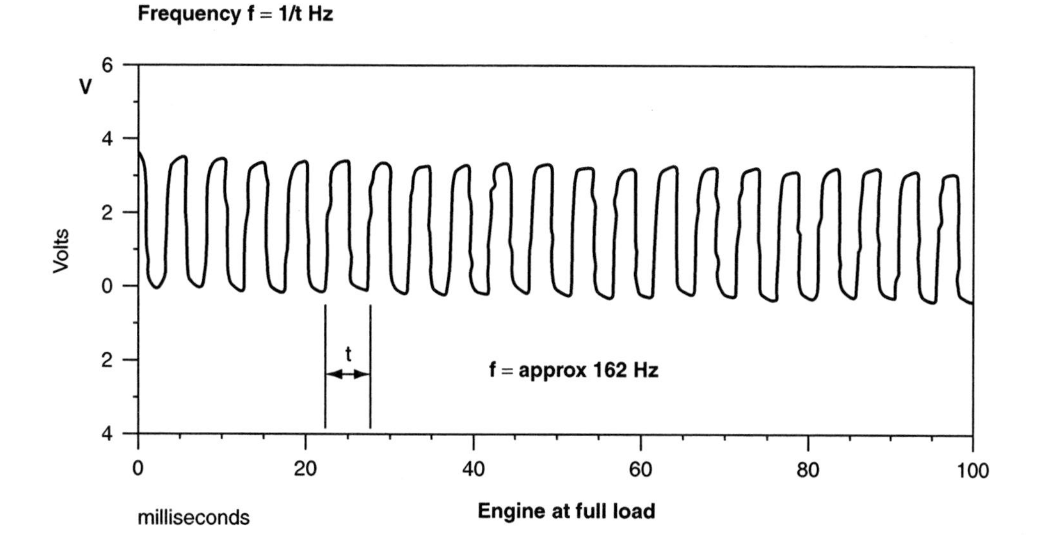

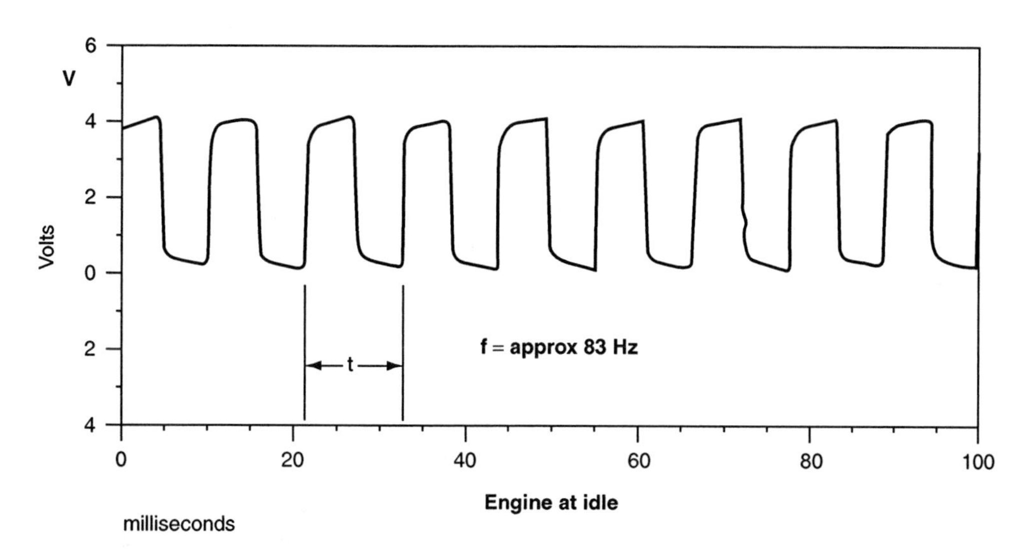

Fig. 5.25 Frequency patterns for a MAP sensor at full load and idle

sensor that assists the ECM to keep the air–fuel ratio within the required limits. The EGO sensor constantly monitors the oxygen content of the exhaust gas, and hence the air–fuel ratio at the engine intake, since the percentage of oxygen in the exhaust gas is an accurate measure of the air–fuel ratio of the mixture entering the engine cylinders. Figure 5.26 shows the relation between the oxygen content of the exhaust gas and the air–fuel ratio of the mixture entering the combustion chambers of the engine.

The information (voltage) from the EGO sensor is fed back to the ECM so that the amount of fuel injected into the engine may be changed to ensure that the

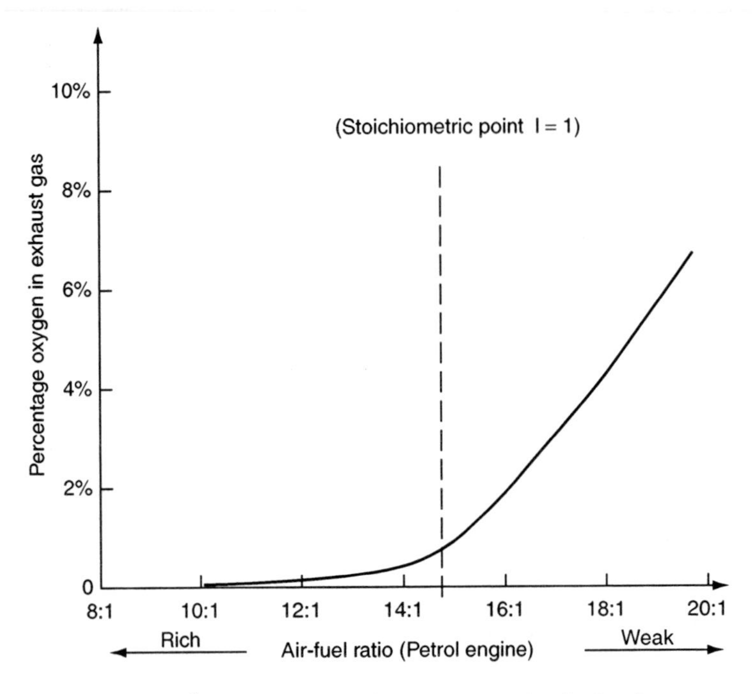

Fig. 5.26 Oxygen in exhaust versus air-fuel ratio

air-fuel ratio is kept within the required limits. It is common practice to refer to the air-fuel ratio that gives chemically correct combustion as lambda = 1. If the mixture is rich, lambda is less than 1 (probably lambda = 0.97), and if the mixture is weak, lambda is greater than 1 (probably lambda = 1.03). For this reason, the exhaust gas oxygen sensor is often referred to as a lambda sensor.

$$\text{Lambda} = \frac{\text{actual air-fuel ratio}}{\text{chemically correct air-fuel ratio}}$$

There are two types of EGO sensor currently in use. One operates on the principle of a voltaic cell, i.e. it is 'chemo-voltaic', and the other relies on the changes in electrical resistance of a material in response oxygen, i.e. it is 'chemo-resistive'.

5.8.1 THE VOLTAIC-TYPE EGO SENSOR

The voltaic, or zirconia (ZrO_2), type oxygen sensor operates on the basis of a difference between the oxygen partial pressure of atmospheric air and the partial pressure of oxygen in the exhaust gas. At sea level, atmospheric air contains approximately 21% oxygen by weight, and this gives the oxygen a partial pressure of approximately 0.2 bar. The oxygen content of exhaust gas varies from zero in a rich mixture, to about 10% in a weak mixture, as shown in Fig.5.26. The partial pressure of the oxygen in the exhaust gas therefore ranges from near zero to approximately 0.01 bar.

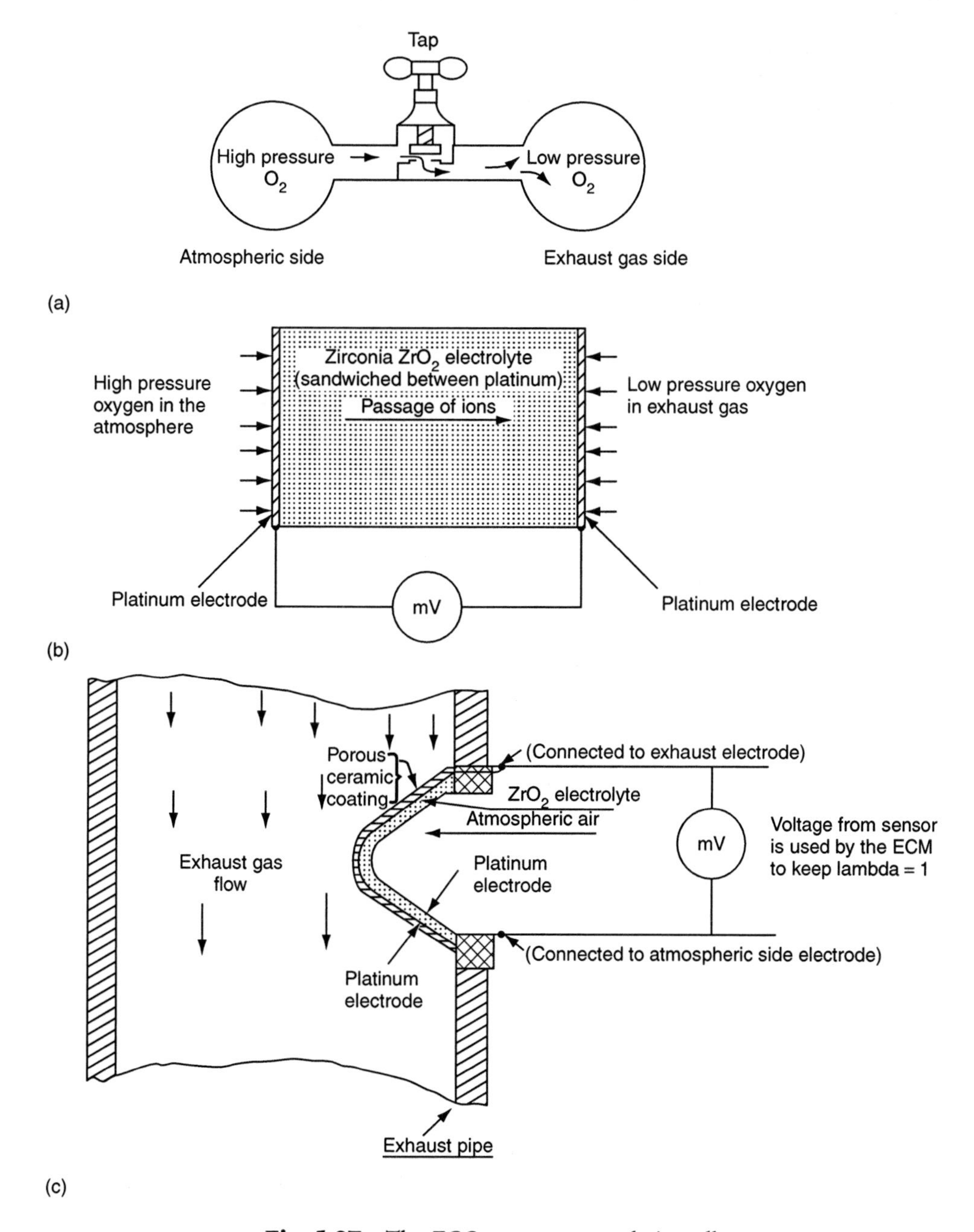

Fig. 5.27 The EGO sensor as a voltaic cell

Figure 5.27 shows that the sensor element is essentially a cell (battery). The plates are made from platinum and they have a layer of ceramic zirconia between them which acts as an electrolyte. The platinum plates act as catalysts for the oxygen which makes contact with them, and they are also used to conduct electricity away from the sensor. The catalyzing action that takes place when oxygen contacts the platinum plates causes the transport of oxygen ions through the electrolyte and this creates the electric current that gives rise to the e.m.f. (voltage) of the sensor. This sensor voltage is an accurate representation of the oxygen content of the exhaust gas.

In practice the sensing element is formed into a thimble shape as shown in Fig. 5.28. This type of construction exposes the maximum area of platinum to the exhaust gas on one side and to the atmospheric air on the other side. The platinum that is exposed to the exhaust gas is covered with a porous ceramic material. This allows the oxygen through to the platinum but protects the platinum against harmful contaminants in the exhaust products.

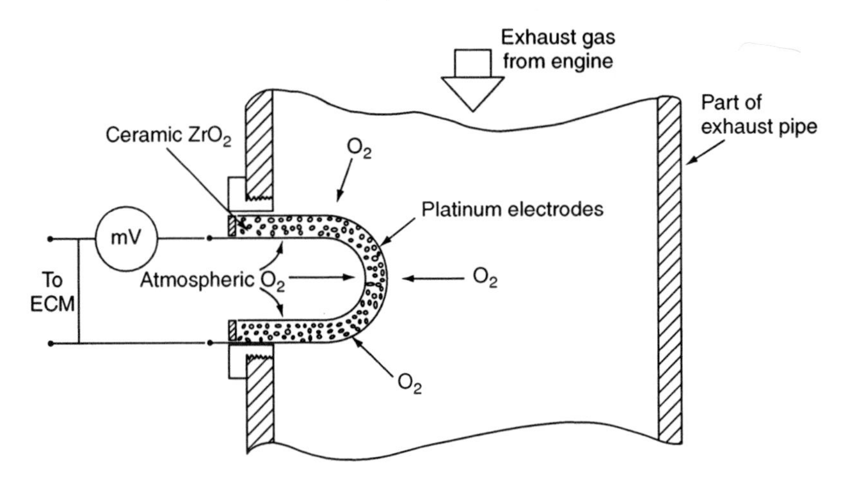

Fig. 5.28 Diagrammatic representation of the oxygen sensor in the exhaust pipe

The greater the difference in oxygen levels between the atmospheric air and the exhaust gas, the greater is the voltage produced by the EGO sensor. When the air–fuel ratio changes from slightly rich, say 14:1 (lambda = 0.93) to slightly weak, 16:1 (lambda = 1.06), there is a marked change in the oxygen partial pressure of the exhaust gas and this leads to a step change in the EGO sensor voltage because the ceramic electrolyte (zirconia) is very sensitive to oxygen levels, as shown in Fig. 5.29.

This sudden change in sensor voltage is used to trigger an action by the ECM, that will alter the fuelling, to maintain lambda = 1 (chemically correct air–fuel ratio). The result of this action is that the EGO sensor output cycles up and down, at a frequency that ensures that the engine runs smoothly and the exhaust catalyst is kept functioning correctly. The actual frequency is determined by the program that the designer places in the ROM of the ECM. All of this means that a voltaic-type EGO produces a standard type of output that can be measured by means of equipment that is readily available to vehicle repairers.

The approximate shape of the voltage waveform from the EGO sensor when in operation is shown in Fig. 5.30. This waveform arises from the way that the ECM alters the amount of fuel injected, i.e. lowering and raising the amount of fuel

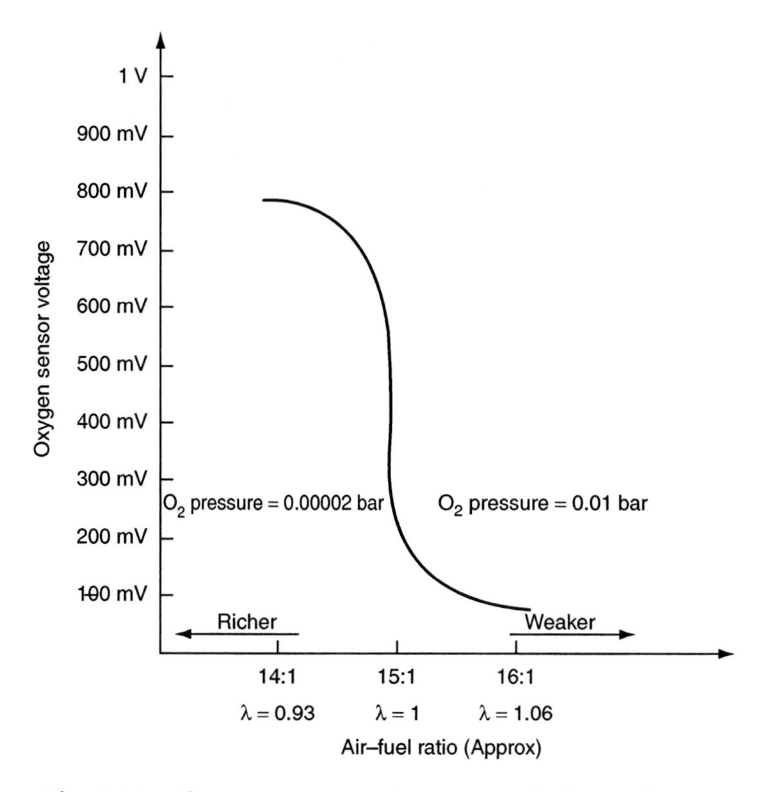

Fig. 5.29 Change in sensor voltage as air-fuel ratio changes

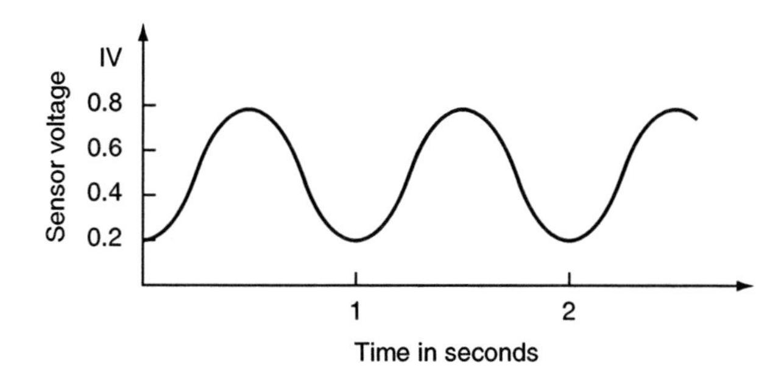

Fig. 5.30 The voltage waveform of an EGO sensor

injected, in an ordered way, so as to keep the air-fuel ratio within the required limits. This means that the time period between peaks and valleys (frequency) of the waveform will vary with engine speed. This time period will also vary according to whether the fuelling system is single-point injection, or multi-point injection. However, the general principle holds good. The type of waveform shown can be expected from any correctly operating oxygen sensor.

The action of the oxygen sensor is dependent on its temperature. The sensor needs to reach a temperature of around 250°C before it starts to function at its best. In order to assist the sensor to reach this temperature as quickly as possible, from a cold start, it is common practice to equip the sensor with a resistive-type heating element as shown in Fig. 5.31.

This means that most oxygen sensors will be equipped with four wires: a signal wire and an earth for the sensor element, and a feed wire and an earth for the heating element. This type of sensor is known as a heated exhaust gas oxygen sensor (HEGO) and is shown in Fig. 5.32.

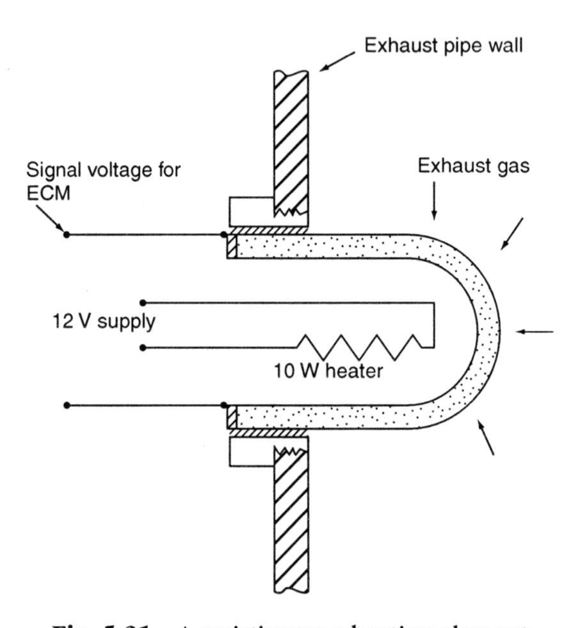

Fig. 5.31 A resistive-type heating element

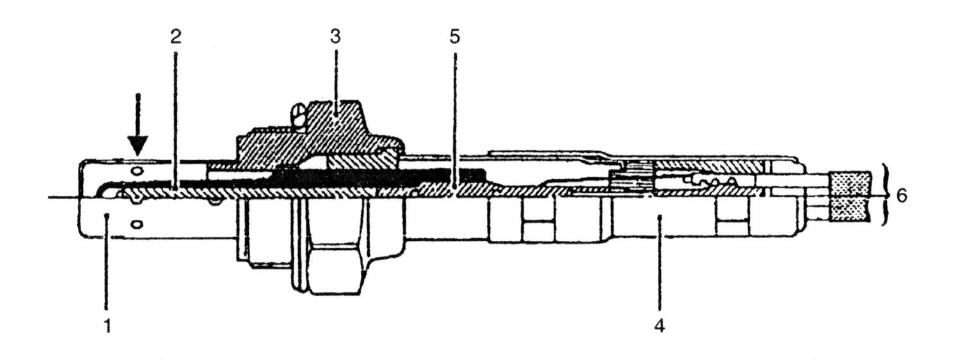

1 Protective sleeve
2 Ceramic sensor
3 Body
4 Protective socket
5 Heating element
6 Electrical connections
→ Exhaust gas flow

Fig. 5.32 A heated exhaust gas oxygen sensor

It must be understood that the EGO sensor is part of a feedback system. If it is disconnected it will cease to function correctly. It must, therefore, be tested while the system is in operation, i.e. when the engine is warmed up and running. When the system is operating correctly the EGO sensor output varies between approximately 200 mV and 800 mV, and the approximate shape of the voltage waveform is shown in Fig. 5.30.

5.8.2 THE RESISTIVE-TYPE EGO SENSOR

The zirconia-type voltaic EGO is somewhat slow in operation and it is claimed that the titanium oxide (titania)-type EGO sensor has a faster response time and is, therefore, better for engine emission control purposes (Fig. 5.33).

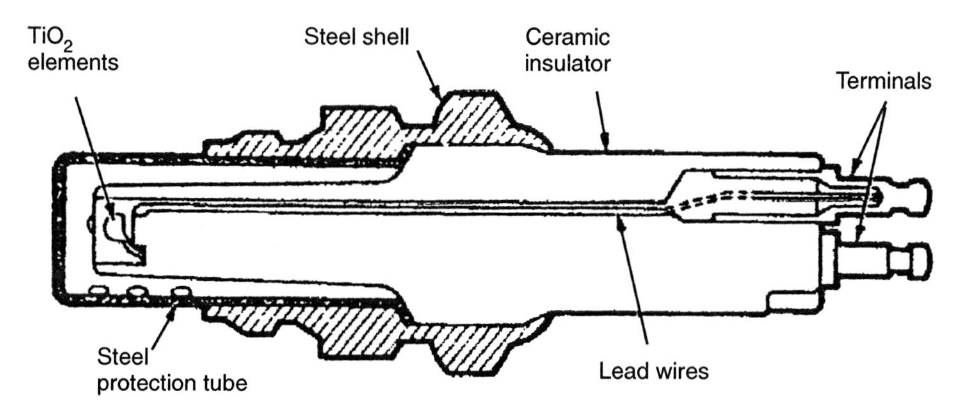

Fig. 5.33 The titanium dioxide (titania) type EGO sensor

The titania sensor reacts to changes in the partial pressure of oxygen in the exhaust gas. Changes in the concentration of oxygen in the exhaust gas cause the resistance of the sensor material to change. When the sensor is supplied with a set voltage from the control unit, the variation in current through the sensing element provides an indication of the oxygen content of the exhaust gas. In the sensing element, the titania is essentially a semiconductor whose resistive properties are affected by the concentration of oxygen that reacts with it. The reaction that occurs affects the resistance of the sensor element and the resultant sensor voltage is an accurate indicator of the partial pressure of the oxygen in the exhaust gas. The main differences between this sensor and the voltaic sensor are that the voltage levels are higher and that there is low voltage for a rich mixture and a high voltage for a weak mixture.

At the critical region, where the air–fuel ratio is chemically correct (lambda = 1), there is a marked change in the resistance of the sensor element which leads to it producing a waveform similar to that of the ZrO_2 sensor, except that the voltage across the sensor is probably higher. The actual value is dependent on the voltage that is applied to the sensor.

5.8.3 ON-BOARD MONITORING OF THE CATALYTIC CONVERTER

The USA OBDII and impending European legislation requires that vehicle emissions systems are equipped with the facilities to illuminate a warning lamp (malfunction indicator lamp or MIL) should the catalytic converter cease to function correctly. In order to meet this requirement it is current practice to fit a second oxygen sensor downstream of the catalyst, as shown in Fig. 5.34.

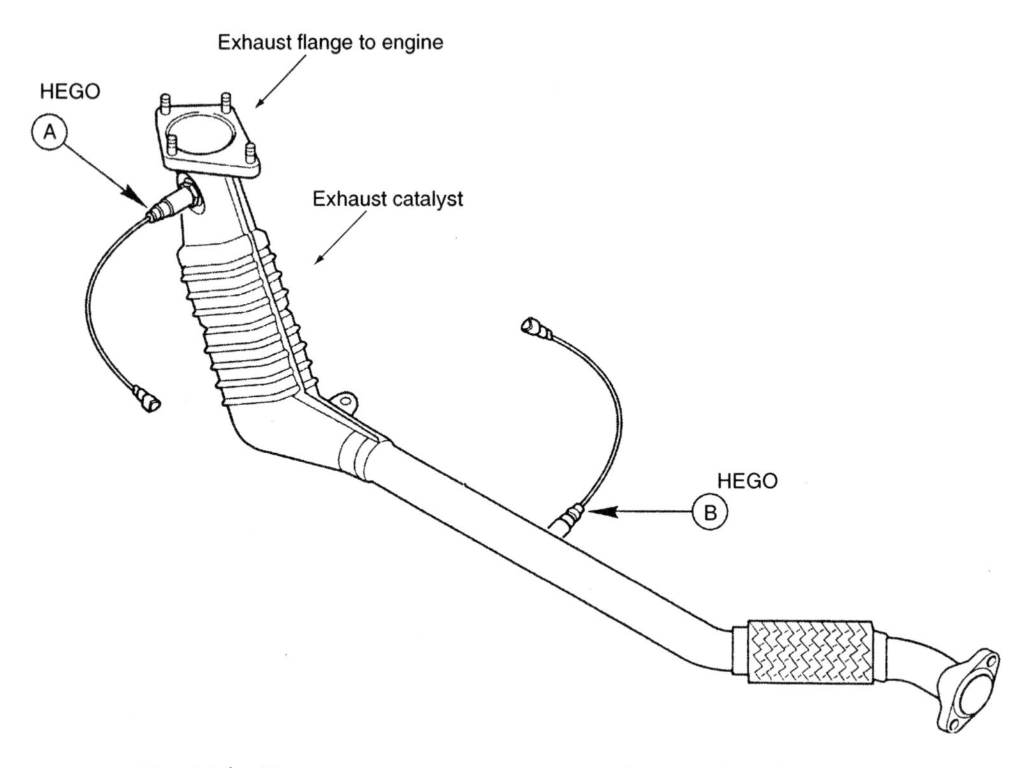

Fig. 5.34 The downstream oxygen sensor that monitors the catalyst

In Fig. 5.34, A represents the upstream oxygen which is on the engine side of the catalyst. It is this sensor that provides the feedback signal that the ECM uses to control the air-fuel ratio within the required limits. The second sensor at B sends a signal to the ECM that is used to determine the efficiency of the catalyst. The voltage amplitude of this second sensor signal is the key to assessing the catalyst efficiency. As the catalyst ages, or is damaged by incorrect fuel etc., the voltage amplitude of this second sensor increases.

5.9 Air flow measurement

An engine requires the correct air-fuel ratio to suit various conditions. With electronic fuel injection the ECM controls the air-fuel ratio and in order to do this it needs a constant flow of information about the amount of air flowing to the engine. With this information, and data stored in its memory, the ECM can then send out a signal to the injectors, so that they provide the correct amount of fuel. Air flow measurement is commonly performed by a 'flap'-type air flow sensor. The air flow sensor shown in Fig. 5.35 uses the principle of the potential divider (potentiometer).

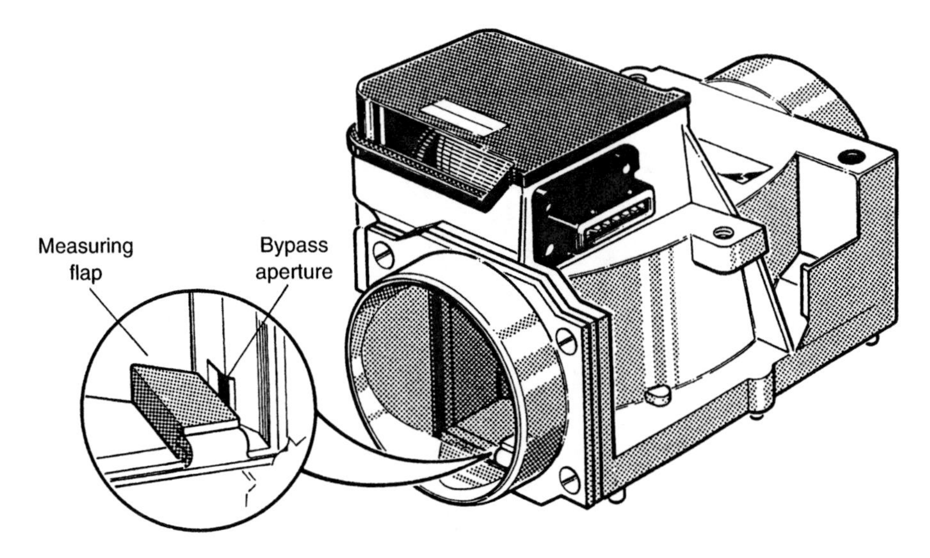

Fig. 5.35 An air flow sensor

Figure 5.36 shows the theoretical form of a simple potential divider. A voltage, say 5 V, is applied across terminals A and B. C is a slider which is in contact with the resistor and a voltmeter is connected between A and C. The voltage V_{AC} is related to the position of the slider C in the form $V_{AC} = V_{AB} \times x/l$.

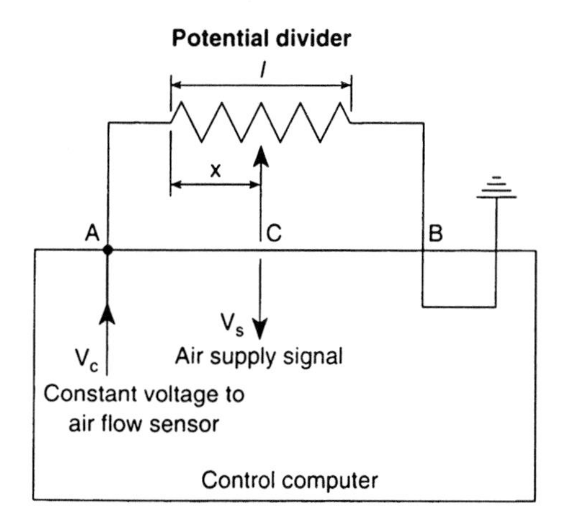

Fig. 5.36 A simple potential divider

In the air flow sensor, the moving probe (wiper) of the potential divider is linked to the pivot of the measuring flap so that angular displacement of the measuring flap is registered as a known voltage at the potentiometer.

Figure 5.37 shows a simplified form of the air flow sensor. The closed position of the measuring flap will give a voltage of approximately zero, and when fully open the voltage will be 5 V. Intermediate positions will give voltages between these values. In practice, it is not quite as simple as this, because allowance must

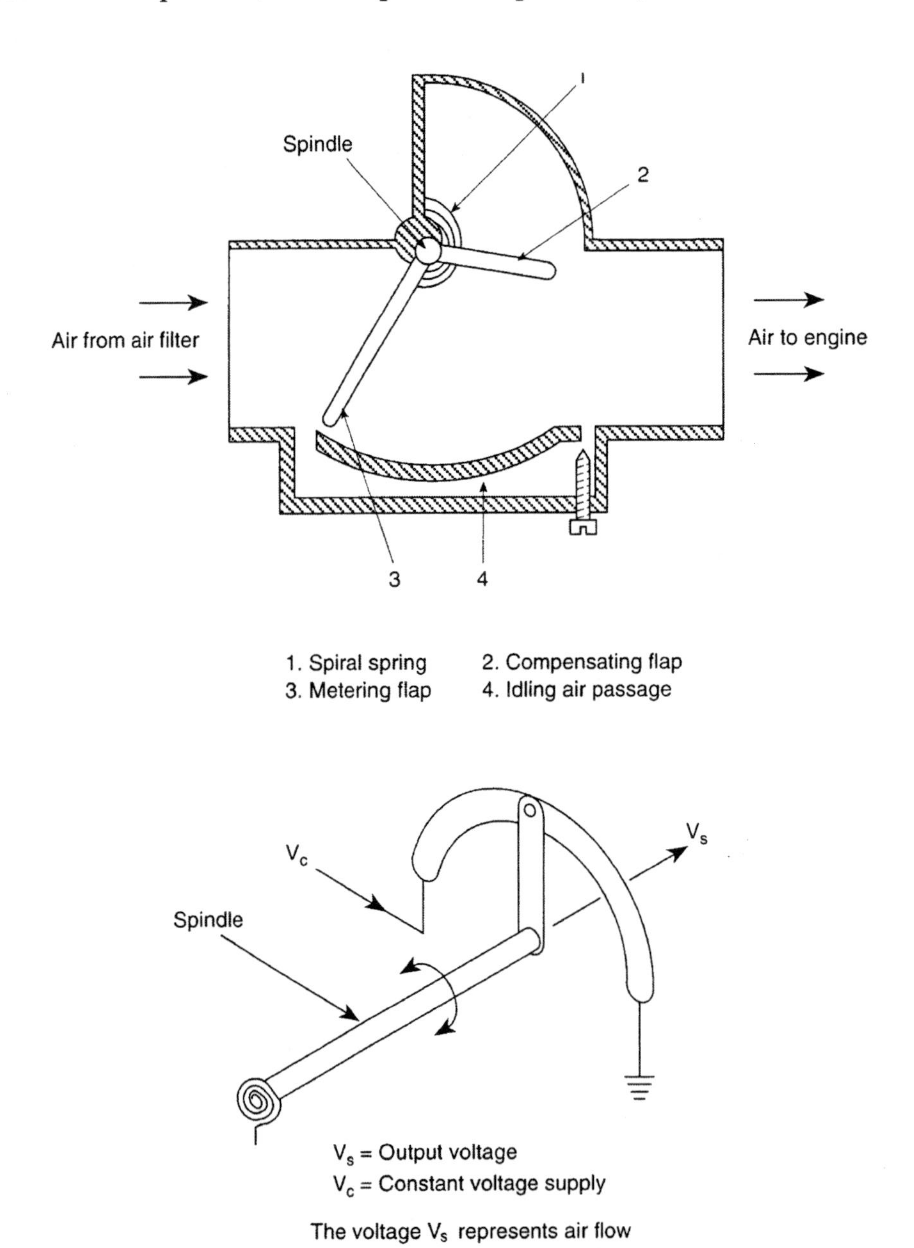

Fig. 5.37 The potential divider applied to an air flow sensor

be made for other contingencies. However, this should not detract from the value of knowing the basic principles because it is these which lead to the diagnostic checks that can be applied. Before we consider some basic checks that can be performed, we will have a more detailed look at the Lucas air flow meter of Fig. 5.35.

Figure 5.38 shows the principle of the Lucas 2AM sensor. The relative position of the measuring flap depends on the air flowing to the engine, and the return torque of the spring. This diagram gives a little more information because it shows a refinement, the idle air bypass, that is required for satisfactory operation of the device.

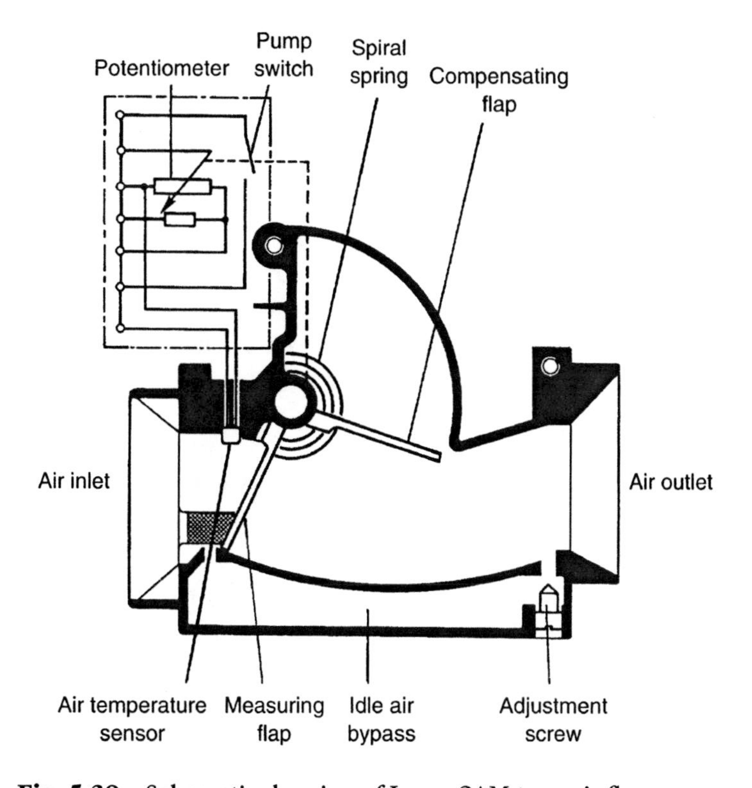

Fig. 5.38 Schematic drawing of Lucas 2AM type air flow sensor

To compensate for production tolerances and different air requirements for similar engines at idling speed, it is desirable that the air-fuel ratio can be adjusted under idling conditions. The idle air bypass together with the adjustment screw permits the idle mixture to be adjusted to suit individual engine requirements. The air flow meter (sensor) incorporates a fuel pump switch. This switch is controlled by the initial (approximately 5°) movement of the measuring flap. This 'free play' is controlled by a very light spring, which is overcome by a very small airflow, i.e. idle speed. This method of switching the fuel pump ensures a minimum fire risk in the event of a collision, when fuel pipes could fracture.

To improve the stability of the measuring plate, when pressure variations in the inlet tract occur, a compensating flap is used. Both the measuring flap, and the compensating flap are part of the same casting and rotate about the same shaft center. Whatever force is felt by the measuring flap is also felt by the compensating flap and this reduces the effect of pressure fluctuations in the air inside the induction manifold. The equilibrium (balance) of the two flaps is only disturbed when the pressure on the engine side of the measuring flap is lower than atmospheric. This pressure is also felt on both sides of the compensating flap.

The air flow sensor also incorporates a temperature sensor which sends a signal to the ECU which uses it to calculate the mass of air. (Mass and volume of air are linked by temperature.) The approximate form of the signal output that can be expected from sensors of this type is shown in Fig. 5.39.

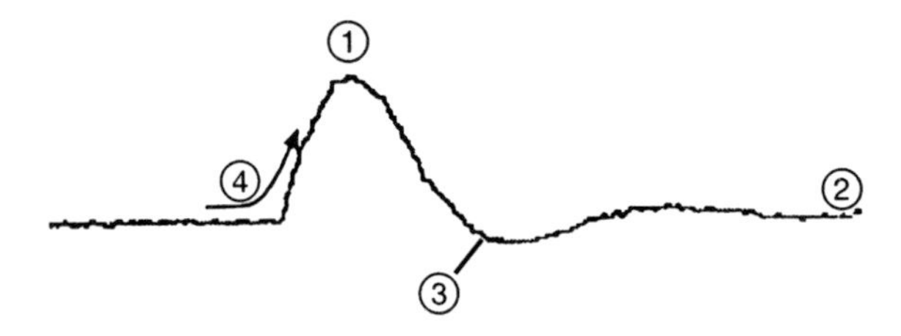

Fig. 5.39 The approximate form of the voltage pattern

The following points should be noted.

1 The voltage is at its highest value when the throttle is wide open and the engine is accelerating.
2 The idling air flow produces a low voltage.
3 This dip is caused by the damping action that is caused by movement of the air flap.
4 The voltage increases as the air flow increases with speed and throttle opening.

5.9.1 HOT WIRE MASS AIR FLOW SENSOR (MAF)

The 'hot wire' air flow meter, shown in Fig. 5.40, incorporates a small orifice inside the main body of the flow meter. The flow meter is positioned between the throttle body and the air cleaner. The main air supply for combustion therefore passes through this meter and a steady flow of the same air passes through the sensing orifice. In the sensing orifice are placed two wires, a compensating wire and a sensing wire. The compensating wire has a small electric current passing through it and the electronic circuit is able to determine the temperature of the incoming air by measuring the resistance of this wire.

The sensing wire is 'hot', about 100°C above the temperature of the compensating wire. This temperature is maintained by varying the current flowing through it. The air flowing to the engine has a cooling effect on the sensing wire so the current in the sensing wire is increased in order to maintain the temperature. As

Fig. 5.40 A 'hot-wire' air flow sensor

air flow increases so the sensing current is increased, and as the air flow decreases so the sensing current decreases. The resultant current flow gives an accurate electrical 'analogue' of air flow which is used in the ECM as an element of the controlling data for the fuel supply.

Modern materials permit the resistive elements of the mass air flow sensor to be built into a metal foil element that is exposed to the intake air stream. Such sensors are frequently built into a housing which is inserted into the air intake system and the principle of operation is similar to that of the hot wire sensor described above.

Figure 5.41 shows the approximate form of the voltage signal that can be expected at the output of a mass flow type of air flow sensor. This pattern shows a voltage that varies as the engine is taken through a range of speeds by operating the throttle.

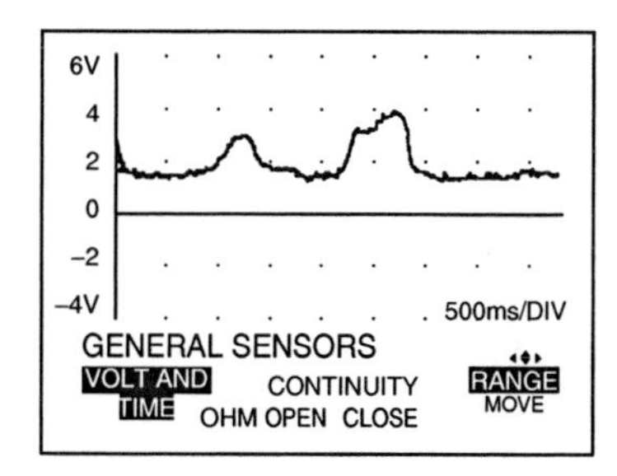

Fig. 5.41 Voltage pattern from a mass flow air sensor as recorded by Bosch PMS 100 oscilloscope

5.10 The practical importance of sensor knowledge

Readers who are familiar with vehicle repair practice will know that, in garage repair shops, most sensors are repaired by replacement. However, it is not unusual to find that parts are replaced in an effort to eliminate a fault, by trial and error. This can prove to be costly and it is not very efficient. In order to be accurate it is essential that the cause of a fault is located before any parts are replaced.

A fault code may just state that a sensor signal is defective. Most sensors are connected to the ECM by a signal wire and this wire may have a number of connectors. Any defect in the circuit between the sensor and the ECM will usually be recorded as a sensor fault. If the sensor output is checked at the ECM (probably via a breakout box) and then checked again, by back probing at the sensor, the connecting circuit can be verified. This point is amplified in Chapter 7 where diagnostic techniques are considered in some detail.

5.11 Review questions (see Appendix 2 for answers)

1. Which sensor plays a major role in the speed density method of air flow sensing?
   (a) The hot wire sensor.
   (b) The knock sensor.
   (c) The cylinder identification sensor.
   (d) The manifold absolute pressure sensor.
2. In OBD II systems a second exhaust gas oxygen sensor is fitted downstream of the catalyst:
   (a) in case the upstream one fails?
   (b) to operate the EGR system?
   (c) to monitor the efficiency of the catalyst?
   (d) to deal with the second bank of cylinders on a vee engine?
3. In the zirconia-type oxygen sensor the difference in oxygen levels:
   (a) changes the resistance?
   (b) causes the sensor to act like a small electric cell and produce a voltage?
   (c) does not affect the sensor output?
   (d) means that the sensor can only be used in diesel engine systems?
4. Coolant temperature sensors often have:
   (a) a negative temperature coefficient?
   (b) resistance that rises with temperature?
   (c) resistance that does not vary?
   (d) screened cables?
5. In some cases, a persistent defect in a sensor reading may cause the ECM to:
   (a) store a fault code and make use of a default value?
   (b) stop the engine?
   (c) make use of a reading from a standby sensor?
   (d) make use of its backup voltage supply?

6. Hall effect sensors:
   (a) generate electricity?
   (b) require a supply of current?
   (c) do not rely on magnetism?
   (d) are only used in ignition systems?
7. An optical sensor:
   (a) uses a light emitter and a light sensitive pick-up?
   (b) operates by switching a bulb on and off at high speed?
   (c) is used to improve the driver's range of vision?
   (d) is only used with fibre optic systems?
8. A potentiometer-type throttle position sensor:
   (a) indicates throttle opening by means of a variable frequency signal?
   (b) cuts off the air supply when stopping a diesel engine?
   (c) indicates throttle opening by means of a varying voltage signal?
   (d) is used to change the octane rating setting of the ECM?

6 Actuators

Actuators are the devices, such as fuel injectors, ignition coils, ABS modulators etc., that are operated by outputs from the ECM. This means that they are normally electromechanical devices and a good deal can be learnt about their operating efficiency through electrical tests conducted by means of multimeters and oscilloscopes. As in the case of sensors, a competent technician can achieve much in the field of diagnosis by knowing the type of electrical behaviour that is to be expected from tests on actuators that are in good condition and, if the test results are not as expected, the types of defect that may be causing an actuator to malfunction. Indeed, modern diagnostic equipment often contains software that enables the user to refer to an ideal scope pattern on a separate window of the scope screen whilst a scope trace is being examined. The alternative to this is that ideal scope patterns are held separately, in a manual, or on CD-ROM or floppy disc.

6.1 Actuator operation

Actuators normally rely on one of two electrical devices for their operation; they are either operated by a solenoid or by an electric motor. Solenoid-operated actuators are normally controlled in one of two ways. One is the duty cycle method, where the solenoid is switched on for a percentage of the time available, e.g. 20 or 80%. This means that pulses of varying width can be used to provide the desired result. The other method of solenoid control is known as pulse width modulation (PWM). Here the solenoid current is switched on and off at frequencies that change to suit operating requirements. Examples of both methods are shown in the oscilloscope patterns that illustrate the injector tests, which are described below.

Electric motors that are used in actuators may be stepper motors, or reversible permanent magnet d.c. motors. A stepper motor can be made to provide small movements of valves by pulsing the current supply. Some stepper motors rotate 7.5° per step, which means that a full rotation of the motor shaft takes 48 steps. A common form of stepper motor uses two sets of windings. Current in one set of windings drives the motor shaft forward and when this is switched off and current is applied to the other set of windings, the motor shaft rotates in the reverse direction. This means that accurate control over the position of a valve can be achieved because the control computer determines the valve position by counting the number of pulses applied to the stepper motor windings.

A selection of commonly used actuators of both types is covered below. Tests on vehicle systems should be approached with caution. It is dangerous to make assumptions, and it is vitally important that you should either be familiar with the product that you are working on, or have to hand the necessary data that relates to the product.

6.2 Petrol engine fuel injectors

In the overview of systems given in Chapter 1, it was shown that there are two basic petrol injection systems in common use. These are single-point (or throttle body) injection and multi-point injection. More recently the concept of direct petrol injection into the combustion chamber has been developed. It is the single-point and multi-point systems only that are covered here.

6.2.1 SINGLE POINT INJECTION

As the name suggests, there is a single injector. This injector is placed at the throttle body, on the atmospheric side of the throttle valve, as shown in Fig. 1.17. The fuel pressure at the injector is controlled by the fuel pressure regulator and the amount of fuel injected is determined by the length of time for which the injector valve is held off its seat. In this particular system, the fuel is injected towards the throttle butterfly where the air velocity helps to mix the fuel spray with the air. Figure 6.1 shows a typical single-point injector.

The injector valve is designed to weigh as little as possible so that it can be opened and closed rapidly. The magnetic field caused by electric current in the solenoid winding opens the valve and when the current is switched off the injector valve spring returns the valve to its seat.

6.2.2 MULTI-POINT PETROL INJECTION

In these systems, there is an injector for each cylinder. The injectors are normally placed so that they spray fuel into the induction tract, near the inlet valve. Figure 6.2 shows the construction of a common type of fuel injector as used for multi-point injection systems.

Multi-point petrol injection systems normally use a fuel gallery to which the fuel pipes of all the injectors are connected. The pressure in this gallery is controlled by the fuel pressure regulator. This means that the quantity of fuel that each injector supplies is regulated by the period of time for which the control computer holds the injector open. This time varies from approximately 1.5 ms at low engine load, up to approximately 10 ms for full engine load. Naturally, these figures will vary

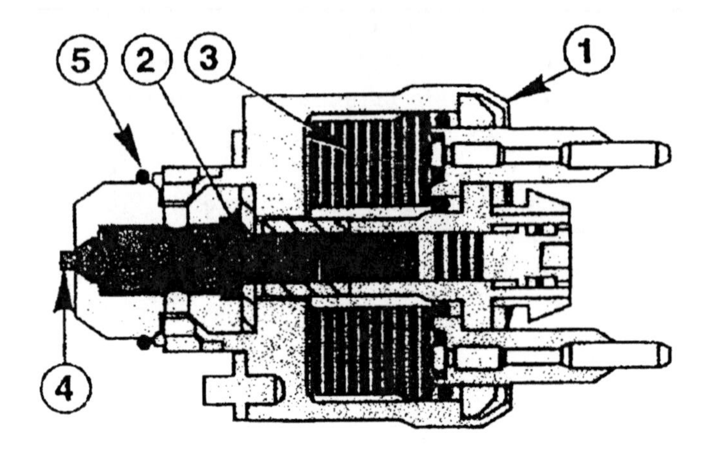

Key
1 Housing
2 Fuel duct
3 Solenoid
4 Jet needle
5 O-ring

Fig. 6.1 A single point, or throttle body injector

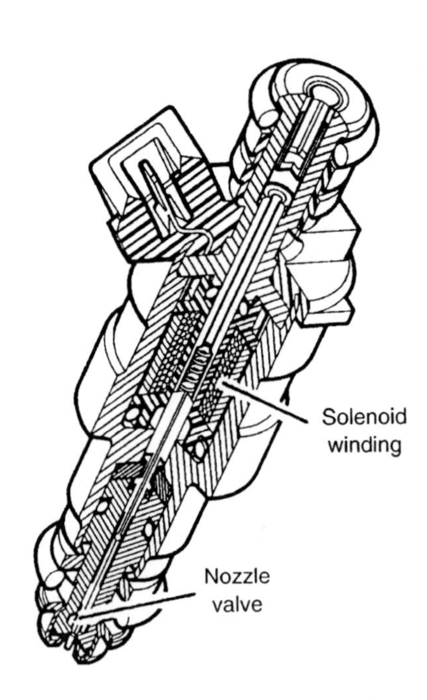

Fig. 6.2 A typical injector for a multi-point injection system

from engine to engine; larger capacity and more powerful engines will require greater amounts of fuel than small capacity and low powered engines.

6.3 Testing of petrol injectors

There are three electrical methods that are commonly used to regulate the operation of petrol injectors and they each aim to keep the injector solenoid winding as cool as possible whilst giving the best possible injector performance. The three methods are:

1. peak and hold;
2. conventional ECM-controlled transistor switching to earth (duty cycle);
3. pulse width modulation.

6.3.1 PEAK AND HOLD

Figure 6.3 shows the current flow in a peak and hold single-point injector. This behaviour is achieved by having two paths to earth that are controlled by the ECM, as shown in Fig. 6.4.

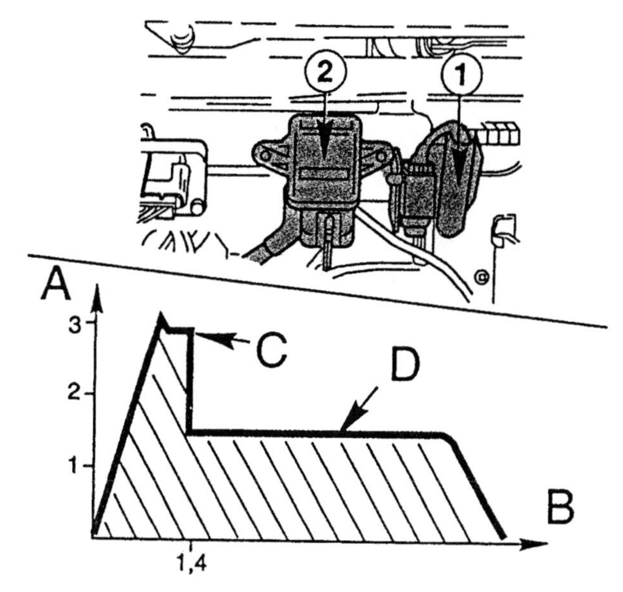

Key
1 Location of ballast resistor
2 MAP sensor
A Current in Amps
B Time in milliseconds
C Pickup current (Phase 1)
D Holding current (Phase 2)

Fig. 6.3 Electric current pattern in injector

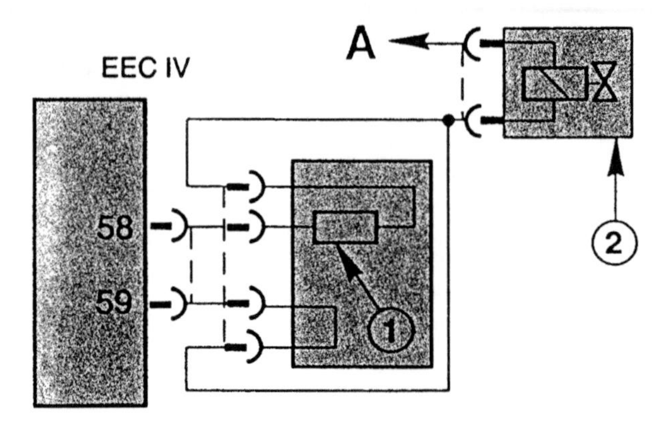

Key
1 Ballast resistor
2 Injector valve
A To power relay

Fig. 6.4 Injector circuit

The peak current occurs when the injector is first earthed through both circuits. After a short period the circuit without the resistor is switched off by the ECM. This effectively inserts the resistor into the injector circuit and leads to a reduction in the current flowing through the solenoid winding. An oscilloscope connected between the ECM side (earth) connection of the injector will show a pattern that reflects the current flow in the circuit. Because the current falls during the hold stage, there will be a rising edge in the voltage trace. This is not the end of the injector period because this comes when the rising edge that accompanies switching off occurs. In order to determine the length of the injector pulse from the scope pattern it is necessary to refer to the time from the first falling edge to the second rising edge of the scope trace.

6.3.2 CONVENTIONAL SWITCHING TO EARTH

Heavy duty transistor switches that are controlled by the ECM (as shown in Fig. 6.5) are used to complete the circuit to earth in order to operate the injectors.

In this case the oscilloscope pattern will be similar to that shown in Fig. 6.6. In order to limit current flow in the injector circuit the earth path may include a series resistor. In some applications the injector solenoid winding is designed to have a higher resistance.

By reference to the trace in Fig. 6.6 it is possible to see the following.

- The high voltage at (1) is caused by the surge that occurs when the current is switched off and the magnetic field of the injector solenoid collapses.
- This is the point at which the ECM switches off the transistor and current ceases to flow. It is the end of the injection pulse.

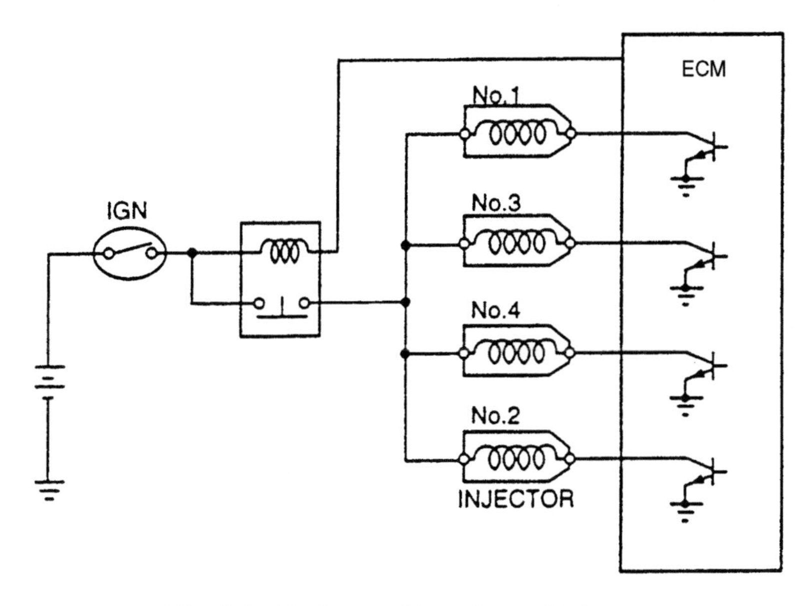

Fig. 6.5 Earth switching of petrol injectors

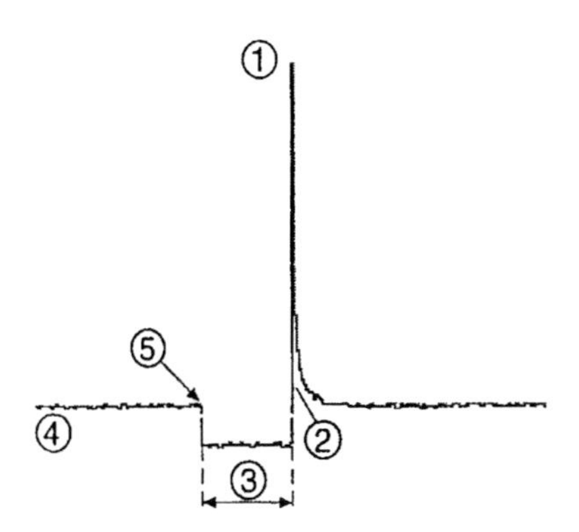

1	Peak voltage caused by the collapse of the injector coil.
2	Driver transistor turnsoff, discontinuing fuel flow.
3	Injector ON-time
4	Battery voltage (or source voltage) supplied to the injector.
5	Driver transistor turns on, pulling the injector pintle away from its seat, starting fuel flow.

Fig. 6.6 Oscilloscope trace of voltage for one pulse of the injector

- This represents the period of time for which the injector is delivering fuel and the actual time (in ms) can be determined by reading off the time base scale of the oscilloscope.
- This represents the voltage supplied to the injector (normally system voltage is 12 V).

- This is the point at which the ECM switches on the driver transistor, to earth the solenoid winding, and it is the point at which injection commences.

6.3.3 PULSE WIDTH MODULATED INJECTORS

Pulse modulated (power control) injectors have a high initial current to provide rapid opening of the injector valve. After the initial pulse, the ECM switches the earthing transistor on and off at high frequency. This permits the injector to remain open for the required time whilst preventing the circuit from overheating. Figure 6.7 shows the type of oscilloscope trace that should be obtained from this type of injector.

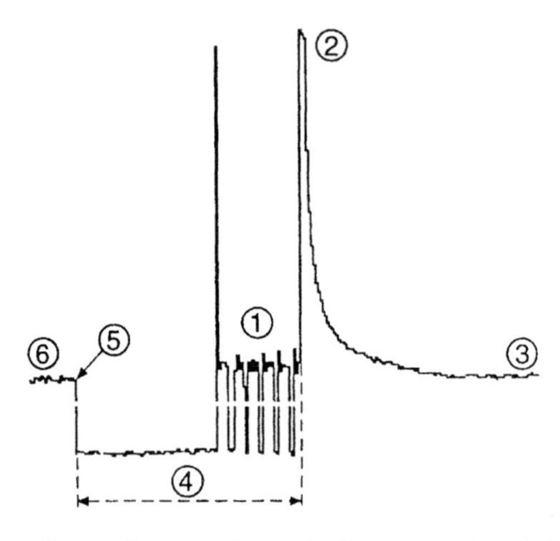

1	Current flow pulsed on and off enough to keep hold in winding activated.
2	Peak voltage caused by collapse of the injector coil, when current is reduced.
3	Return to battery (or source) voltage.
4	Injector ON-time.
5	Driver transistor turns on, pulling the injector pintle away from its seat, starting fuel flow.
6	Battery voltage (or source voltage) supplied to the injector.

Fig. 6.7 Oscilloscope trace for a pulse width modulated injector

Referring to the Fig. 6.7, please note the following.

1. Current flow pulses on and off at a frequency that holds the injector valve open.
2. Voltage peak is caused by the voltage surge that occurs when the injector solenoid current is switched off and the magnetic field collapses. (In some systems this peak may be shortened because the system is provided with surge protection.)
3. Here the injector voltage is returned to battery voltage ready for the next injector action.

4 This represents the injector 'on-time'. (The sharp rise in voltage as the injector starts to pulse is caused by the sudden interruption of current as the high frequency pulsing commences.)
5 This is where the ECM switches on the driver transistor and starts the injection period.
6 This is the supply voltage, probably battery voltage.

These explanations of the methods of operation of petrol injectors, together with the oscilloscope traces, should give a good insight into the types of test that

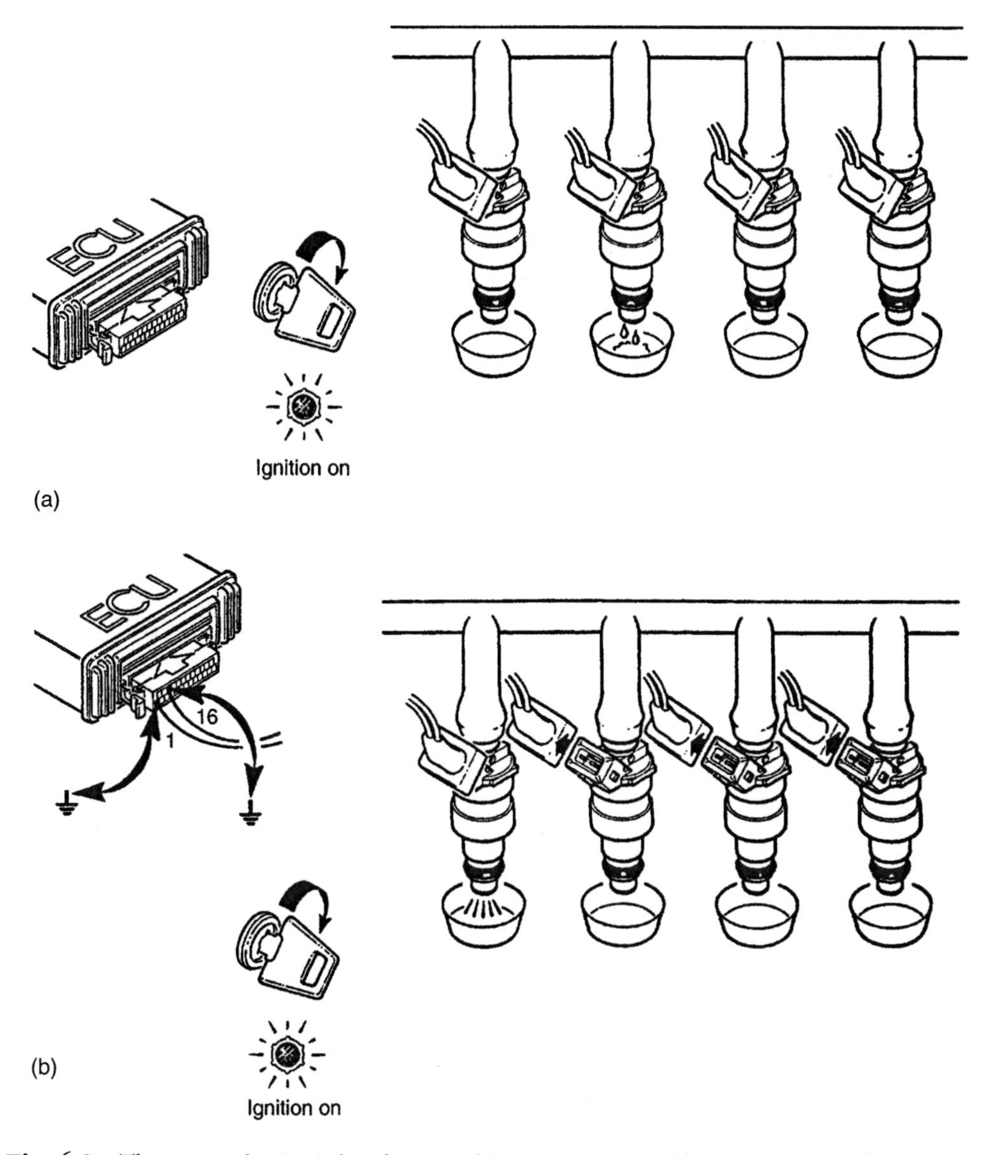

Fig. 6.8 The general principle of a petrol injector test. (a) Testing injectors for leakage. (b) Testing the amount of fuel per injector

can be performed in order to assess the effectiveness of injectors and thus assist technicians in their diagnostic work. However, it is important to remember that the function of an injector is to provide the correct amount of fuel, in a combustible condition, as and when the engine requires it. This means that complete testing of injectors requires examination of other aspects of injector performance, such as the amount of fuel delivered and the spray pattern.

6.3.4 FURTHER INJECTOR TESTS

Figure 6.8 shows examples of two tests that are possible on a vehicle. Again these should only be undertaken by competent personnel, in possession of the necessary instructions and who are careful to take all necessary safety precautions.

Figure 6.8(a) illustrates the type of test that can be performed to check for injector leakage. The injectors are removed from the manifold and securely placed over drip trays. Note that they remain connected to the fuel and electrical system. The approved procedure is then applied to energize the injectors and a careful visual inspection will reveal if there are any leaks. In the case shown it is recommended that any injector that leaks more than 2 drops of fuel per minute should be replaced.

Figure 6.8(b) shows the same injectors. In this case three of them are disconnected and the amount of fuel delivered by the injector which is still connected, is carefully measured. It is possible to obtain special injector test benches on which these tests can be performed in ideal conditions.

In many applications, the petrol injectors can be activated by the diagnostic tester through the serial port. This makes it possible to hear the injectors 'click' as they are pulsed. If the fuel supply is active during this process, a certain amount of fuel will enter the intake manifold. In order to prevent damage to the exhaust catalyst, steps must be taken to prevent the fuel passing through the engine. One procedure advocates cranking the engine over with the spark plugs removed, to expel the unburnt fuel, on completion of the injector test.

6.4 Exhaust gas recirculation

The principle of EGR is described in Chapter 1. In this chapter a test that can be performed on some EGR valves is considered. The exhaust gas recirculation valve permits the passage of a regulated amount of exhaust gas, into the intake side of the engine, under the control of the engine ECM. The effect is to reduce temperature in the combustion chamber below approximately 1800°C and this prevents the formation of the oxides of nitrogen. In order to function at its best, EGR is used only under certain conditions, e.g. at cruising speeds and during reasonable acceleration. In order to determine the state of the EGR valve, i.e. is it on or off?, the system is equipped with a sensor. A common form of EGR sensor is the voltage divider (variable resistor) type that is mounted on top of the EGR valve so that the position of the EGR valve is represented by a voltage.

6.4.1 TESTING THE EGR SENSOR

In order to avoid injury it is recommended that the test is performed on a cold engine. This is only possible where the EGR valve can be lifted from its seat manually. The type of voltage trace that can be expected from this type of sensor is shown in Fig. 6.9. In order to obtain a voltage trace the sensor must be energized and it is always wise to check that the supply voltage to the sensor is correct.

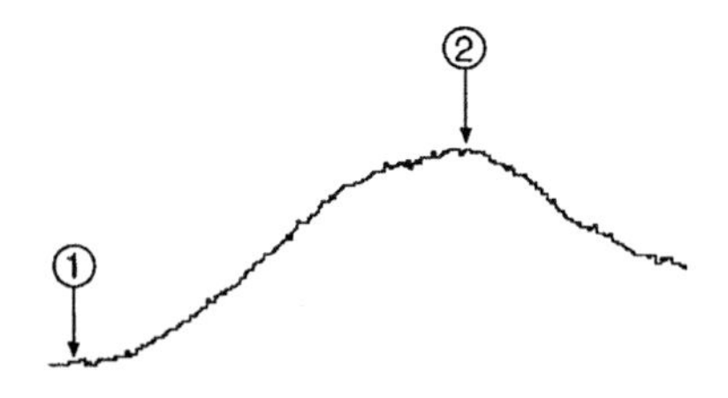

1 EGR valve closed restricting the flow of exhaust gas.
2 EGR valve open nearly all the way allowing flow of exhaust gas.

Fig. 6.9 An impression of the voltage trace from an EGR sensor

If it is necessary to perform the test under load, the scope connections should be made secure and the test instrument placed in a secure position so that the pattern can be observed without risk of personal injury from either the dynamometer, or the moving parts of the vehicle. If the test is performed during a road test it is essential that an assistant accompanies the test driver.

The valve that controls the vacuum that energizes the EGR valve is often controlled by a solenoid which, in turn, is controlled by the ECM. Figure 6.10 shows the connections for the oscilloscope for this test and Fig. 6.11 shows a typical duty cycle trace for the EGR valve.

As the duty cycle may be increased or decreased by the ECM, the operation of the EGR valve is controlled to permit varying quantities of exhaust gas to be transferred from the exhaust system into the intake system.

6.5 Petrol engine idle speed control

Idle speed control is an important element of the control strategy for any engine management system. The control strategy for engine idling must take account of factors such as engine coolant temperature, engine load, power assisted steering, alternator load, etc. Many systems are fitted with an idle speed control valve that provides a supply of air that by-passes the throttle valve, whilst other systems

● MultiScope key sequence

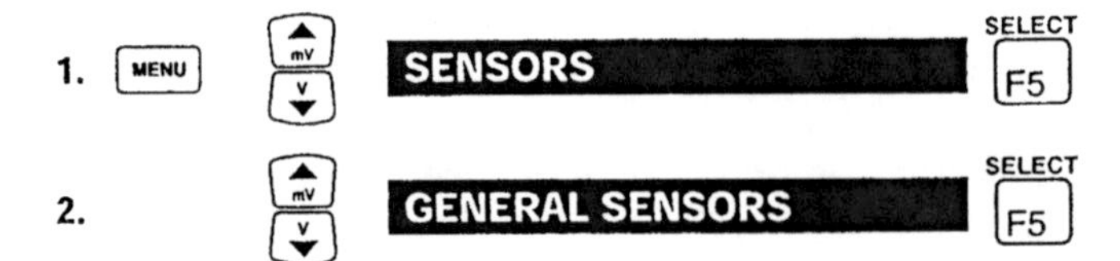

3. Connect the test leads as displayed by the MultiScope's **Connection Help** as shown below.

4. OK F1 Starts the EGR test.

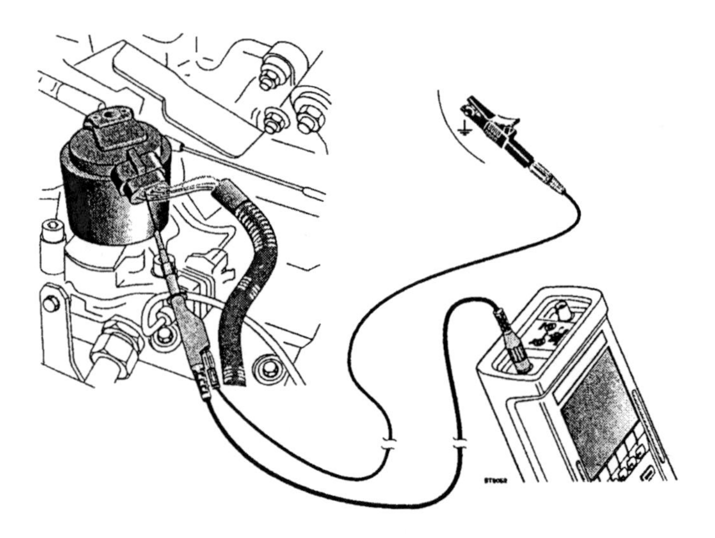

Fig. 6.10 Making the connections for the test on the EGR valve

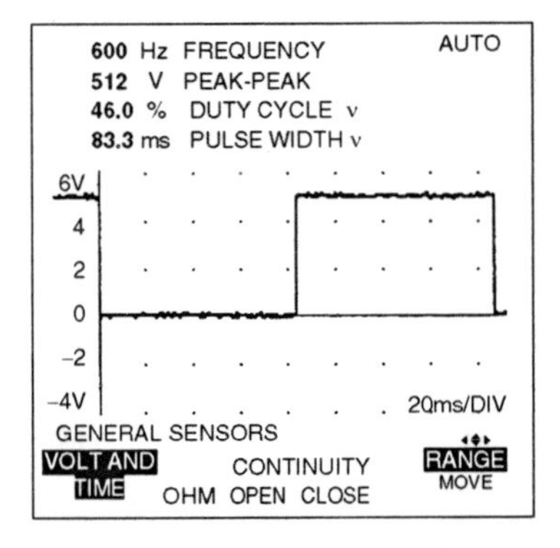

Fig. 6.11 Oscilloscope trace from an EGR valve operated by duty cycle

may make use of the electronic throttle control. Two types of valve are used to provide a computer controlled idle air supply. One makes use of a stepper motor, as shown in Fig. 6.12, and the other uses a solenoid operated valve as shown in Fig. 6.15.

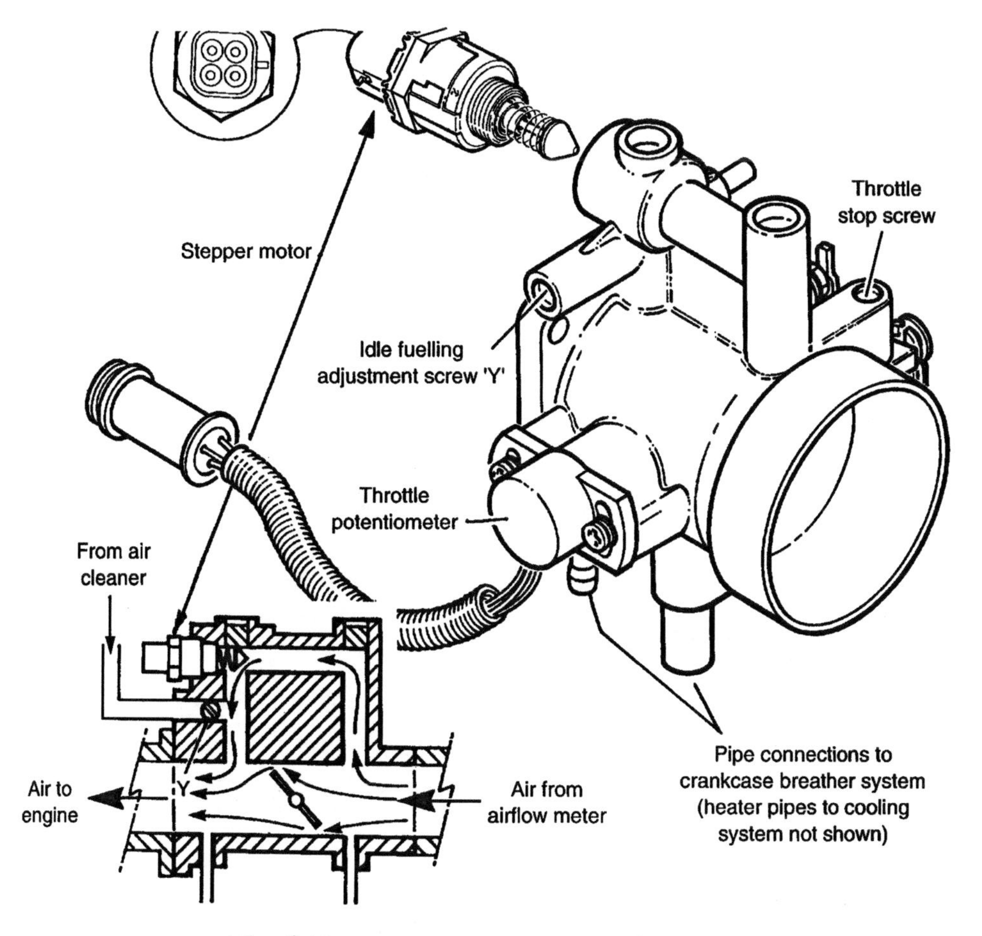

Fig. 6.12 A stepper motor operated air valve

6.5.1 *STEPPER MOTOR-OPERATED VALVE*

Figure 6.12 shows a simplified arrangement of the extra air (air by-pass) valve that is built into the throttle body of some petrol injection systems.

The ECU pulses the transistor bases, in the correct sequence, so that the stepper motor moves the air valve to provide the correct air supply, for any given condition. In addition, other sensor signals will enable the ECU to provide the correct amount of fuel to ensure that the engine continues to run smoothly.

Figure 6.13 shows the stepper motor with the air valve attached. The multiple pin connection is typical of the type of connection that is used to electrically connect the stepper motor to the ECU. The stepper motor can normally be checked

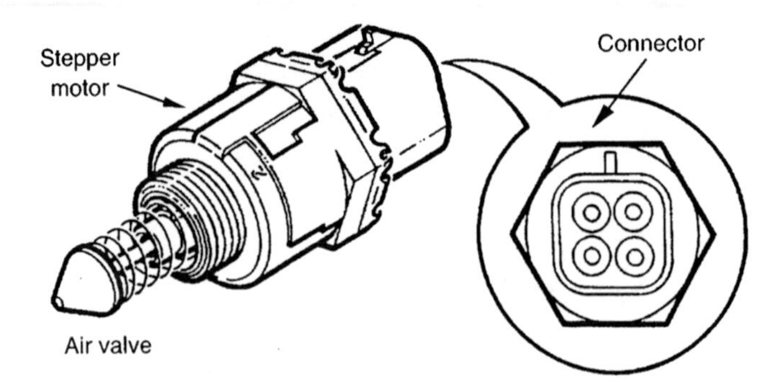

Fig. 6.13 The stepper motor and extra air valve (Lucas)

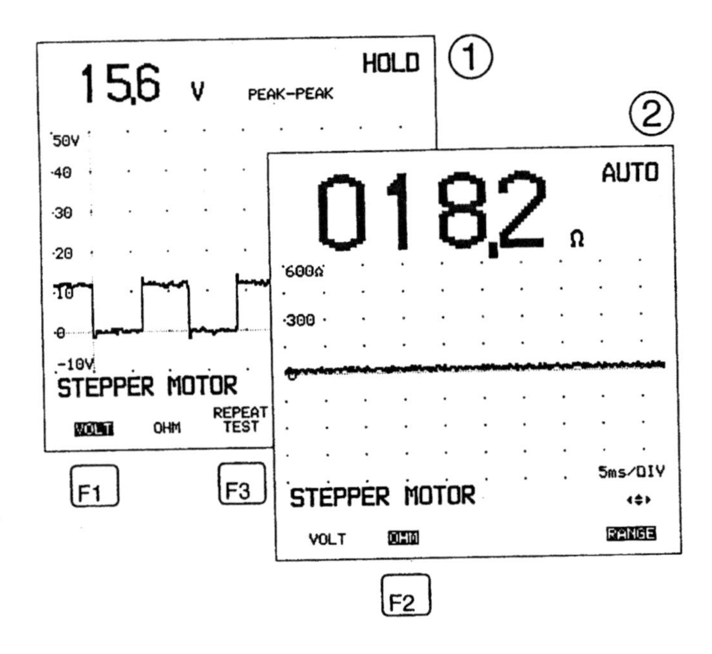

(1)	**PEAK-PEAK**	Indicates the difference in voltage between the lowest and the highest value of the displayed waveform.
(2)	This screen displays the measured resistance in ohms (Ω).	

Fig. 6.14 Voltage trace from a stepper motor test

by operating it with the diagnostic tool connected to the serial communication port of the ECM. An oscilloscope can also be used to check the pulses that are sent to the motor from the ECM. Figure 6.14 gives an impression of the type of result that is to be expected from the PMS 100 oscilloscope when used to test a stepper motor.

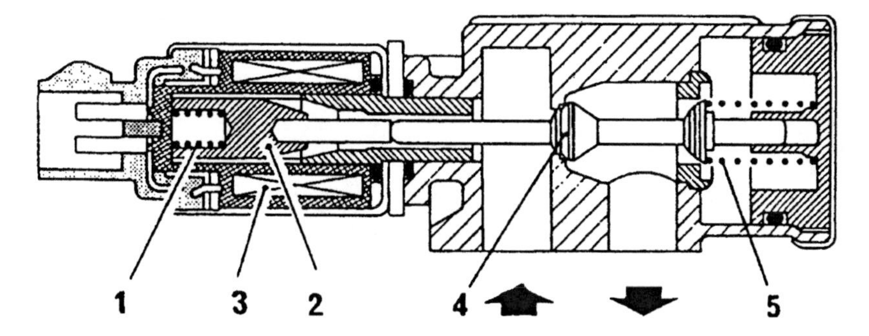

Fig. 6.15 A solenoid operated idle speed control valve

● Measurement conditions

- Engine RUNNING after you have connected the MultiScope to the Idle Air Control valve. Monitor the valve's operation with the engine cold, warming up, and hot.
- Introduce a small vacuum (false air) leak, and watch the signal from the ECU as it adjusts the valve's opening.

● MultiScope key sequence

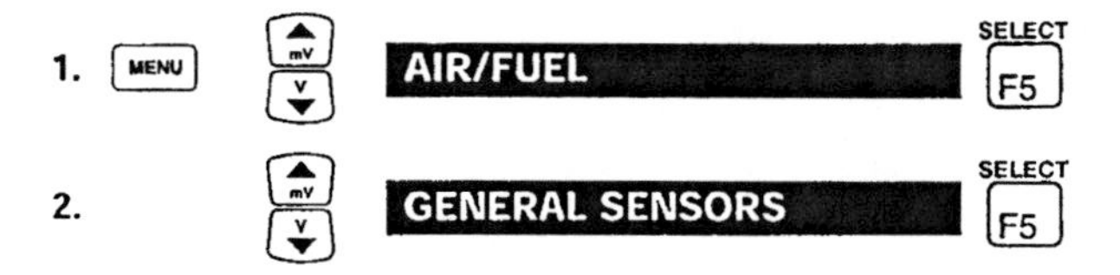

3. Connect the test leads as displayed by the MultiScope's **Connection Help** as shown below.

4. OK F1 Starts the idle air control test.

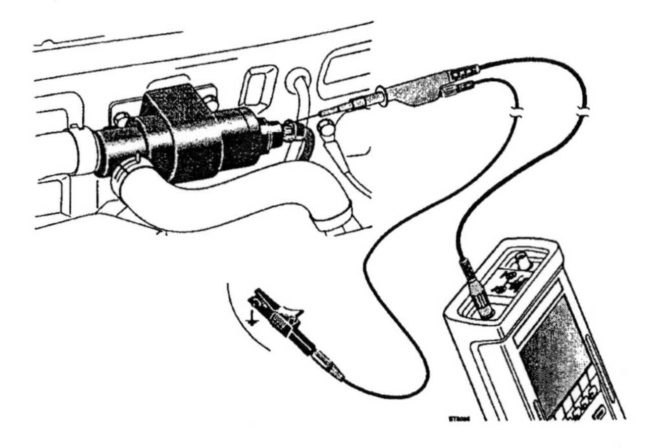

Fig. 6.16 Checking an idle speed air valve

6.5.2 SOLENOID-OPERATED VALVE

This type of valve regulates the amount of air that by-passes the throttle valve through the medium of a solenoid-operated valve of the type shown in Fig. 6.15.

In the rest position shown, the valve (4) is closed by the spring (5) and the armature of the solenoid (2) is pushed back inside the solenoid coil (3). When operating, the energized solenoid opens the valve (4) and admits air to the induction system. The quantity of air admitted is controlled by duty cycle pulses that are sent from the ECM.

Testing a solenoid-operated valve

Figure 6.16 shows the portable oscilloscope being used to check the performance of a solenoid-operated idle speed control valve and Fig. 6.17 shows the pattern that can be expected as the ECM pulses the solenoid.

The suggested test conditions are that the valve's operation should be monitored when the engine is cold, when it is warming up, and when it is hot. It should be noted that different types of valve will produce different shaped patterns. The different shapes of the waveforms arise from the behaviour of inductive-type circuits. The pronounced saw tooth effect in Fig. 6.17(c) arises from the inductive reactance which is a feature of many inductive devices. Once again, we see the need to have good information about the product being worked on in order to make informed judgements about condition.

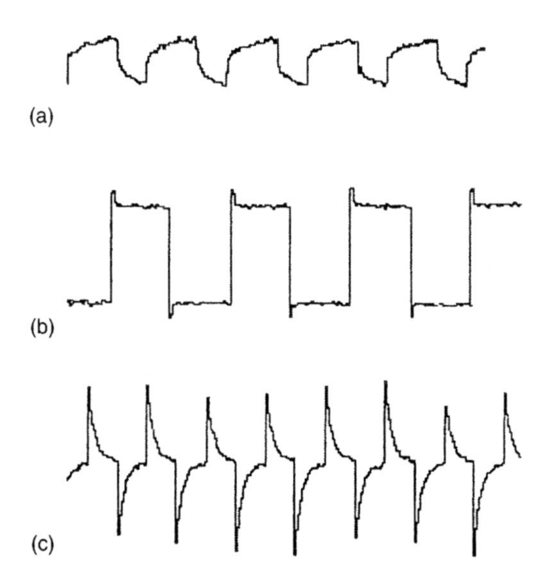

Fig. 6.17 Idle air bypass waveforms

Computer controlled purging of the charcoal canister is frequently performed by the operation of a (duty cycle) valve that is under the control of the ECM. The length of the 'on pulse' of the solenoid varies according to operating conditions.

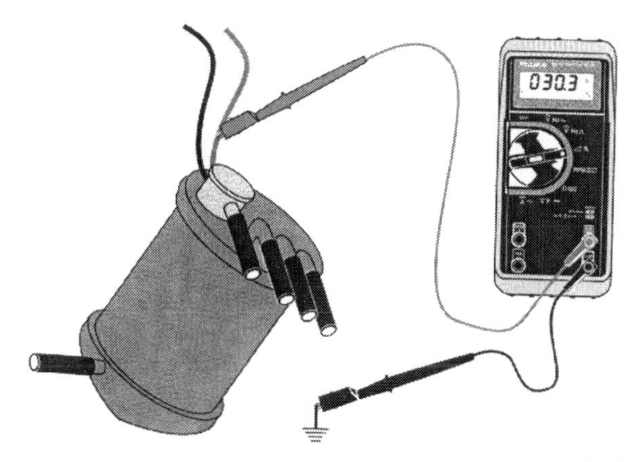

To measure the duty cycle of a solenoid, attach the red lead to the signal wire and the black lead to a good engine ground. Select duty cycle and read the value directly.

Fig. 6.18 Testing the operation of the solenoid valve on a charcoal canister

Figure 6.18 shows the Fluke meter connected to the purge valve solenoid signal wire and earth. The meter shows a 30.3% duty cycle.

Were the oscilloscope to be used for this test the connections would be made in an identical way to the Fluke meter connections shown above. The pattern then obtained would show the characteristic shape similar to that shown for the test on the solenoid-operated idle valve. The 'on-pulse' length will vary as the ambient temperature and vehicle operating conditions change.

6.6 Ignition system

Computer controlled engine management systems frequently use distributorless ignition systems that rely on the lost spark principle. In the system shown in Fig. 6.19 there are two secondary windings that generate the HT sparks.

Both ends of each HT coil are connected to sparking plugs as shown in Fig. 6.20. One plug, say cylinder number 1, sparks near TDC on the compression stroke and this means that the sparking plug in cylinder number 4 sparks near the end of its exhaust stroke. The control computer activates the primary side of the ignition system. Because of the direction of current flow in the secondary winding of the coil, the ignition voltage polarity is different at the spark plug placed at either end of each secondary winding. In one, the current is flowing into the spark plug, and in the other, the current is flowing away from the spark plug.

6.7 ABS actuators

One of the main actuators on an anti-lock braking system is the unit known as the 'modulator'. The modulator contains a pump driven by an electric motor and

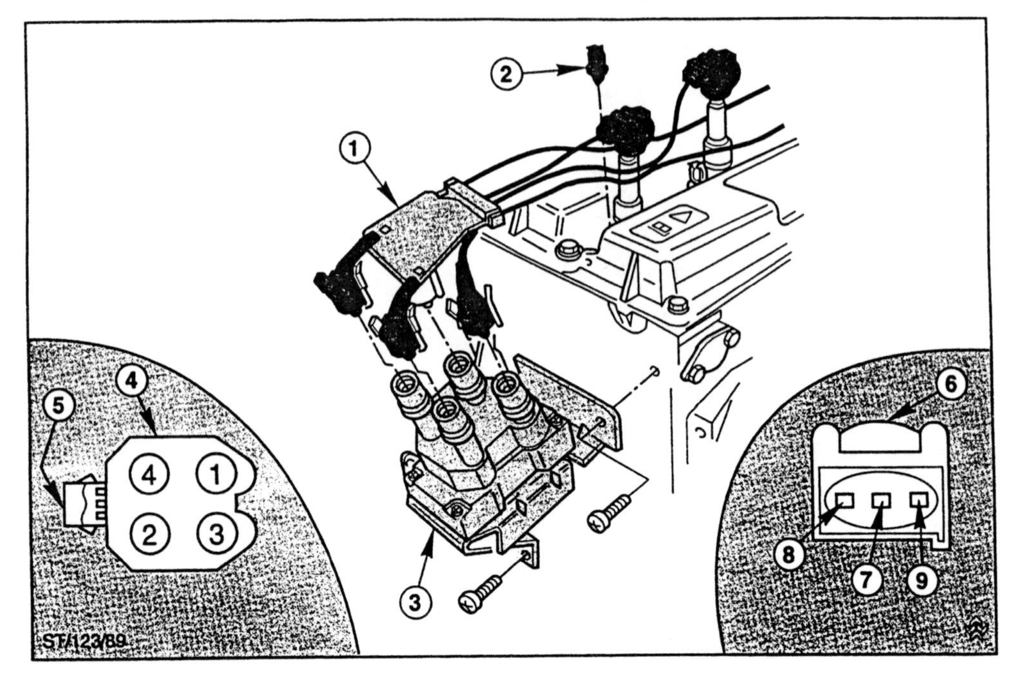

1 HT lead - cylinders 1 to 4 with protection cap for DIS coil
2 Securing clip for HT lead, secured in camshaft cover
3 DIS ignition coil
4 Marking of HT lead connections at underside of coil housing
5 Connection - engine wiring loom
6 Loom connector
7 From ignition switch, terminal 15 (Battery +)
8 From E-DIS module, Pin 10 (cylinder 1-4)
9 From E-DIS module, Pin 12 (cylinder 2-3)

Fig. 6.19 Distributorless ignition

various solenoid-operated valves. The actual working conditions of ABS are difficult to simulate for test purposes and it is normal practice for the ABS computer to have quite a large self-diagnosis capacity. Figure 6.21 shows the extent of the fault analysis data that is available through the diagnostic link of a fairly recent Lexus vehicle.

6.8 A clamping diode

In several of the oscilloscope tests on actuators, mention is made of the voltage 'spike' that occurs when the device is switched off. In cases such as relays, horns, electric fan motors etc., the circuit that operates the device is equipped with a diode that 'blocks' current. This is known as a clamping diode. Figure 6.22 shows the type of waveform that can be expected from a test on these devices.

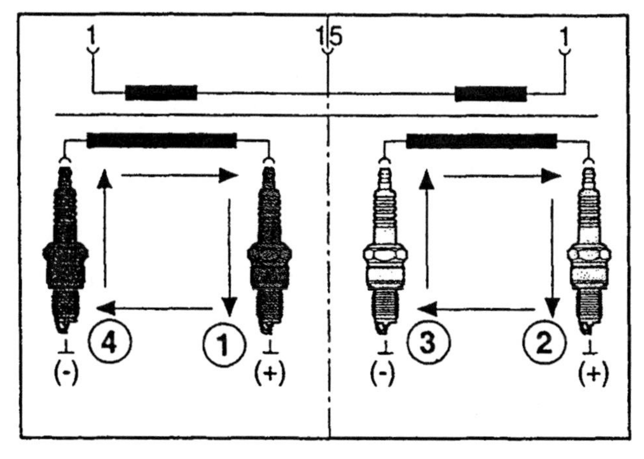

Secondary current : ignition circuit 1–4

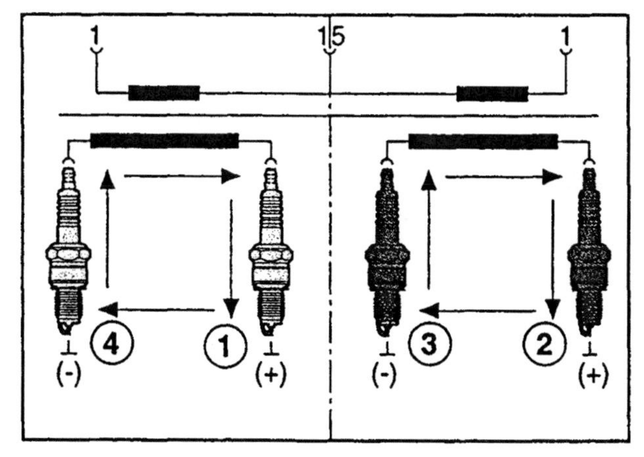

Secondary current : ignition circuit 2–3

Fig. 6.20 The HT coil connections

6.9 Electronic unit injectors

Electronic unit injectors (EUI) are used on engine management systems for large diesel engines. The map that provides the data for the system to work on is stored in the ROM of the ECM. Each injector combines a cam-operated high pressure pump with an electrically-operated spill valve and there is one EUI for each cylinder of the engine. The solenoid that operates the spill valve is called a colenoid, it has a short travel and operates at high speed. The principle of operation is as follows (Fig. 6.23).

1. Fuel enters the injector from a fuel gallery in the cylinder head of the engine. The plunger is pushed up by the spring and the spill valve is opened. The injector valve is on its seat.

Self-Diagnosis

In case of any sensor or actuator malfunction, the ABS warning light located in the combination meter will light up to alert the driver that a malfunction has occurred. The ECU will also keep the codes of the malfunctions in memory. For details on the check procedure, see the GS300 Repair Manual (Pub. No. RM 352E1). The description of the diagnostic trouble codes is identical to that of the LS400.

▶ **Diagnostic Trouble Code** ◀

Code No.	Diagnosis
11	Open circuit in solenoid relay circuit.
12	Short circuit in solenoid relay circuit.
13	Open circuit in pump motor relay circuit.
14	Short circuit in pump motor relay circuit.
21	Open or short circuit in 3-position solenoid valve of front right wheel.
22	Open or short circuit in 3-position solenoid valve of front left wheel.
23	Open or short circuit in 3-position solenoid valve of rear wheels.
31	Front right wheel speed sensor signal malfunction.
32	Front left wheel speed sensor signal malfunction.
33	Rear right wheel speed sensor signal malfunction.
34	Rear left wheel speed sensor signal malfunction.
35	Open circuit in front left and rear right speed sensors.
36	Open circuit in front right and rear left speed sensors.
41	Low battery voltage (9.5 V or lower) or abnormally high battery voltage (17 V or higher).
51	Pump motor locked or open circuit.
Always ON	Malfunction in ECU.

Fail-Safe

In the event of a malfunction in an input signal to the ECU, the ECU cuts off its current to the actuator. As a result, the brake system operates in the same way as in a vehicle without ABS, and normal braking function is assured.

Fig. 6.21 Self-diagnosis data from an ABS computer

2. The cam rotates and the pump plunger covers the feed port. The spill valve remains open and the injector valve is still on its seat.
3. Further rotation of the cam pushes the plunger towards the end of its stroke. The colenoid now closes the spill valve. This causes the high pressure fuel to lift the injector valve from its seat and injection takes place.
4. The plunger has now reached the end of its stroke. The colenoid has opened the spill valve and the injection process is completed.

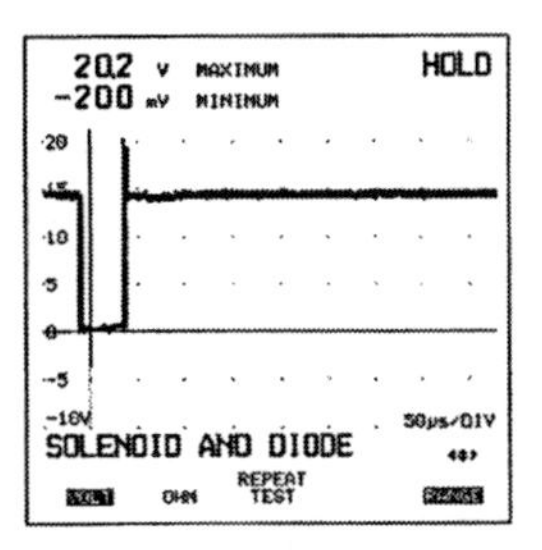

Result display from a solenoid and clamping diode test

● Measurement conditions

Activate the item under test and watch the MultiScope's display

● MultiScope key sequence

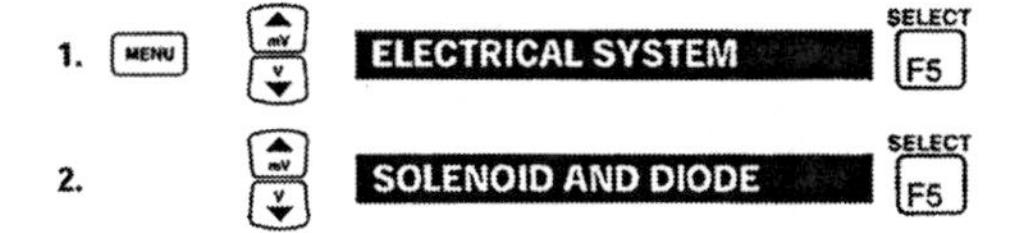

3. Connect the test leads as displayed by the MultiScope's **Connection Help** as shown below.
4. OK F1 Starts the solenoid and clamping diode test.

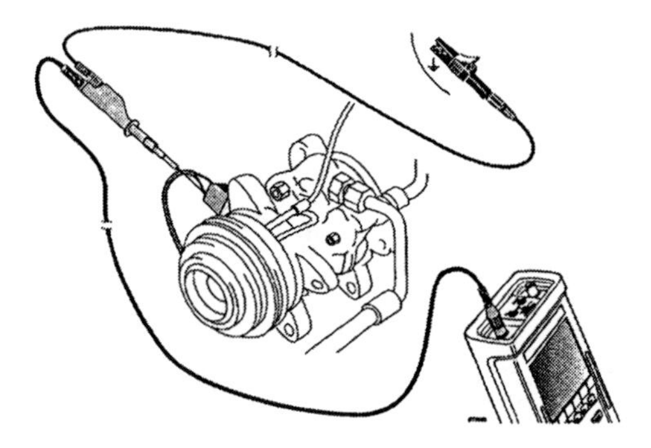

● Clamping diodes

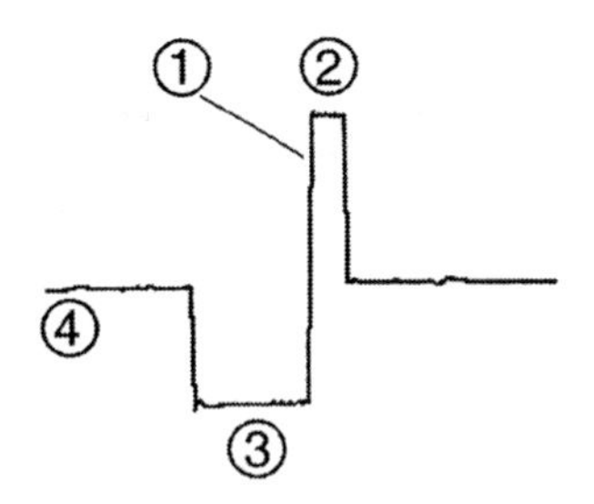

1 Voltage increases due to the collapse of coil voltage.
2 Peak is clamped against the diode threshold voltage.
3 Ground.
4 Battery voltage.

! Specifications may vary. Consult manufacturer's specifications.

Fig. 6.22 Test result from a circuit that includes a clamping diode

The commencement and ending of injection can be varied by the computer control of the colenoid and spill valve.

6.10 Review questions (see Appendix 2 for answers)

1. ABS computers have a good self-diagnosis capacity because:
   (a) they work at high speed?
   (b) their actual working conditions are difficult to simulate for test purposes?

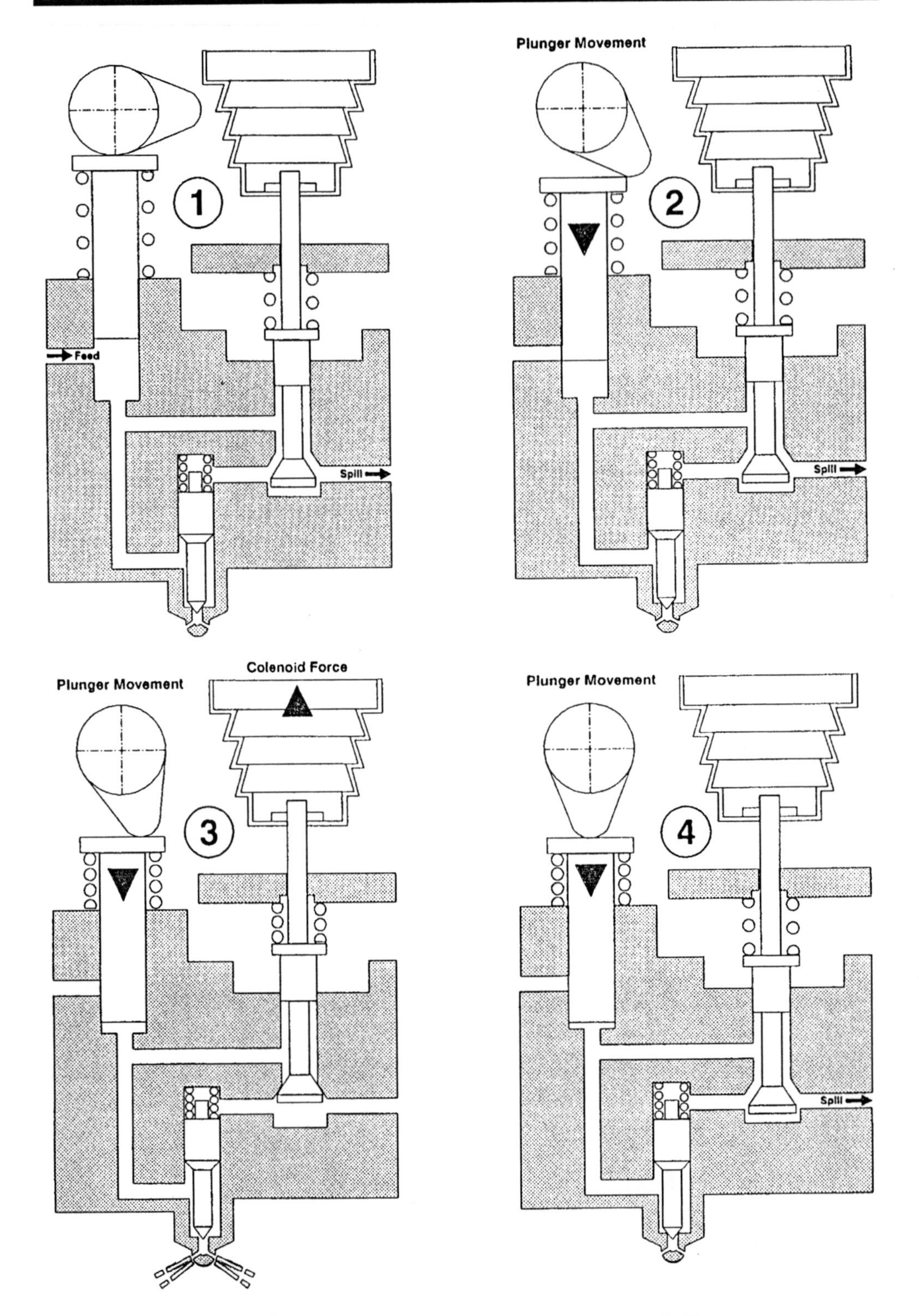

Fig. 6.23 The electronic unit injector (Lucas CAV)

(c) the braking system stops working if the ABS fails?
(d) the modulator is not under the control of the ECM?

2. A clamping diode is used on some solenoid-operated devices:
   (a) to restrict current flow?
   (b) to block the voltage spike that occurs when the magnetic field collapses?
   (c) to prevent the solenoid connections being reversed?
   (d) to prevent overheating of the solenoid winding?
3. Duty cycle:
   (a) is the number of times per second that the solenoid operates?
   (b) refers to the 'on time' compared with the total time available to operate a solenoid?
   (c) is only used on the evaporative purge control solenoid?
   (d) is the length of time allowed for the transmission of an ECM command?
4. The electronic unit injector:
   (a) has a high powered solenoid that controls the spill valve?
   (b) is a single-fuel injector that is placed at the throttle body of a diesel engine and one injector serves all of the engine cylinders?
   (c) is supplied with high pressure fuel from a 'common rail'?
   (d) is only suitable for small engines?
5. Sequential petrol injection:
   (a) uses a single throttle body injector?
   (b) has two injectors for each cylinder?
   (c) delivers fuel to each cylinder on each induction stroke?
   (d) must operate the injectors by pulse width modulation?
6. In the 'lost spark' ignition system:
   (a) there is a single ignition coil for each cylinder?
   (b) the cycle of operations is completed in 540° of crank rotation?
   (c) there is one coil for each pair of cylinders?
   (d) the polarity of the spark is the same at all sparking plugs?
7. Stepper motors:
   (a) open a valve which is then closed by a spring?
   (b) can open and close valves because they can be reversed?
   (c) are not suitable for use with computer controlled systems?
   (d) operate at very high speed?
8. Exhaust gas recirculation valves:
   (a) blow hot gas into the exhaust system to heat up the catalyst?
   (b) control the movement of exhaust gas from the exhaust system into the intake system to help reduce NO_x?
   (c) are not used on diesel engines?
   (d) are used to operate exhaust brakes on heavy vehicles?

7
Diagnostic techniques

In Chapter 5 several examples of sensor performance details are given which show how an oscilloscope can be used to obtain readings from sensors while they are in operation. Similarly, in Chapter 6 the types of test results that are obtained when actuators are operating are also shown. This shows that there is a 'body' of knowledge which is applicable to fault diagnosis across a wide range of automotive systems. This chapter goes more deeply into the general principles of fault diagnosis techniques that are applicable to computer controlled automotive systems in general. For example, in the testing of circuits which often follows the reading out of a fault code, the descriptions cover the general principles of circuit testing, such as continuity, voltage drop, resistance and current flow. These principles, like the oscilloscope patterns for sensors and actuators, apply to virtually any vehicle system and the ability to perform them will assist vehicle service technicians in their work.

In keeping with the aim of concentrating on factors that have general application across a range of systems, it is not intended to provide details of individual systems. Such treatment requires a good deal of repetition of basic knowledge, and the variations that exist across vehicle types and makes means that the amount of information that is needed for accurate diagnosis is vast. In some cases the diagnostic data is freely available, in other cases it is restricted to authorized repairers. In most cases the makers of diagnostic equipment supply diagnostic support services and these services provide a valuable source of information for independent garages and others.

7.1 Circuit testing

The test shown in Fig. 7.1(b) is useful for determining whether the circuit is complete between the points at which the meter leads are applied. The circuit itself must be switched off and the electrical supply for the test derived from the battery of the multimeter. Some meters are equipped with a buzzer which sounds if the circuit is complete and others use the ohm-meter scale of the multimeter.

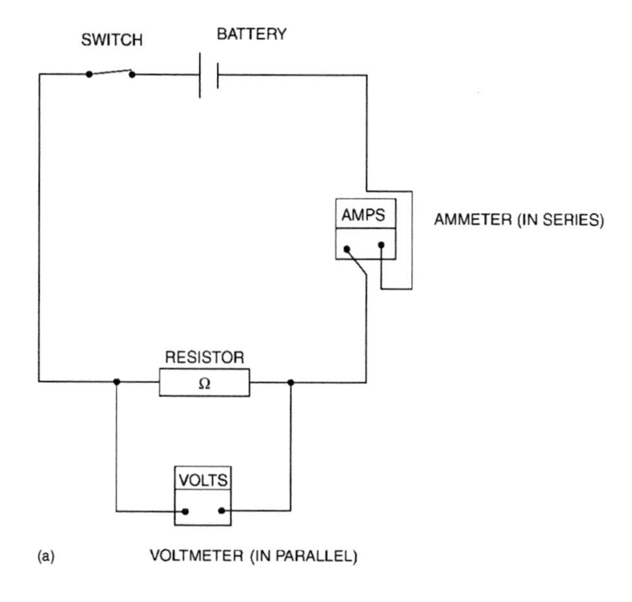

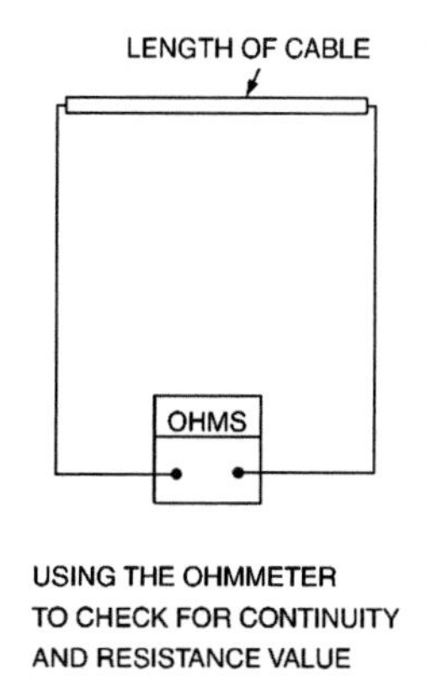

Fig. 7.1 Some uses of a digital multimeter

If there is continuity, i.e. a complete circuit, the ohm-meter will show zero (or very nearly) resistance, whereas an open (incomplete) circuit will show infinite resistance. The test is also useful for locating the two ends of a wire. The ohm-meter scale can also be used to check the resistance value of an HT lead, or any component where the resistance value is important.

Factors such as dirty or corroded connections and poorly made earthing points can lead to voltage drop in a circuit. The results of these voltage drops can be quite unpredictable. Whenever current flows through a resistor it causes a voltage drop across the resistor. Figure 7.1(a) shows the multimeter being used for voltage drop testing. In this case the circuit must be switched on and, depending on the device being tested, be in operation.

Figure 7.1(a) also shows the position of an ammeter for measuring current flow; the meter is in series.

Charging system problems often come to you as a „no-start" complaint. The battery will have discharged and the starter will not crank the engine. The first step is to test the battery and charge it if necessary.

● Measuring system voltage

Bleed the surface charge from the battery by turning on the headlights for a minute. Now turn the lights off and measure the voltage across the battery terminals. When possible, individual cell specific gravity should be checked with a hydrometer. A load test should be done to indicate battery performance under load. Voltage tests only tell the state of charge, not the battery condition.

● Measurement conditions

- Connect the MultiScope to the vehicle's battery as described on the MultiScope's help screen.
- Crank the engine while watching the instrument's display.

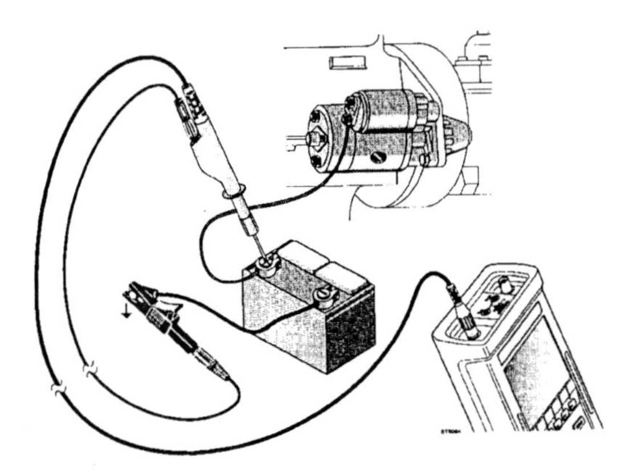

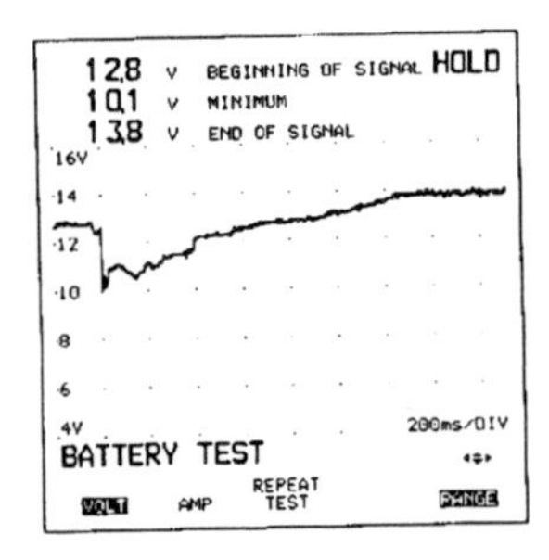

Fig. 7.2(a) Testing battery condition

Checking battery voltage is an important part of diagnostic work, as is the need to check the charging system voltage. These checks are shown in Fig. 7.2.

In many cases, sensors are supplied with a voltage via the ECM. This may be a 5 V supply and is known as the source voltage. If the sensor is to function

● Measurement conditions for the charging output test

- Connect the MultiScope to the vehicle's alternator.
- Engine RUNNING. Test the alternator at idle and under load Slowly increase engine speed.
- Load the charging system by turning on vehicle accessories such as the headlights, heater blower motor fan, and wind shield wipers.

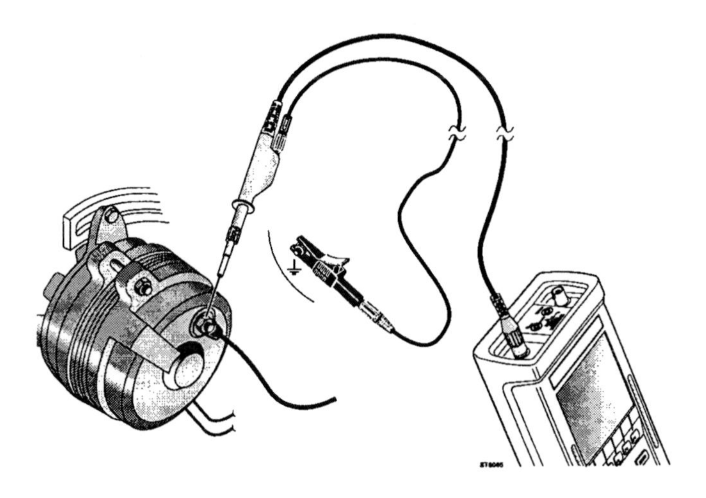

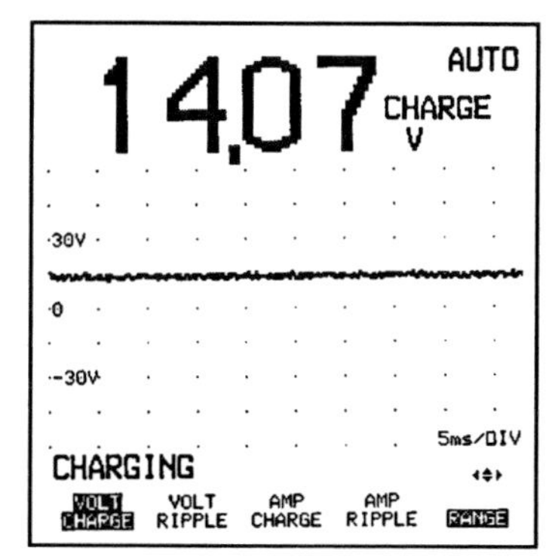

Fig. 7.2(b) Charging system voltage tests

correctly, this source voltage must be accurate within certain limits. This requires careful voltage measurement. Figure 7.3 shows a backprobe test to check source voltage. In this case the Bosch PMS 100 scope is being used. The reading obtained is shown as 5.09 V.

In all of the above tests it is important to ensure that good quality leads and clips, or probes are used. It is essential that 100% effective electrical connections are made between the test leads and the electrical connections that are being tested. If this is not achieved, a great deal of time and effort can be lost because of false readings.

● Measurement conditions

Refer to the vehicle manufacturer's wiring diagram for additional information on pin location and circuit descriptions.

● MultiScope key sequence

3. Connect the test leads as displayed by the MultiScope's **Connection Help** as shown below.

4. OK F1 Starts the Voltage test.

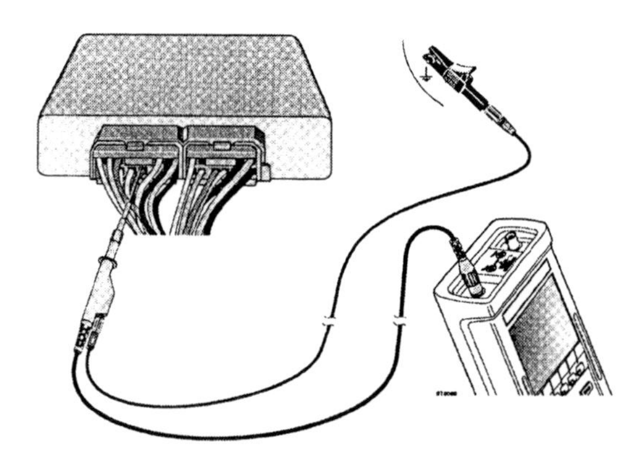

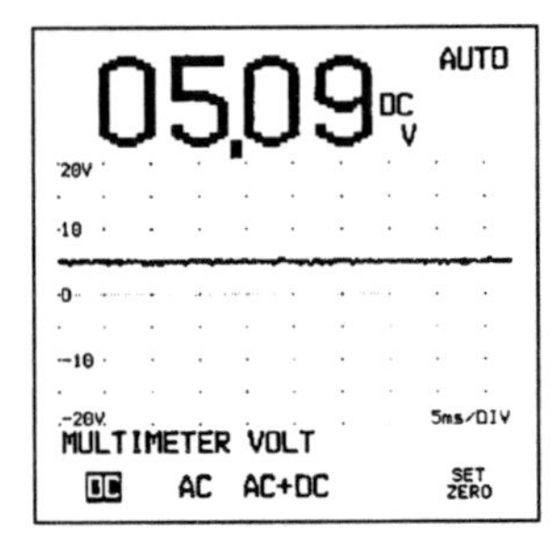

Fig. 7.3 Checking source voltage

7.2 Vehicle specific details

Those readers with experience of working on different makes of vehicles will know that the range of differences that are to be found within computer controlled systems makes it virtually impossible to cover all different systems in one volume. Changes in computing technology and electronics occur frequently and this means that the design of computer controlled systems on vehicles also changes quite frequently as new designs incorporate the new technology. This means that data and information is frequently updated and it is a major task to keep abreast of these changes. Technicians in franchised dealerships are normally notified of changes and, with modern technology, this updating can be done rapidly. In the case of the general garage workshop, technicians will probably have to rely on data and information from specialized companies. An updating and information back-up service is offered by companies such as Robert Bosch Ltd and Lucas Aftermarket Operations etc. Whatever the case, it is essential that the information that relates exactly to the system that is being worked on should be readily to hand and that all procedures are properly understood before work commences.

7.3 The 'six-steps' approach

At this stage it is important to emphasize the need to be methodical. A simple, but effective approach to diagnostic work is known as the 'six-steps' approach. This six-step approach may be recognized as an organized approach to problem solving, in general. As quoted here it may be seen that certain steps are recursive. That is to say that it may be necessary to refer back to previous steps as one proceeds to a solution. Nevertheless, it does provide a proven method of ensuring that vital steps are not omitted in the fault tracing and rectification process. The six steps are:

1. collect evidence;
2. analyze evidence;
3. locate the fault;
4. find the cause of the fault and remedy it;
5. rectify the fault (if different from 4);
6. test the system to verify that repair is correct.

Collect evidence

Collecting evidence means looking for all the symptoms that relate to the fault and not jumping to conclusions, e.g. because the system is controlled by an ECU it must be the ECU that is at fault. In order to collect the evidence it is necessary to know which components on the vehicle actually form part of the faulty system. This is where sound basic skills comes in. If an engine control system is malfunctioning because one cylinder has poor compression it is important to discover this at an early stage of the diagnostic process.

Analyze the evidence

In the case of poor compression on one cylinder, given above as an example, the analysis would take the form of tests to determine the cause of low compression, e.g. burnt valve, blown head gasket etc. The analysis of evidence that is performed will vary according to the system under investigation. But these steps are obviously important otherwise one may embark on a drawn out electronics test procedure which will prove unproductive.

Locate the fault

The procedure for doing this on an electronics system varies according to the type of test equipment available. It may be the case that the system has some self-diagnostics which will lead you to the area of the system which is defective. Let us assume that this is the case and the self-diagnostics report that an engine coolant temperature sensor is defective. How do you know whether it is the sensor, or the wiring between it and the remainder of the system? Again this is where a good basic knowledge of the make-up of the system is invaluable.

Find the cause of the fault and remedy it

With electronic system repair it is often the case that a replacement unit must be fitted. However, this may not be the end of the matter. If the unit has failed because of some fault external to it, it is important that this cause of failure is found and remedied before fitting the new unit. It is often not just a matter of fitting a new unit.

Give the system a thorough test

Testing after repair is an important aspect of vehicle work and especially so where electronically controlled systems are concerned. In the case of intermittent faults, such testing may need to be extended because the fault may only occur when the engine is hot and the vehicle is being used in a particular way.

7.4 Skills required for effective diagnosis

Studies, over a period of time, show that the skills possessed by vehicle technicians, who are successful in diagnostic work on vehicle electronic systems, consist of many elements. The most important of these may be classed as 'key skills'. These key skills may be summarized as follows.

- Use appropriate 'dedicated' test equipment effectively.
- Make suitable visual inspection (assessment) of the system under investigation.
- Make effective use of wiring diagrams.
- Use instruction manuals effectively.
- Use multimeters and other (non-dedicated) equipment effectively.
- Interpret symptoms of defective operation of a system and, by suitable processes, trace the fault and its cause.
- Work in a safe manner and avoid damage to sensitive electronic components.
- Fit new units and make correct adjustments and calibrations.
- Test the system, and the vehicle for correctness of performance.

7.5 An approach to fault finding

Having dealt with tools and technology in previous chapters and with skills and method in the sections above, it is appropriate to turn to more specific detail. Whilst the self-diagnosis capacity of the on-board computer (ECM) and its code storing capacity is a powerful aid to fault diagnosis and repair, it is by no means a complete solution. By taking an example of an engine management system, such as that shown in Fig. 7.4, it may be seen that there are many factors that relate to the operation of the engine.

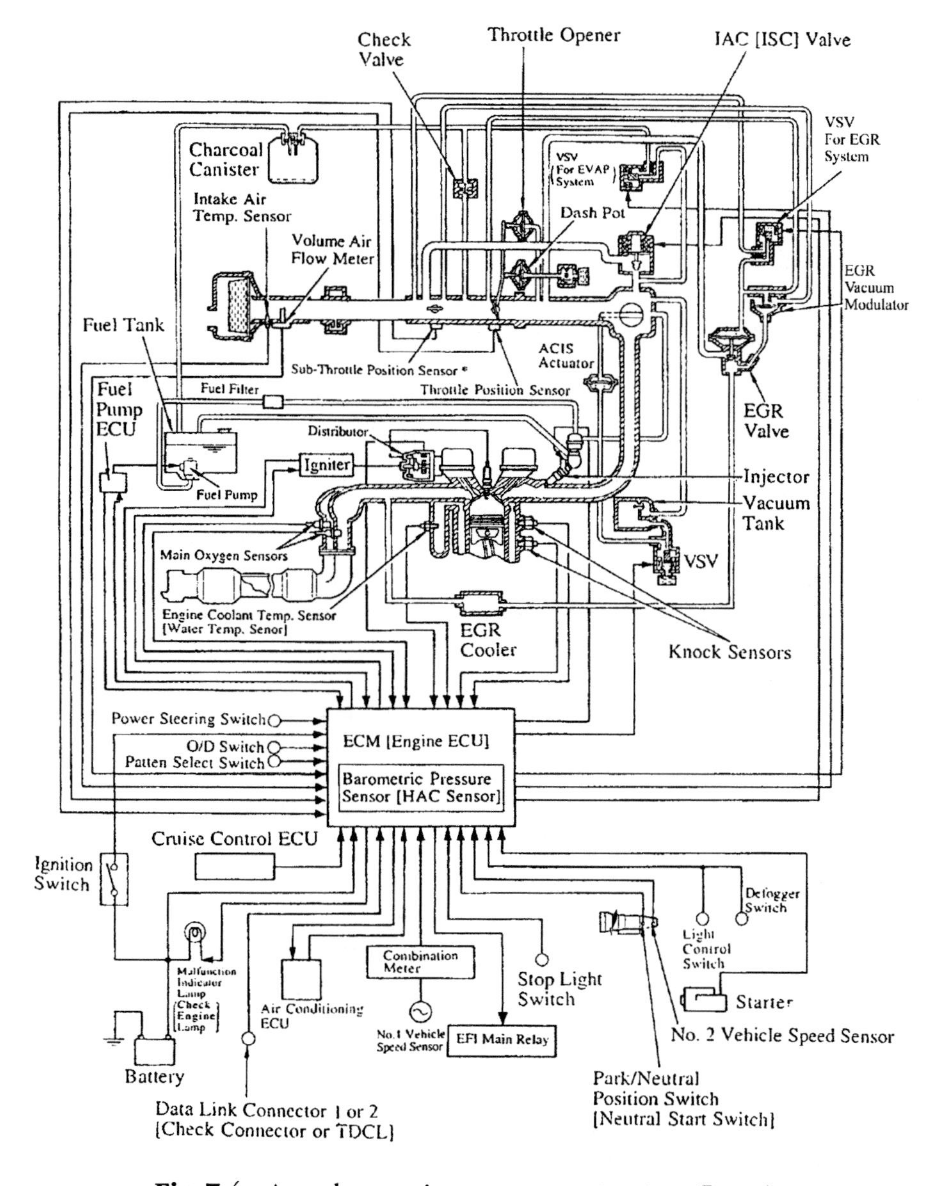

Fig. 7.4 A modern engine management system (Lexus)

At the 'heart' of the diagram (Fig. 7.4) and the engine system itself, is the engine. In order for this engine to operate it must have:

1. fuel and air;
2. a spark;
3. compression in the cylinder;
4. a cooling system to prevent overheating;
5. a lubrication system to deal with friction and assist with cooling;
6. the mechanism of the engine, i.e. pistons and piston rings, exhaust and inlet valves, connecting rods and bearings, crankshaft, bearings and a flywheel. These are required to convert the energy of combustion into mechanical energy at the flywheel.

Whilst failures do occur in areas (4), (5) and (6), on modern vehicles, they happen far less frequently than they once did. Most failures in automotive computer controlled systems are likely to be related to the areas (1), (2) and (3). In other words, faults are more likely to be found in some areas than others. In fact the term used is probability. When we are looking for causes of faults we can safely say that some possible causes are more likely to be the reasons for the failure than others. These likely causes have a high probability of being the reason for failure.

To take a simple case: some time ago, a certain vehicle had its ignition distributor situated behind the radiator and facing directly into oncoming rain. In very wet weather it was quite common for these vehicles to suffer breakdowns. The cause was often found to be that water had affected the distributor. This became common knowledge and one had a good chance of getting the diagnosis right first time by going direct to the most probable cause, wetness at the distributor. This type of knowledge (based on experience) is based on empirical evidence, i.e. we know that it is so because that is what happens.

Evidence gained by experience is gathered by motoring organizations, such as the AA and the RAC, and also by technicians who are called upon to recover and repair vehicles. This evidence often points to common causes of failure, such as discharged batteries, lack of fuel, poor maintenance etc.

Whilst urging technicians to remain aware of the value of knowledge gained by experience, we now turn to the instruments and tools aspects of computer controlled systems diagnosis.

Firstly, remember the six steps; one of these is 'carry out a thorough visual inspection'. This is a step that is often overlooked until one has got well into the fault finding procedure. Carrying out this relatively simple step can often avoid much unnecessary work. By taking an example of a system failure and by working through a diagnostic procedure it is possible to establish a 'methodology' for fault diagnosis. The example chosen is a vehicle that is not 'pulling' well, the engine check lamp shows a malfunction and the workshop documentation provides the following information which is also shown on the flow chart in Fig. 7.5.

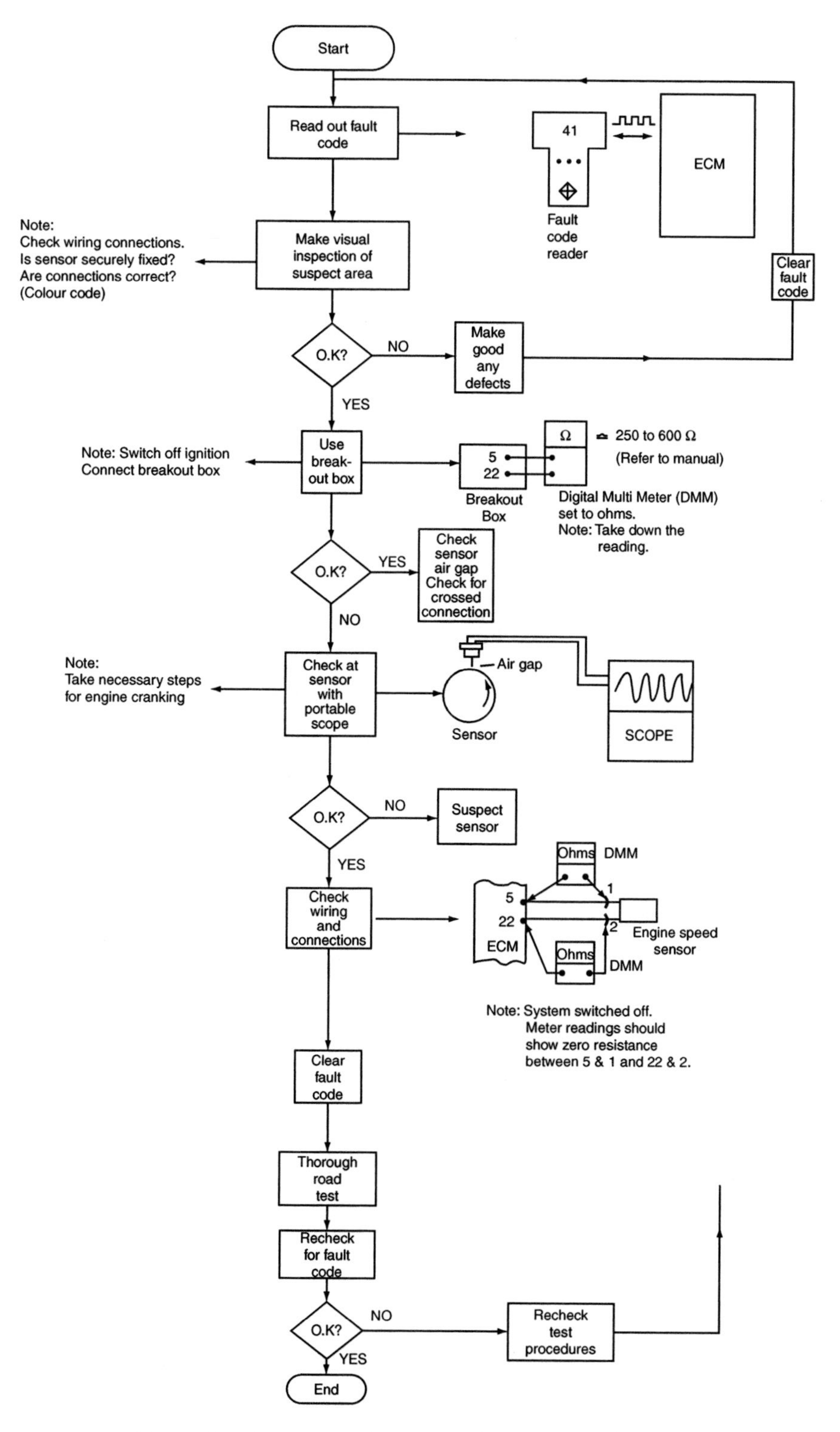

Fig. 7.5 Flow chart showing a diagnostic technique

Fault:
Engine speed sensor signal out of limits.

Symptom:
Low engine power.

Fault code:
41.
Checks to find possible cause:
Circuit between ECM pin 5 and sensor pin 1.
Circuit between ECM pin 22 and sensor pin 2.
(Either of these may be open circuit, short circuit or have high resistance due to corrosion at connectors.)
Cables connected to sensor pins 1 and 2 may be cross-connected.

Engine speed sensor:
The air gap may be too large.
The sensor coil may or its insulation may be damaged.

Note that the possible causes include a number of options other than 'defective sensor'. The flow chart in Fig. 7.5 shows a sequence of procedures and tests that can be applied by using some of the equipment that is described in Chapter 4.

Assuming that a thorough visual inspection of the system has been performed, the vehicle is prepared for test and the diagnostic test tool (scan tool) is connected to the serial port so that the fault codes can be read out. In this particular case, the fault code 41 is shown and the details of this code are given above. Following the steps down the flow chart a series of tests are shown.

The first of these is a resistance check that is performed at the breakout box connections. Resistance checks are commonly used and they provide a good guide to the condition of a component. However, a resistance check may not be sufficient because it is possible to get a resistance value that is within the limits - in this case 250-600 ohms - and yet to find that the sensor does not perform correctly under operating conditions. This is why a dynamic test, such as that performed when an oscilloscope is used with the sensor in operation, is a more satisfactory guide to condition. The oscilloscope (dynamic) test can be conducted at the ECM connectors and at the point where the sensor is connected into the main wiring harness. The checks performed at each end of the circuit that connects the sensor to the ECM thus provide a check on the conditioning of the circuit. Should the reading at the sensor be different from the reading at the ECM it is a fair indication that there is a fault in the circuit, possibly a bad connection. The multimeter can then be used to verify the condition of the connections, as shown on the flow chart.

7.6 Emissions related testing

From time to time, reports circulate about vehicles that have failed an exhaust emissions test. Often, it seems that an assumption has been made that the exhaust catalyst is defective, a new catalyst has been fitted and when retested the vehicle still fails the exhaust emissions test. From this point things can often get worse. It is, therefore, useful to make an examination of the techniques that should prevent such misdiagnosis from happening.

7.6.1 OXYGEN SENSOR

The pre-catalyst oxygen sensor is an important element in the control system that regulates the mixture strength (air-fuel ratio) in spark ignition engines. As explained in Chapter 5, the oxygen sensor samples the exhaust gas before it enters the catalytic converter and produces a signal that tells the ECM the air-fuel ratio of the mixture that is entering the combustion chambers. It is a feedback system. The oxygen sensor signal is used by the ECM to change the amount of fuel injected so that lambda is kept in the range of approximately 0.97-1.03. Figure 7.6 should remind you of the principle.

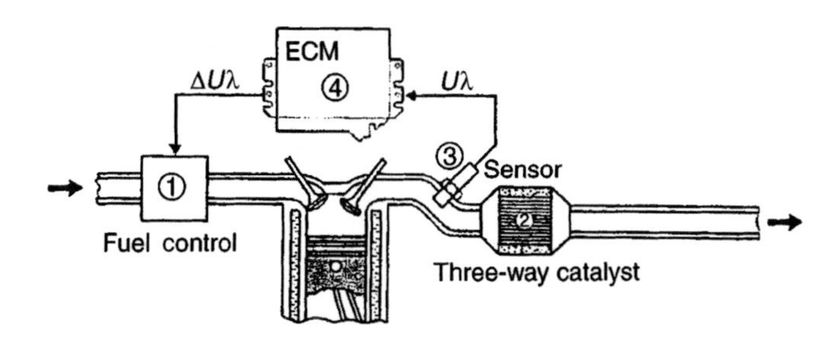

Fig. 7.6 Oxygen sensor - feedback principle

Should the oxygen sensor be disconnected, there will be no feedback to the ECM, which will probably have been programmed to use a substitute value to cause the 'limp home' mode to operate and to minimize the possibility of damage to the catalyst. Add to this the fact that, like the catalyst, the oxygen sensor needs to be at an operating temperature of over 300°C for it to operate efficiently, and it becomes evident that electrical testing of the oxygen sensor is work that requires care and attention to detail.

If we now take an example of an oxygen sensor fault code we can explore some of the details that should be taken into account when attempting diagnosis and rectification work.

On a particular vehicle, the fault code 51 means a fault at the 'oxygen sensor or in the oxygen sensor circuit'. Assuming that there are no other faults recorded that may affect the performance of the oxygen sensor, an *in situ* voltage test with the oscilloscope should assist in locating the problem. It is advisable to remember that the ECM has registered a fault code because the value received at the ECM is not within the programmed limits. It may be the case that the sensor is working correctly but the signal is not reaching the ECM because of a defect between the sensor output terminals and the connecting pins at the ECM. This indicates that the oxygen sensor should be checked:

1 at the ECM diagnostic port, using equipment such as the Bosch KTS 500 tester. This will produce a trace of the oxygen sensor voltage, similar to the one shown in Fig. 7.7. This particular trace was taken at idle speed and the frequency is quite low.
2 by backprobing at the sensor, as shown in Fig. 7.8.

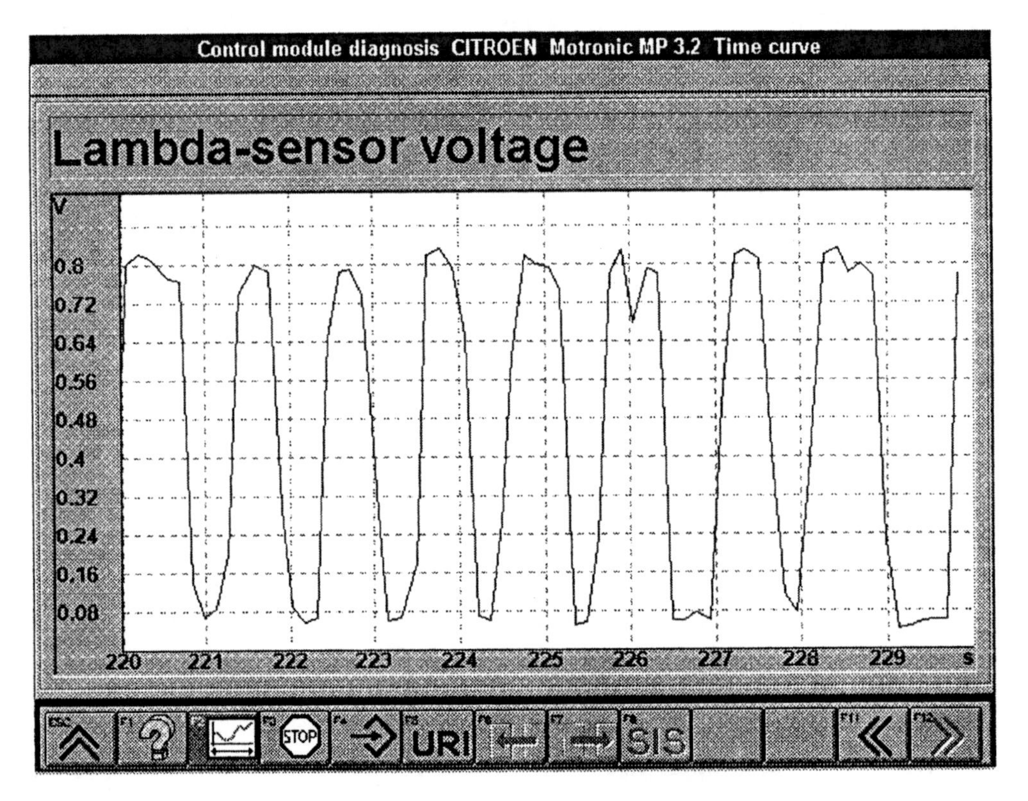

Fig. 7.7 Voltage test at the ECM diagnostic connector using the Bosch KTS 500 tester

When the system is at its operating temperature and the sensor and circuit are in good order the two sets of readings should be identical. However, if the reading at the sensor is correct and the one at the ECM is not, it is reasonable to assume that there is a defect in the circuit between the sensor and the ECM. The cables and connectors should be examined for signs of damage, looseness and corrosion. With the system switched off, it should be possible to test for continuity

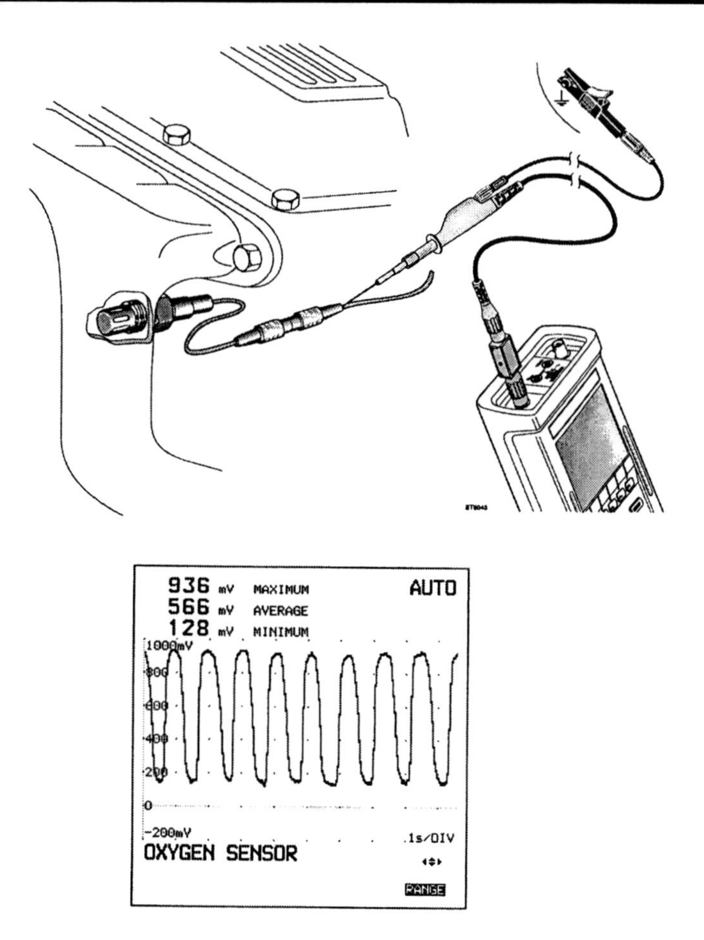

Fig. 7.8 Voltage test at the sensor using the Bosch PMS 100 portable oscilloscope

between the ends of the signal cable and also the condition of the sensor's earth connection.

If the checks at the sensor and the ECM both produce similar defective signals, a sensor defect is indicated and an analysis of the scope patterns should give useful clues about the cause. Considerable amounts of information about the performance of the zirconia-type oxygen sensor is contained in the oscilloscope trace and the principal features are described below.

Switching characteristics

As the air-fuel ratio moves from perfect combustion (lambda = 1) to slightly rich (lambda = 0.98), the sensor voltage rises rapidly, and when the air-fuel ratio moves from lambda = 1 to lambda = 1.02, which is slightly weak, the sensor voltage drops rapidly.

This characteristic is used by the ECM to regulate the amount of fuel injected and thus to control lambda and provide the conditions for the catalyst to function, i.e. maintaining the air-fuel ratio as near as possible to the chemically correct

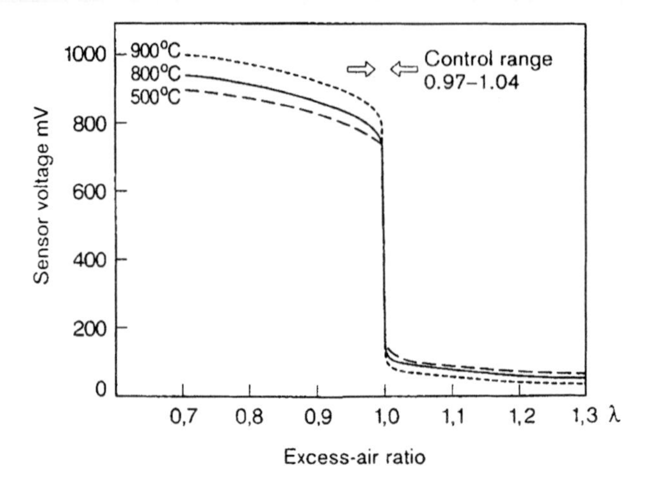

Fig. 7.9 The zirconia oxygen sensor voltage in the region of lambda = 1

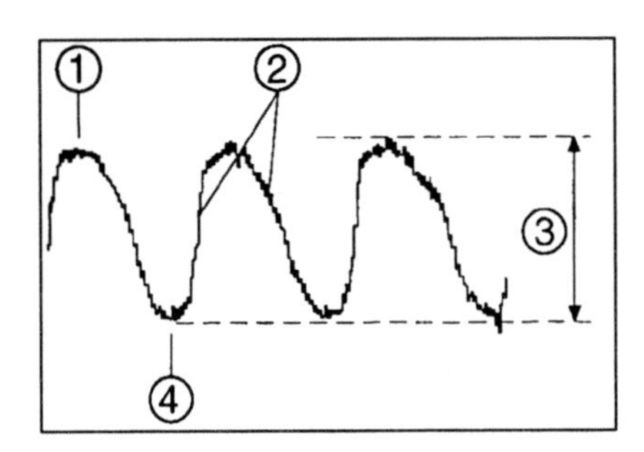

Fig. 7.10 Voltage trace for a zirconia oxygen sensor

value which gives lambda = 1. The result is that the output trace of the zirconia oxygen sensor is of the form shown in Fig. 7.10.

Points to note are as follows.

1 The maximum voltage should be between 800 mV and 1 V.
2 The slope on the rise and fall sides of the trace becomes less steep as the sensor ages, or is damaged by use of incorrect fuel or lubricants.
3 The peak-to-peak voltages should be at least 600 mV with an average of 450 mV.
4 The minimum voltage should be approximately 200 mV.

As stated above, the sensor pattern changes as the sensor ages or is damaged in some way (poisoned) by leaded fuel or other contaminants. Figure 7.11 (P16 KTS 300) shows a comparison between the voltage trace for a good sensor

The ageing of the lambda sensor upstream of the catalytic converter is monitored by analysis of the sensor signal.
An aged lambda sensor (dotted line) reacts more slowly to changes in the oxygen in the exhaust than a new lambda sensor (unbroken line).

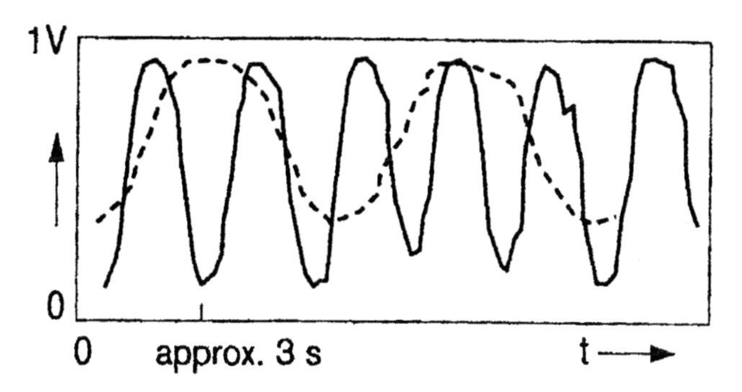

- The frequency of the control speed of the lambda sensor thus becomes smaller as the sensor ages.

- The amplitude of the sensor signal becomes smaller. This becomes evident in the case of a lean mixture, where the 100 mV of an intact sensor (difference in voltage approx. 900 mV - 100 mV = 800 mV) is exceeded by several hundred mV (difference in voltage approx. 900 mV - 400 mV = 500 mV).

Fig. 7.11 Comparison between a good sensor and a sluggish one

and a sluggish, or aged, one. The 'good' sensor pattern is the continuous line and the sluggish one is the broken line.

Obtaining these scope patterns is but part of the diagnostic process. It is in fact, the collecting evidence step of the six-steps approach. Analyzing the evidence is another step and in this case the main features to note are the lower peak voltages and lower frequency of the sluggish one. Among the factors to consider when seeking a cause of the sluggishness would be the age of the vehicle, the hours of running, type of fuel used, the maintenance record, and the condition of the exhaust system, for example, is it damaged? are there leaks or obstructed pipes?

Dual oxygen sensors for catalyst monitoring

Figure 7.12 shows the general arrangement of the sensors and catalyst in a system that uses a second oxygen sensor to monitor the performance of the catalyst. The second oxygen sensor downstream of the catalyst is a requirement of the OBD II standard.

If the catalyst is operating efficiently, the control voltage of the downstream oxygen sensor is smoothed, as shown in Fig. 7.13. This difference in voltage patterns is the method by which the efficiency of the catalyst is monitored by

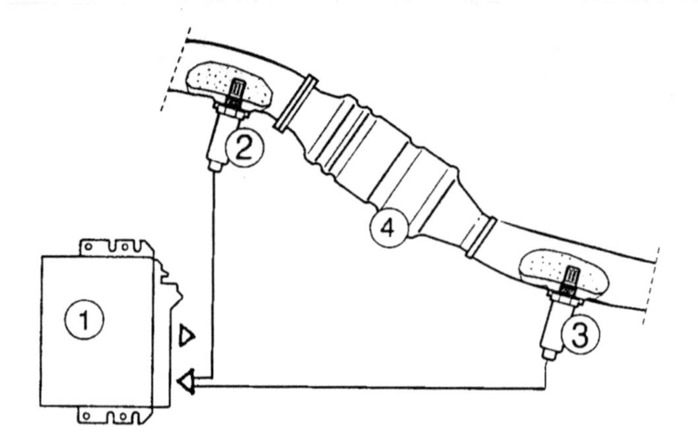

1	*Control unit*
2	*Lambda sensor upstream of catalytic converter*
3	*Lambda sensor downstream of catalytic converter*
4	*Catalytic converter*

Fig. 7.12 Dual oxygen sensors as used on OBD II type systems

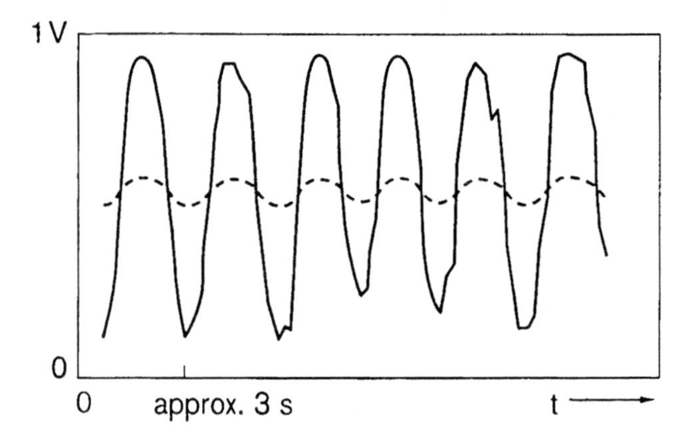

Fig. 7.13 Comparison of voltage patterns for upstream and downstream oxygen sensors

the ECM. If the catalytic converter is not working, the voltage patterns of the two sensors will not show the marked difference and this is the basis of the fault monitoring procedure.

The downstream oxygen sensor is less likely to age than the oxygen sensor upstream of the catalyst. This permits the downstream sensor to be used as a 'guide' signal to allow the fuelling ECM to compensate for any ageing in the upstream sensor.

Features of oxygen sensor performance that are monitored by the ECM include:

- output voltage
- short circuits
- internal resistance

- speed of change from rich to weak
- speed of change from weak to rich.

Testing the upstream and downstream oxygen sensors

Figure 7.14 shows the principle for the use of dual-trace portable oscilloscope. The leads are equipped with electrical filters for this test and the test allows the two sensor patterns to be observed simultaneously. The general form of the voltage traces is shown in Fig. 7.15.

● Measurement conditions for the oxygen sensor test

- Run the engine until the oxygen sensor is warmed to at least 600 °F (315 °C), in closed loop. Use jumper leads or back probe to make connection at the sensor wiring connector.

● MultiScope key sequence

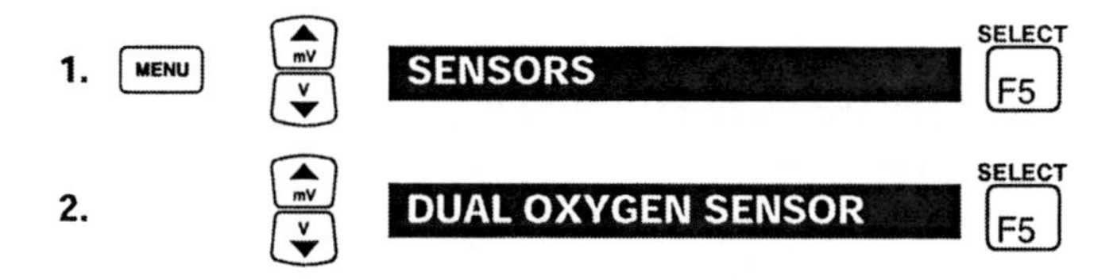

3. Connect the test leads as displayed by the MultiScope's **Connection Help** as shown below.

4. OK F1 Start the dual oxygen sensor test.

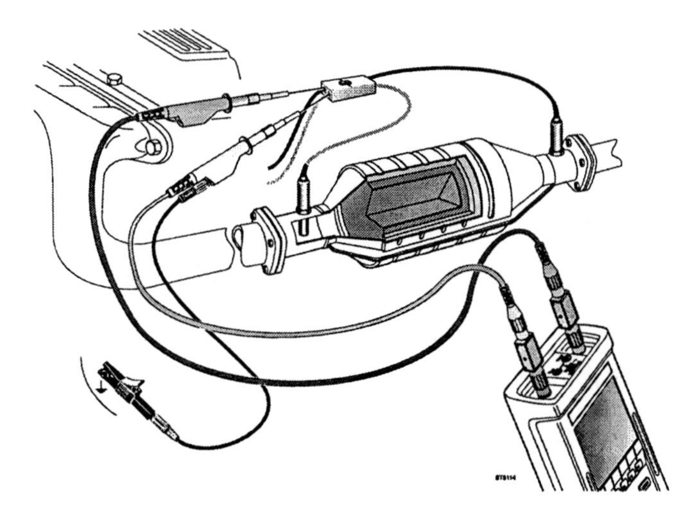

Fig. 7.14 The set-up for testing the two oxygen sensors

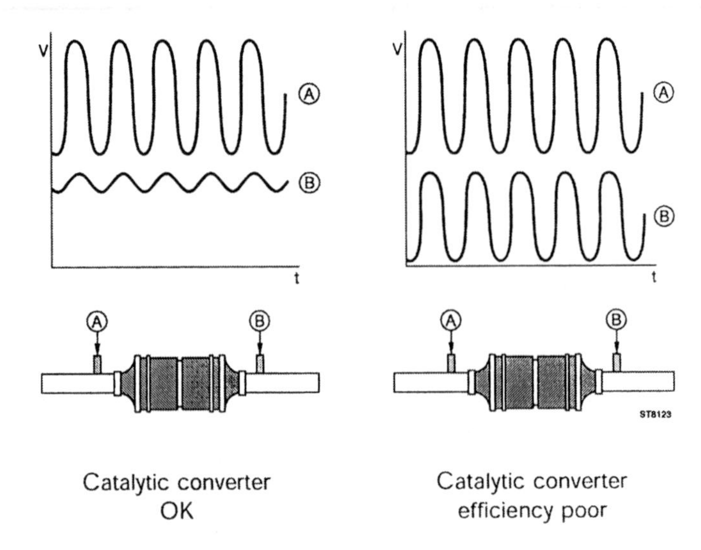

Fig. 7.15 The voltage traces from the upstream and downstream oxygen sensors

A similar test could be conducted with the aid of two digital voltmeters but, as you will appreciate, this test would be more difficult to perform. It must be noted that the test must be performed when the system is hot and this, coupled with the fact that the exhaust system is often not in the most accessible place, means that care must be taken in preparing for the test. A vehicle lift is virtually essential, precautions must be taken to prevent personal injury through burns etc., and the instruments and leads must be kept away from the hot exhaust.

Titania oxygen sensors

These are described in Chapter 5. Because they operate on the voltage changes that are produced by changes of resistance which in turn are caused by changes in oxygen levels in the exhaust, they produce a high voltage when the mixture is weak and a low output when the mixture is rich. Quite the opposite of the zirconia sensor. The response time is somewhat faster, the voltage trace is similar, but the voltages can be measured in volts rather than millivolts.

7.6.2 KNOCK SENSORS

If a knock sensor of the type described in Chapter 5 is correctly fitted, i.e. not overtightened and the electrical connections are good, the sensor is likely to be very reliable. However, if 'pinking' should occur it is probable that the knock sensor may be failing to produce a signal. In order to test the knock sensor, the sensor can be activated by tapping the cylinder block with a small spanner, or similar object, close to the sensor. It should be possible to observe the effect on an oscilloscope screen.

Figure 7.16(a) shows the oscilloscope probe connected to the signal lead of the sensor, the type of pattern, together with a maximum voltage and frequency that

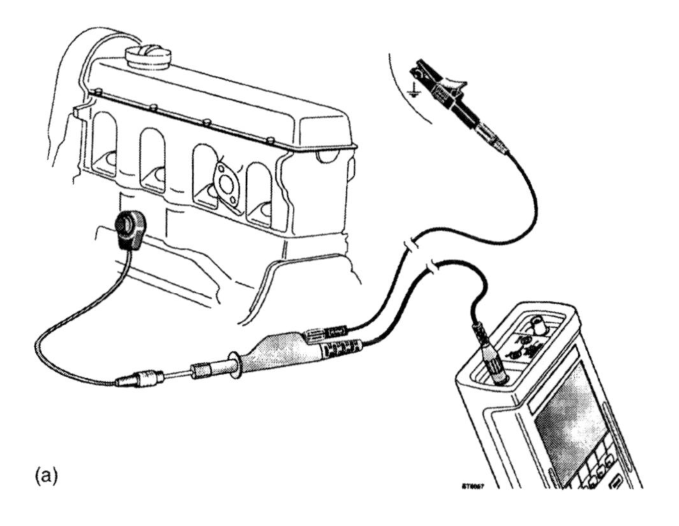

(a)

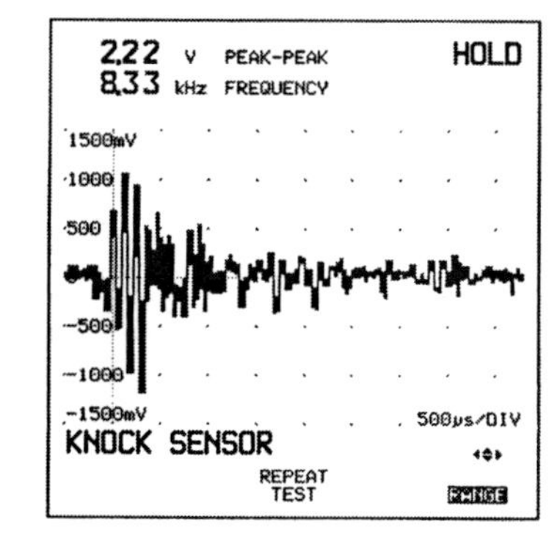

(b)

The pattern from this sensor is directly related to the cause and severity of the knock. For this reason each signal looks slightly different.
The main thing is to check for the presence of a signal.
On most vehicles, when the ECU receives a knock signal from the knock sensor, it retards ignition timing until the knock disappears.

! Specifications may vary. Consult manufacturer's specifications.

Fig. 7.16 Testing a knock sensor (Robert Bosch Ltd)

should be obtained. It should be noted that the voltage and frequency depend on the severity of the knock. Extreme care must be exercised when tapping the engine casting in order not to cause any damage.

7.6.3 AIR FLOW METERS

The air flow meter provides the ECM with a number of signals that are used in engine management. These signals include: measuring the air flow so that fuel injection matches it to give the correct air fuel ratio, indicating when the throttle valve is in the idle position, giving an indication of load on the engine etc. The commonly used types of air flow sensor have been described in Chapter 5 and the information given here is to show some practical tests that can be performed to assess the performance of air flow sensors.

The flap type (potentiometer) air flow sensor

In Fig. 7.17 the test probe is making contact with the signal connection on the air flow sensor. The time base of the oscilloscope must be set to a suitable value. If the throttle is moved steadily from fully closed to fully open and then released slowly, an oscilloscope trace similar to that shown in Fig. 7.18 should be obtained for the potentiometer type of sensor.

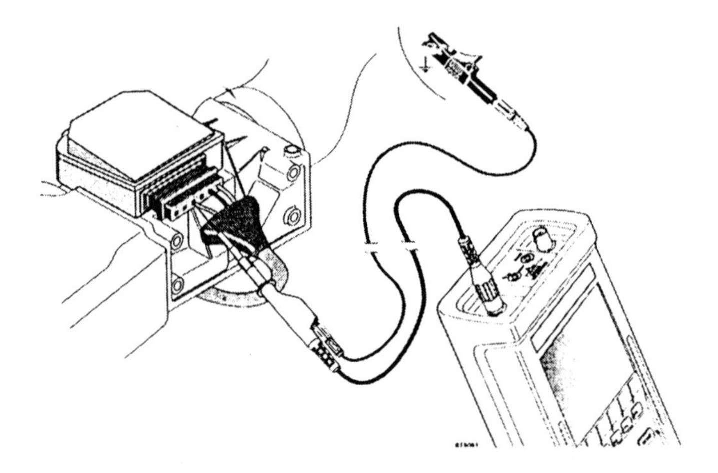

Fig. 7.17 Test connections for a MAF

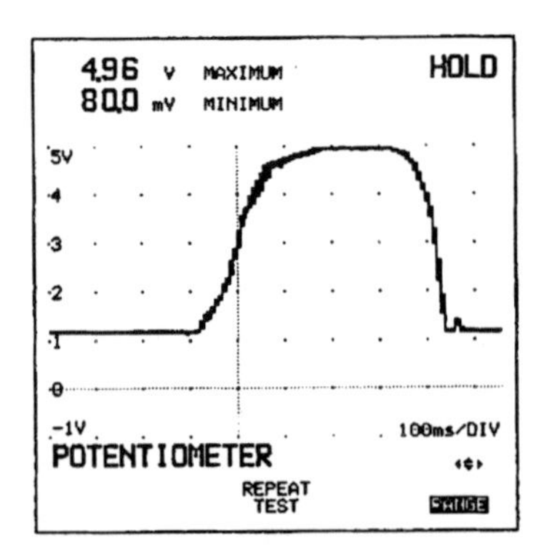

Fig. 7.18 Voltage pattern for the full movement of the accelerator pedal

In Fig. 7.19(a) the arrow at (5) indicates a rise in sensor voltage which shows that air flow into the manifold is increasing, as would be expected with an increase in throttle opening and engine speed. At section (2) the maximum air flow is entering the engine and this is the type of flow that happens when the engine is running at steady load high speed. At (3) the voltage is falling, showing a decrease in air flow into the manifold. At (4) the sensor voltage is at its minimum and this happens when the throttle valve is closed. In this position a switch, incorporated into the air flow sensor, normally causes the ECM to switch to idle control. Figure 7.19(b) shows the type of pattern that may be obtained from a defective potentiometer type of air flow meter. The spikes indicate that parts of the potentiometer track are defective and these would cause a problem whenever the sensor wiper is located in these positions.

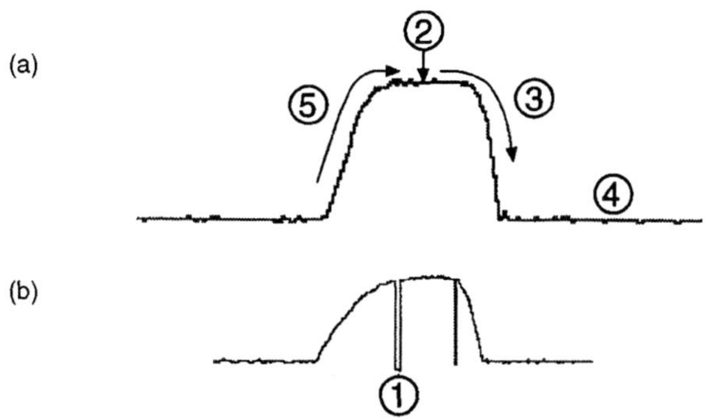

1 Spikes in downward direction indicate a short to ground or an intermittent open in the resistive carbon strips.
2 Peak voltage indicates maximum air flow entering intake manifold.
3 A voltage decrease indicates less air is flowing into the intake manifold.
4 Minimum voltage indicates a closed throttle plate.
5 A voltage increase identifies an increase of airflow into the intake manifold.

Fig. 7.19 Potentiometer-type air flow sensor voltage signal

Mass air flow sensors

As described in Chapter 5, these sensors normally rely on changes in resistance in the sensing element that are brought about by changes in air flow. The signal output may be of analogue or digital form depending on the type of sensor. One obviously needs to check this sort of detail before attempting any tests. Figure 7.20 shows the oscilloscope connected to the signal wire of a sensor that gives an analogue signal.

When the scope is securely connected and safely positioned, the engine is started and allowed to idle. The engine speed is slowly increased whilst observing the scope screen. If the sensor casing is very lightly tapped with a small screwdriver it should show up any poor connections by breaks, or blips in the scope pattern. Figure 7.21 shows a signal pattern for an analogue MAF in good condition.

- Measurement conditions

- Connect the MultiScope to the output signal from the air flow sensor (or meter).
- Start the engine and allow the engine to idle. Slowly accelerate the engine while watching the display.
- Use a screwdriver handle and gently tap on the sensor while performing this test. Loose connections in the sensor can cause momentary hesitations and flat spots.

- MultiScope key sequence for a mass air flow sensor test

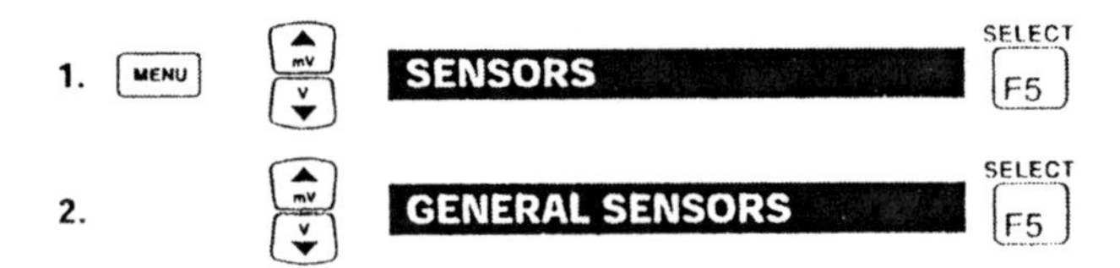

3. Connect the test leads as displayed by the MultiScope's **Connection Help** as shown below.

4. OK F1 — Starts the mass air fllow sensor test. If necessary, use the arrow keys to range.

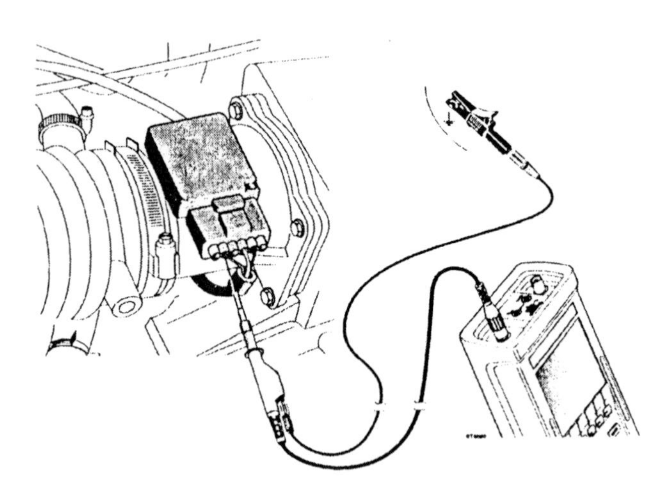

Fig. 7.20 Testing an analogue air flow sensor

7.6.4 THROTTLE POSITION SWITCHES

The ECM needs to know the position that the throttle valve is in at any given time, in order that it knows when the throttle is closed and the idle control is required to operate. The throttle position sensor also gives a very good indication of engine load and also the fully open position of the throttle butterfly valve (plate). The throttle position sensor is normally positioned on the throttle butterfly spindle at the opposite end to the throttle lever.

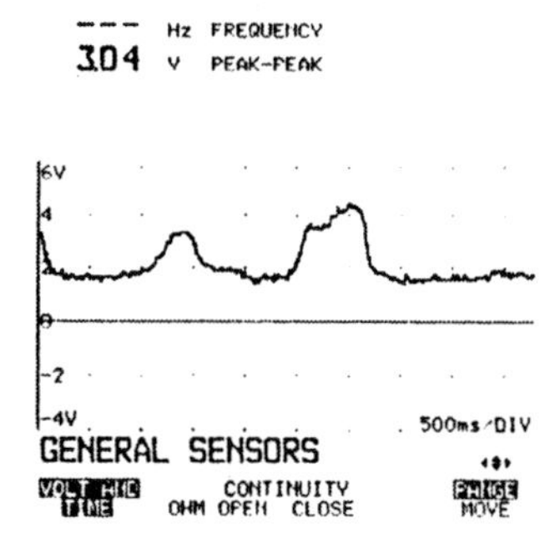

Fig. 7.21 Oscillscope voltage pattern for an analogue MAF

● MultiScope key sequence

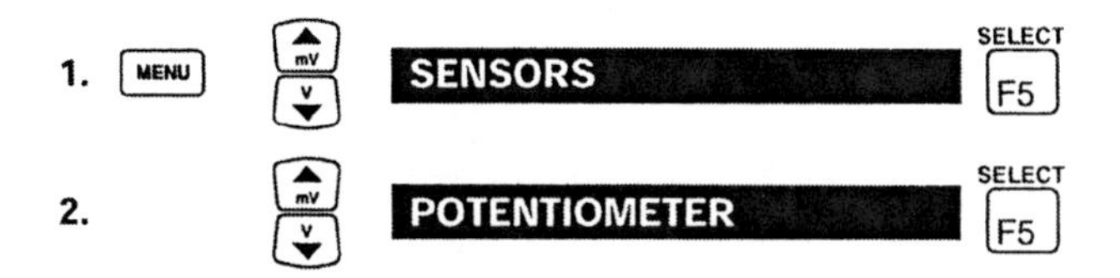

3. Connect the test leads as displayed by the MultiScope's **Connection Help** as shown below.

4. OK F1 Starts the potentiometer sweep test.

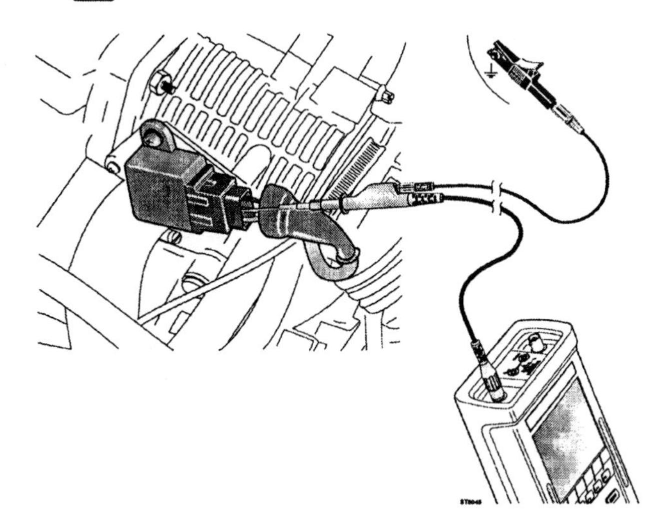

Fig. 7.22 Testing a throttle position sensor (switch)

Figure 7.22 shows a throttle position sensor being backprobed in order to obtain a scope pattern of the voltage signal.

With the time base of the oscilloscope set to a suitable value, the ignition is switched on. The scope screen should then be observed while the accelerator

(a)

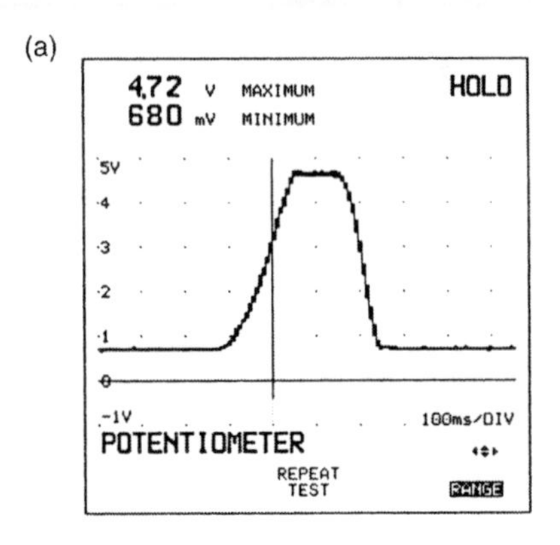

(b)

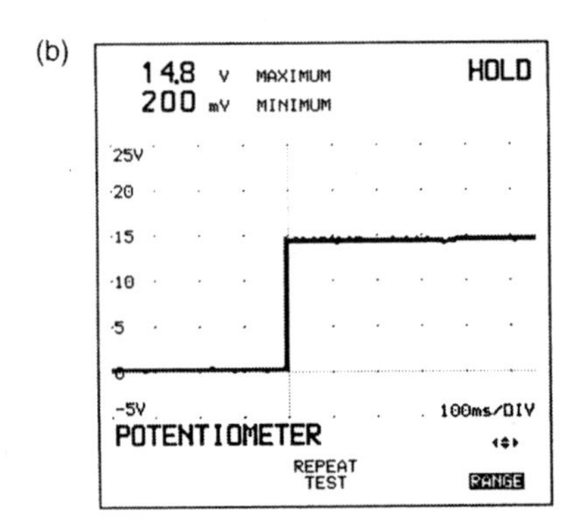

Fig. 7.23 The voltage trace for (a) a potentiometer-type and (b) a switched-type throttle position sensor

pedal is gradually moved from the fully closed to fully open position and then back again. The resulting pattern should then look like Fig. 7.23.

An analysis of the voltage trace is given in Fig. 7.24. You will note that this is very similar to the trace from the flap type air flow meter. In addition to the variable voltage that shows throttle position, a throttle switch contains switch contacts that give a step signal at the throttle-closed and throttle fully-open positions. This type of voltage trace is shown in Fig. 7.25.

7.6.5 A COOLANT TEMPERATURE SENSOR

The coolant temperature sensor (CTS) operates on signal voltages that range from approximately 0.2–4.8 V. Generally, a fault code will only be registered if the voltage falls outside these values. This means that an engine may operate with a defective coolant temperature sensor. A likely outcome of this is that the CTS may indicate a cold engine reading to the ECU when in fact the engine is hot. A probable outcome could be that the engine may attempt to operate on a mixture that is too rich for the conditions. Tests of the type shown below will verify the sensor through its operating range and the CTS can then be eliminated from the possible causes of a problem.

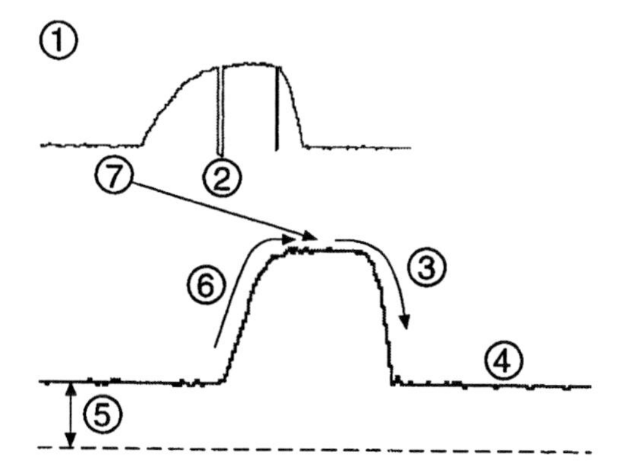

1 Defective TPS pattern.
2 Spikes in a downward direction indicate a short to ground or an intermittent open in the resistive carbon strips.
3 Voltage decrease identifies enleanment (throttle plate closing).
4 Minimum voltage indicates closed throttle plate.
5 DC offset indicates voltage at key on, throttle closed.
6 Voltage increase identifies enrichment.
7 Peak voltage indicates wide open throttle (WOT).

Fig. 7.24 Analysis of voltage trace for a potentiometer-type throttle position sensor

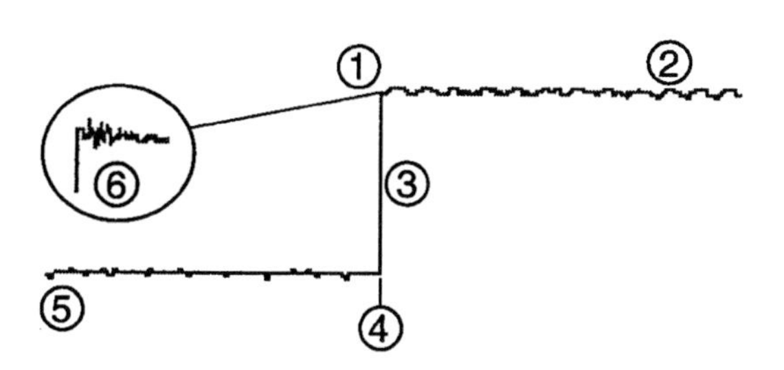

1 Throttle at position other than closed. Not necessarily wide open throttle.
2 Reference voltage.
3 Transitions should be straight and vertical.
4 Throttle opening and voltage transitioning.
5 Throttle plate closed.
6 Ringing may indicate worn contacts or loose throttle return spring.

Fig. 7.25 The throttle switch type of voltage pattern

It is likely that the ECM will detect voltage as measured at the sensor terminals and this means that a voltmeter or oscilloscope can be used for sensor checks. Figure 7.26 shows the Bosch portable oscilloscope being used to examine coolant temperature sensor performance.

- Measurement conditions for the temperature sensor test

- Turn the key ON, Engine OFF. With the sensor wiring harness connected, measure the output voltage (engine COLD)
  or
- Run the engine and monitor the voltage decrease (NTC) as the engine warms.
- This same test sequence can be performed while monitoring the resistance value of the sensor. The sensor must then be disconnected.

- MultiScope key sequence for the temperature sensor test

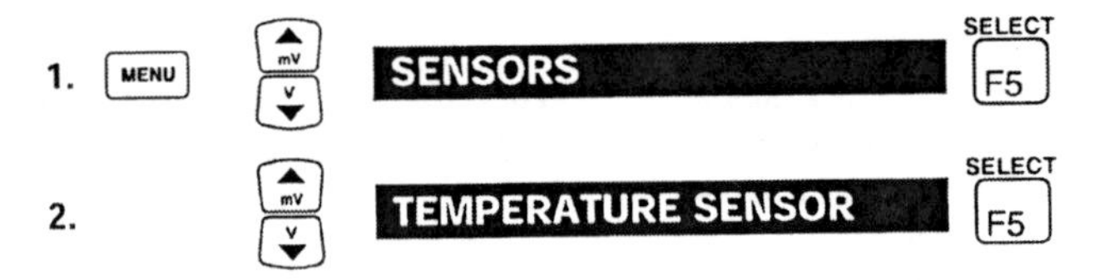

3. Connect the test leads as displayed by the MultiScope's **Connection Help** as shown below.

4. OK F1 — Start the temperature sensor test now

   or

   F2 — You should disconnect the sensor before you press **F2** to test the sensor resistance.

Use the optional temperature probe to measure the actual coolant or intake air temperature.

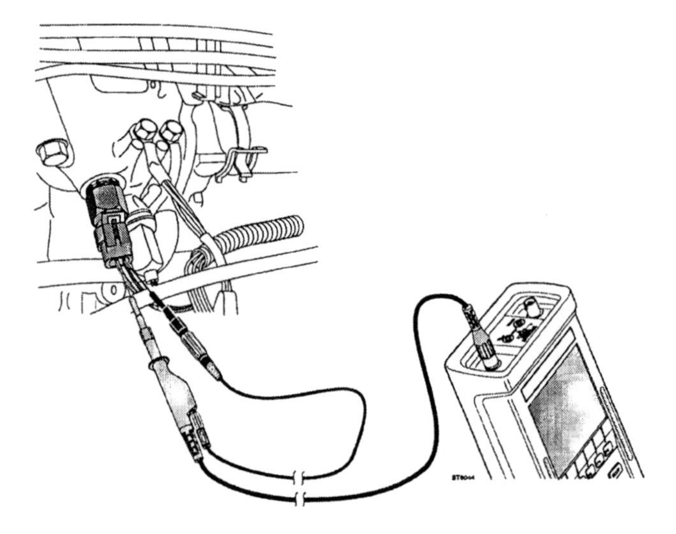

Fig. 7.26 Testing the coolant temperature sensor

In this test the voltage drop across the sensor terminals is being measured. For this test the ignition is switched on, but the engine is not running. The sensor wiring remains connected and the signal voltage will be high because the sensor element is cold. Next the engine is started and as it warms up to operating temperature the sensor voltage will change as shown in Fig. 7.27. To check the sensor performance across the operating range, from engine cold to engine hot, is likely to take some time. This means that the time base of the scope must be set accordingly.

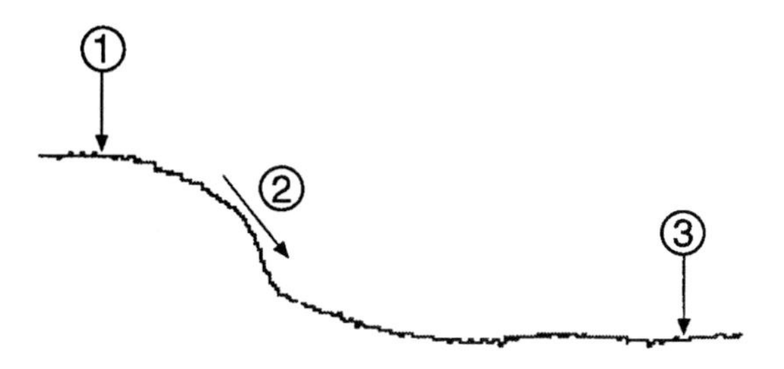

1 Temperature HOT
2 Temperature is decreasing, causing the resistance to increase
3 Temperature COLD

Fig. 7.27 A voltage trace from a coolant temperature sensor

7.6.6 MANIFOLD ABSOLUTE PRESSURE SENSOR (MAP) TESTS

In this case the preliminary visual inspection is very important because it is sometimes the case that the vacuum pipe from the intake manifold to the MAP sensor has become loose or damaged. Figure 7.28 shows the oscilloscope connections for this test.

Figure 7.29(a) gives an impression of the voltage trace from an analogue MAP as observed over a period of 3 s or so. In Fig. 7.29(b) a section of this trace is enlarged and the following points should be noted:

- low engine load
- high engine load
- high voltage (low manifold vacuum)
- as the throttle opens the vacuum falls and the voltage rises
- a low voltage indicates high vacuum.

Figure 7.30 shows the voltage trace for a MAP sensor that gives a variable frequency digital signal output. The signal frequency changes with manifold vacuum and the range of frequency is approximately 50–110 Hz. Points to note are as follows.

- The top line should be very close to the reference voltage that is supplied to the sensor.

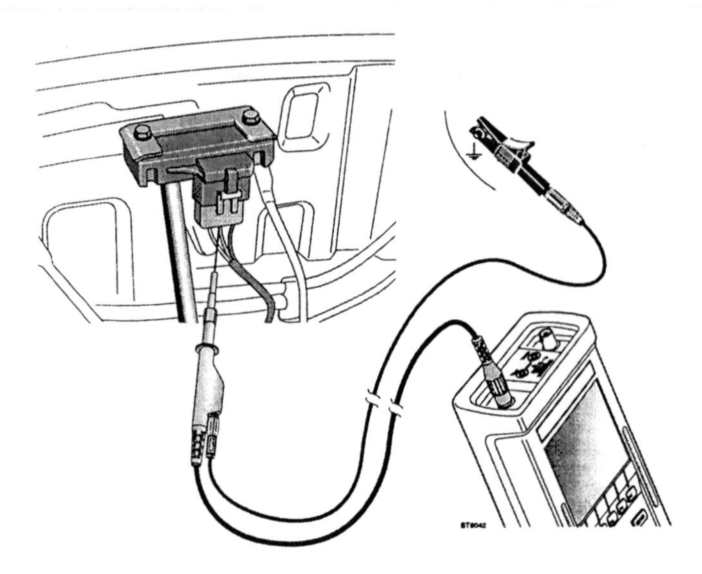

Fig. 7.28 Testing a manifold absolute pressure sensor

(a)

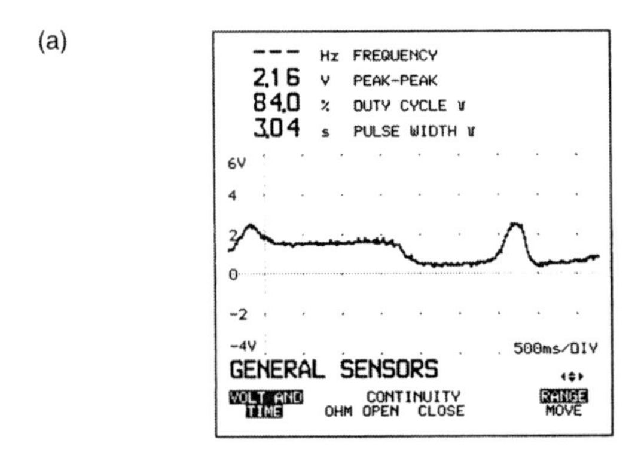

(b)

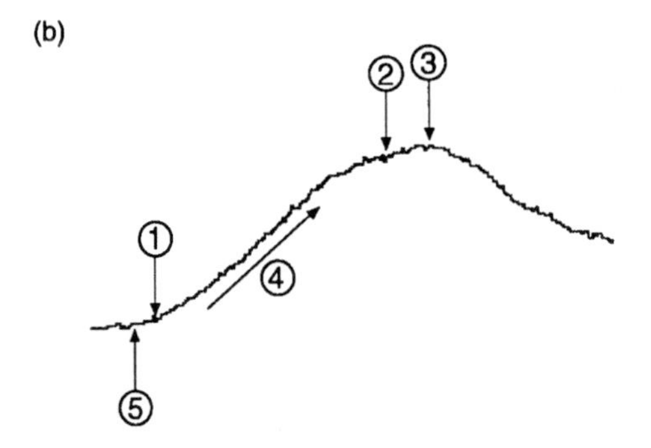

The intermittent record function is a powerful means to watch the signal over time. This function gives you time to activate the sensor while recording is in progress, and then stop recording to display the result.

Fig. 7.29 Signal voltage from an analogue MAP

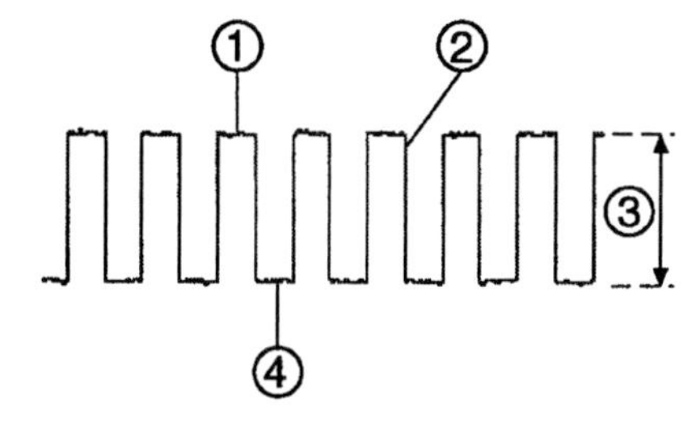

1 The upper horizontal lines should reach reference voltage.
2 Voltage transitions should be straight and vertical.
3 **Peak-Peak** voltage should equal reference voltage.
4 The lower horizontal lines should almost reach ground.

Voltage drop to ground should not exceed 400 mV.
If the voltage is greater than 400 mV, look for a bad ground at the sensor or ECU.
Signal frequence increase as the throttle is opened (vavuum decreases). As the throttle closes the frequency decreases.
The 'f' ranges from approx. eq 50 Hz to 120 Hz.

Fig. 7.30 The voltage pattern for a variable frequency (digital) type MAP sensor

- The rise and fall lines should be near to vertical.
- The peak-to-peak voltage should be very close to the reference voltage.
- The lower lines should be very close to the earth voltage level. Any voltage difference greater that 400 mV at this point requires investigating.

The MAP sensor may be tested with the aid of a vacuum pump, to simulate manifold vacuum, and a voltmeter as shown in Fig. 7.31.

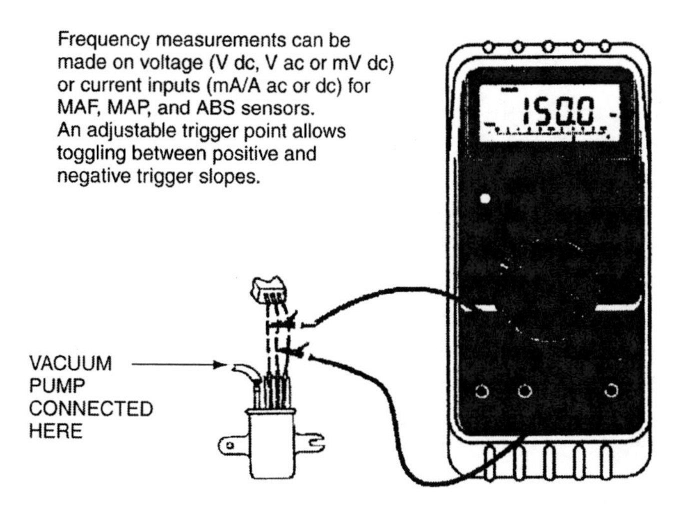

Fig. 7.31 Testing a MAP sensor with the aid of a vacuum pump

This test requires that the vacuum pipe from the manifold to the sensor should be disconnected and the manifold end of the pipe must be connected to the vacuum pump. The positive lead of the voltmeter must be connected to the signal output from the sensor and the negative lead connected to a good earth. The actual means of making a good electrical contact between the signal cable of the sensor and the voltmeter lead is dependent on the means available to the operator. However, any form of connector that impairs insulation or connector reliability must only be used with caution. The advantage of using the vacuum pump is that the sensor output can be checked accurately against measured vacuum and this has advantages over other methods.

7.7 Ignition system tests

Without a good quality spark, in the right place at the right time, the engine performance will be affected, as will the operation of the emissions control system. A misfire can lead to unburnt fuel reaching the exhaust and this will quickly harm the catalyst, often irreparably. For this reason, modern systems monitor the performance of each cylinder, in relation to combustion. One method of doing this is to 'sense' the angular acceleration of the engine flywheel; a firing cylinder will produce more acceleration than a misfiring one. In order to identify the cylinder that is misfiring the ECM requires a reference signal and this is often provided by the camshaft position sensor.

On modern systems, the ECM has the ability to detect misfires because the unburnt fuel that results can cause serious damage to the exhaust catalyst. The ECM achieves this diagnosis by reading the time interval between pulses from the crankshaft speed sensor. Persistent misfires will activate the MIL and a fault code (DTC) will be recorded. Urgent remedial work will then be required if serious catalyst damage is to be avoided.

7.7.1 TESTS ON DISTRIBUTORLESS IGNITION DIS

Firstly, it is important to observe 'Key skill 7: work in a safe manner'. Electric shocks from ignition systems must be avoided - not only is the shock dangerous in itself but it can also cause involuntary muscular actions which can cause limbs to be thrown into contact with moving or hot parts.

The secondary voltage trace is often used in the analysis of ignition system performance. Figure 7.32 shows the portable oscilloscope set-up for this test. The inductive pick-up is clamped to the HT lead, as close as possible to the spark plug.

Note that the screen display shows data about engine speed and burn time. These and other details may more easily be seen by taking an enlargement of the trace for a single cylinder, as shown in Fig. 7.33. With reference to this figure the details are as follows.

1. Firing line. This represents the high voltage needed to cause the spark to bridge the plug gap.

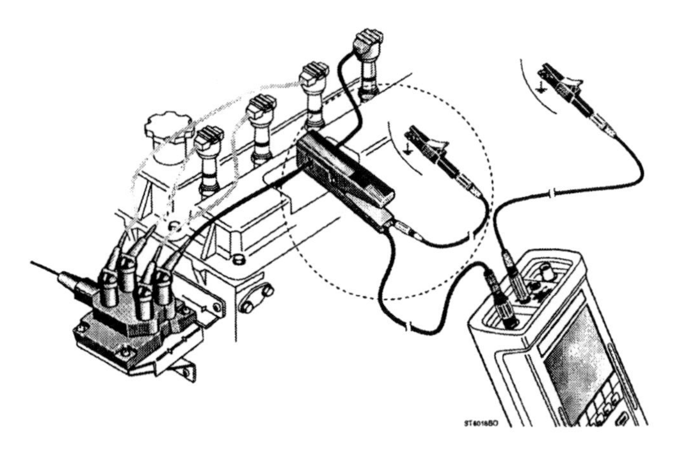

Fig. 7.32 The oscilloscope set-up for obtaining a secondary voltage trace from a DIS

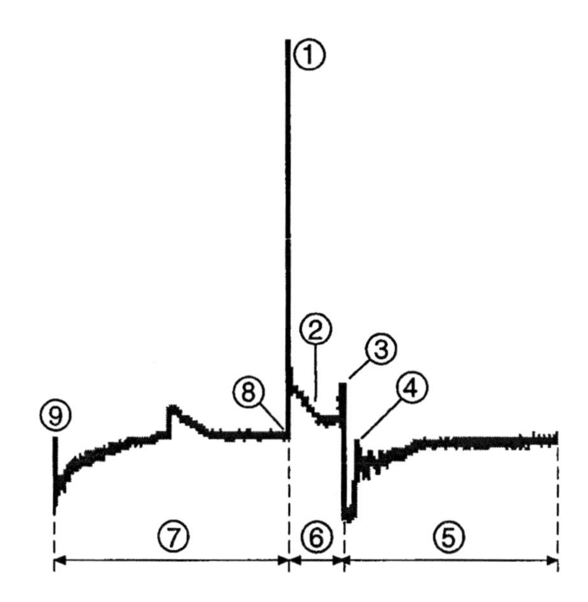

Fig. 7.33 Details of the HT voltage trace for a single cylinder

2. The spark line.
3. Spark ceases.
4. Coil oscillations.
5. Intermediate section (any remaining energy is dissipated prior to the next spark).
6. Firing section (represents burn time).
7. Dwell section.
8. Primary winding current is interrupted by transistor controlled by the ECM.
9. Primary winding current is switched on to energize the primary. The dwell period is important because of the time required for the current to reach its maximum value.

A comparison of the secondary voltage traces for each of the cylinders should show them to be broadly similar. If there are major differences between the patterns then it is an indication of a defect. For example, a low firing voltage (1) indicates low resistance in the HT cable or at the spark plug. The low resistance could be attributed to several factors, including oil- or carbon-fouled spark plug, incorrect plug gap, low cylinder pressure or defective HT cable. A high voltage at (1) indicates high resistance in the HT cable, or at the spark plug. Factors to consider here include a loose HT lead, wide plug gap, or an excessive amount of resistance which has developed in the HT cable. Table 7.1 summarizes the major points.

Table 7.1 Factors affecting firing voltage for high and low firing voltages

Factor	High firing voltage	Low firing voltage
Spark plug gap	Wide	Small
Compression pressure	Good	Low
Air-fuel ratio	Weak	Correct
Ignition point	Late	Early

Sparking plugs can be removed and examined, HT leads can also be examined for tightness in their fittings and their resistance can be checked with an ohm-meter. Resistive HT leads are used for electrical interference purposes and they should have a resistance of approximately 15 000-25 000 ohms/meter length.

Here, as in all cases, it is important to have to hand the information and data that relates to the system being worked on.

7.8 Diesel injection

The diesel injection system together with heat generated by compression provides the ignition system in a diesel engine. The pipes that convey the fuel from the pump to the injectors operate at very high pressures of several hundred bar. This causes the pipes to expand with each injection pulse. By clamping a piezoelectric pick-up to the injection pipe, this expansion of the fuel injection pipe is converted into an electrical signal which is displayed on an oscilloscope screen. The principle is illustrated in Fig. 7.34.

For best results, the pipes should be thoroughly cleaned and the piezo-adaptor placed as close as possible to the injector, on a straight section of pipe. It will not work if it is placed on a bend in the pipe. When the instrument is set up correctly the engine should be started and allowed to idle. A pattern of the type shown in Fig. 7.35 should then be seen.

● MultiScope key sequence

Set **DIESEL** in the **VEHICLE DATA MENU** as follows:

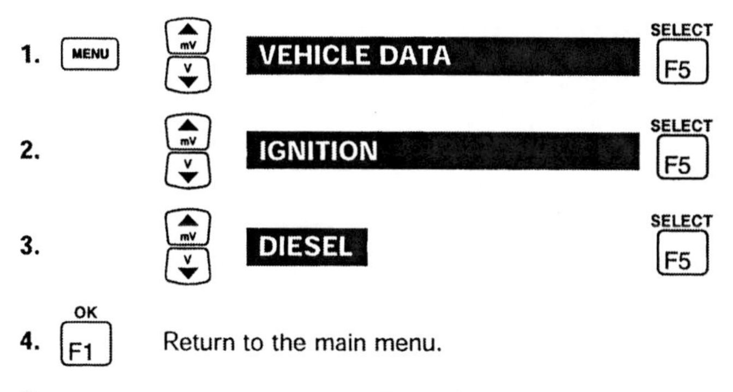

4. OK F1 Return to the main menu.

Proceed as follows to select the Diesel Injector test:

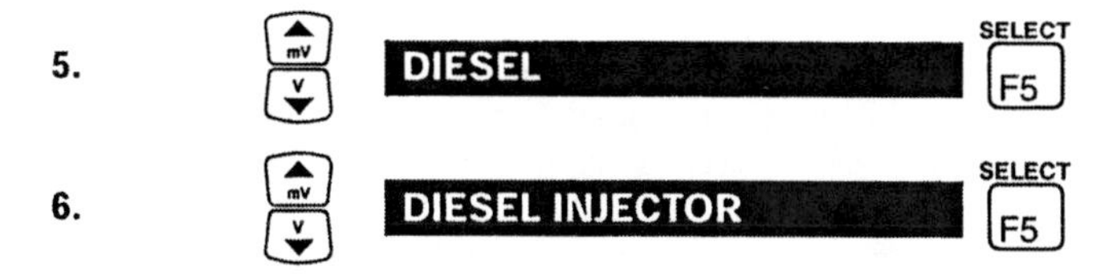

7. Connect the test leads as displayed by the MultiScope's **Connection Help** as shown below.
8. OK F1 Starts the diesel injector test.

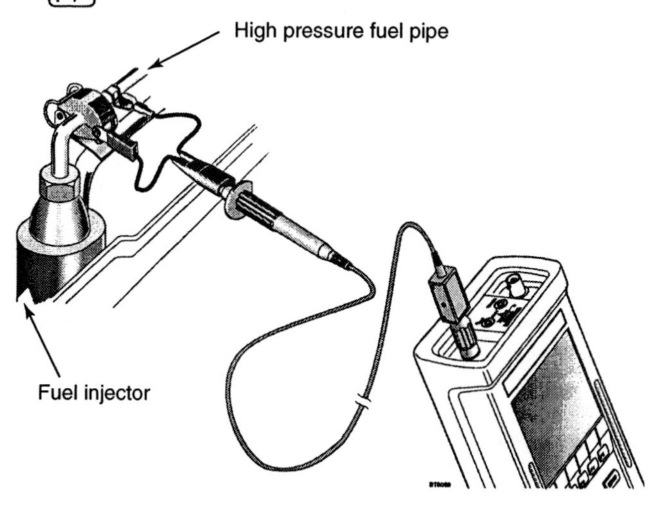

Fig. 7.34 Using the piezoelectric pick-up and the Bosch PMS 100 to examine diesel injection details

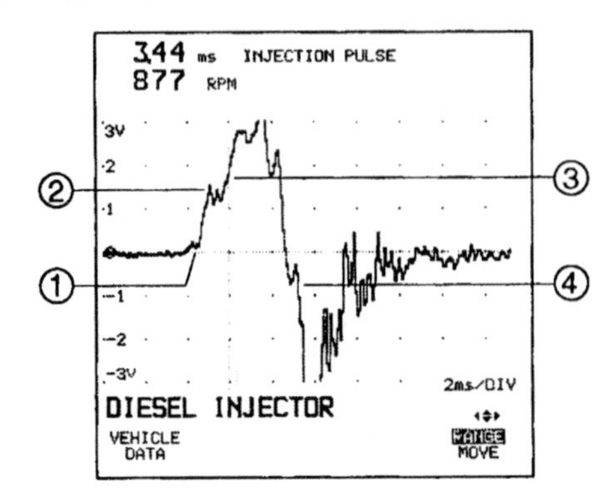

1. The injection pumps plunger moves in the supply direction thus generating a high pressure in the pressure gallery.
2. The delivery valve opens and a pressure wave proceeds toward the injector.
3. When the injector opening pressure is reached, the needle valve overcomes its needle spring force and lifts.
4. The injection process ends, the delivery valve closes and the pressure in the line drops. This quick drop causes the nozzle to close instantly, preventing the nozzle from opening again, and preventing backflow of combustion gases.

Fig. 7.35 Diesel injection pressure pattern at idle speed

7.8.1 TESTING THE INJECTION POINT ADVANCE

When other facilities on the PMS 100 scope are brought into use it is possible to obtain a trace of the TDC sensor signal and the pressure signal simultaneously. The set-up is shown in Fig. 7.36.

If the engine is now run at idle speed, the advance period can be observed, as shown by the two vertical lines on the trace shown in Fig. 7.37(a). When the speed is increased to 1700 rpm the effect on the advance period is seen in Fig. 7.37(b). The main benefit of this test is it enables one to check the injection point in relation to TDC and this gives an indication that the timing control on the pump is functioning correctly.

7.9 Sensor tests on other systems

The survey of computer controlled systems that is the subject of Chapter 1 shows that the systems rely on sensor inputs for the live data that permits the systems to operate. The diagnostic techniques that are described in the above sections are equally applicable to virtually any computer controlled system on a vehicle. The following examples show other tests that can be applied as part of a diagnostic procedure.

The piezo pickup is clamped on the fuel line of the first cylinder, close to the injector and connected via the blue filter probe to **INPUT A**.(See figure below).The TDC sensor signal is connected to **INPUT B**.
Do not use the ground lead of Channel B, since the instrument is already grounded through the pickup adapter to the fuel line.

● MultiScope key sequence

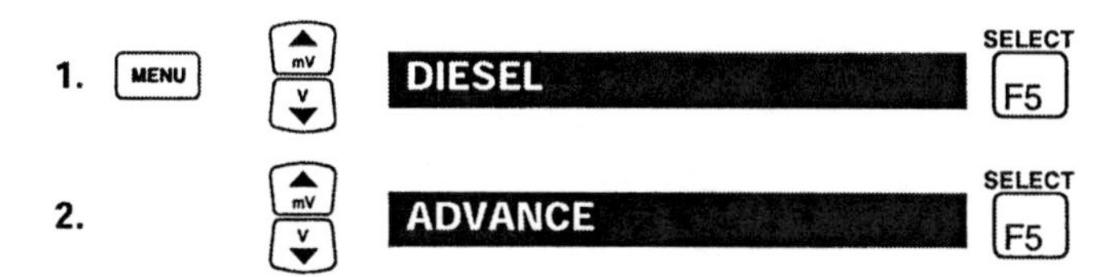

3. Connect the test leads as displayed by the MultiScope's **Connection Help** as shown below.
4. OK F1 Starts the Diesel Advance test.
5. Position the cursors as shown

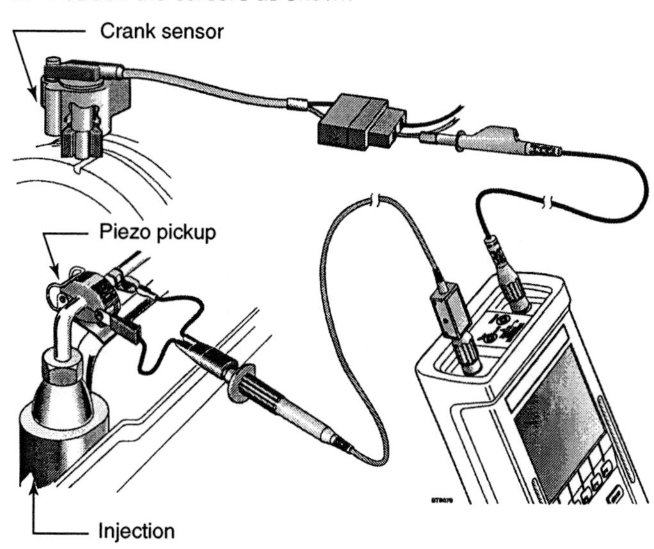

Fig. 7.36 Connecting to the crank position sensor and the piezo pick-up

7.9.1 ABS WHEEL SPEED SENSORS

The anti-lock braking wheel speed sensors are an essential signal for operation of the system and, as explained in Chapter 1, they are also used in other systems such as traction control and stability control. A commonly used ABS sensor operates on the variable reluctance principle and Fig. 7.38 gives an indication of the set-up for a wheel speed sensor test.

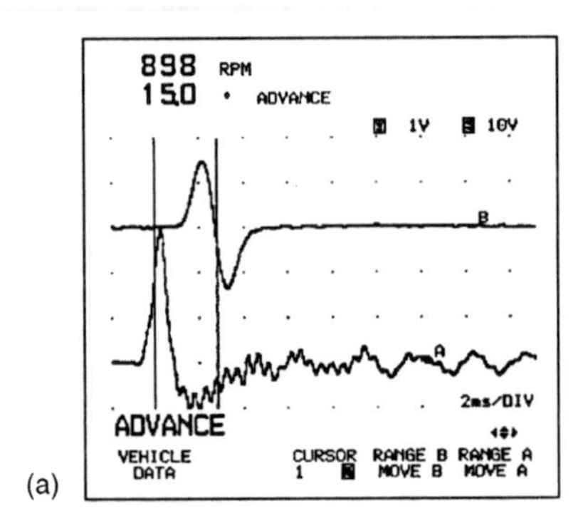

(a)

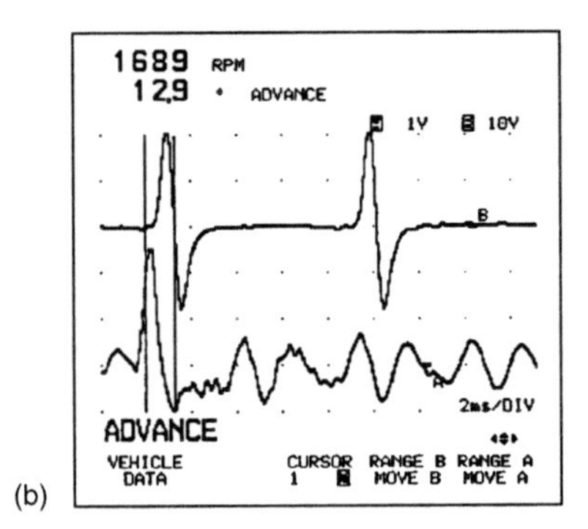

(b)

Fig. 7.37 Diesel injection advance

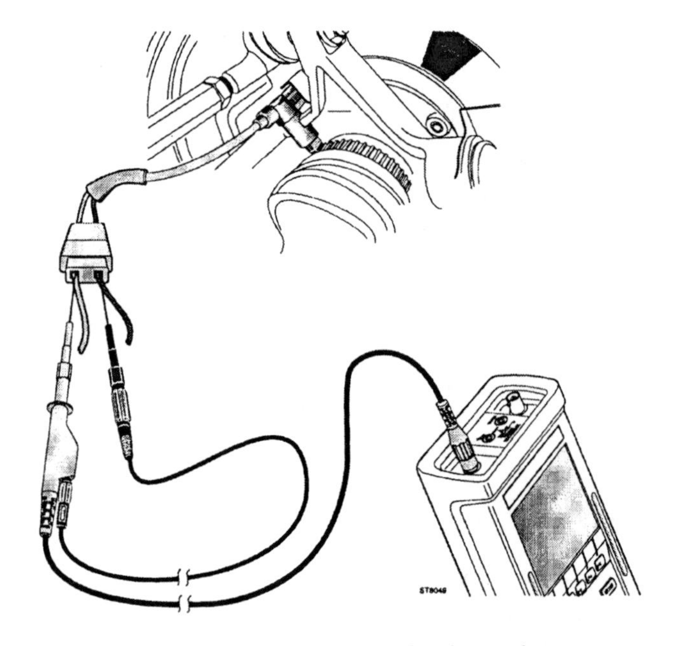

Fig. 7.38 A test on an ABS wheel speed sensor

The test requires that the road wheel be rotated at sufficient speed to generate a signal. This can normally be achieved by safely jacking up the vehicle and rotating the wheel by hand. [Any other method of rotating the wheel must only be in accordance with the vehicle manufacturer's recommended procedure.] The sensor output should be of the form shown in Fig. 7.39. The voltage will vary with speed, as will the frequency.

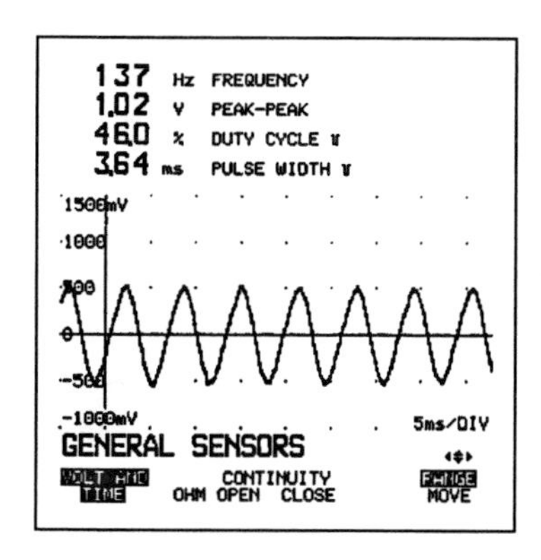

Fig. 7.39 Testing an ABS wheel sensor

The main points to look for on the voltage trace are:

- a regular waveform
- any gaps in the waveform that may indicate a missing tooth or displaced reluctor ring
- low voltage.

Some ABS and traction control systems make use of Hall-type sensors. The Hall sensor requires a power supply and testing this type of sensor will require special attention to be given to this aspect. The remainder of the test would be similar to that shown above and, once again, special attention must be given to safe jacking up of the vehicle and rotation of the wheel. A Hall-type sensor will produce a pattern, of varying frequency, of the form shown in Fig. 7.40.

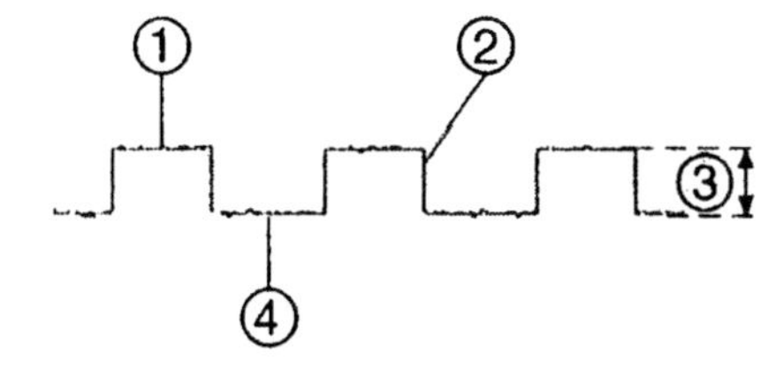

1 The upper horizontal lines should reach reference voltage.
2 Voltage transitions should be straight and vertical.
3 **Peak-Peak** voltages should equal reference voltage.
4 The lower horizontal lines should almost reach ground.

Fig. 7.40 The pattern of the signal from a Hall type ABS sensor

7.9.2 TESTING THE RIDE HEIGHT CONTROL SENSOR

The computer controlled suspension system uses a variable resistance-type sensor to provide the computer with a voltage signal that represents ride height. Figure 7.41 shows the portable oscilloscope being used to obtain a voltage

● Measurement conditions

- Turn the key ON, engine OFF. Backprobe the sensor's connector or use jumper wires. Disconnect the moveable arm of sensor (attached to the rear axle.) Move the arm from stop to stop to monitor the full
- Turn the key OFF, engine OFF. Test the sensor's resistance by carefully disconnecting the sensor from its associated wiring harness. Use the resistance mode to determine if there is an open or short in the potentiometer.
- Reconnect the movable arm to the rear axle and adjust the ride height sensor to the specifications found in the vehicle's service manual.

● MultiScope key sequence

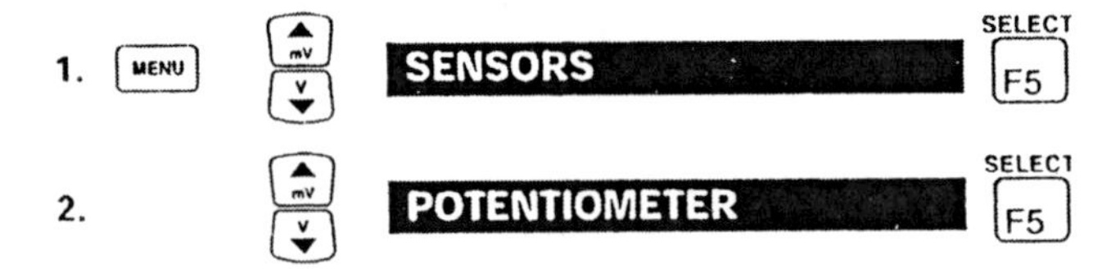

3. Connect the test leads as displayed by the MultiScope's **Connection Help** as shown below.
4. OK F1 Starts the sweep test.

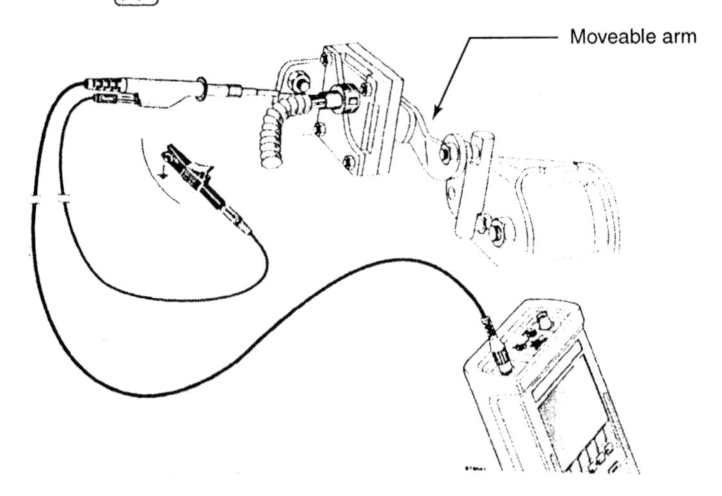

Fig. 7.41 Testing a ride height sensor

trace from the ride height sensor. The same test can be performed by using a digital voltmeter set to a suitable range.

When the electrical supply to the sensor has been verified and the signal cable has been located, the sensor's moveable arm should be disconnected from the axle or suspension unit so that the sensor can be activated manually. When the oscilloscope or voltmeter has been securely connected to the signal lead (probably by backprobing), the sensor arm should be moved through its full operating range. A scope pattern similar to that shown in Fig. 7.42 should result. Failure to produce a voltage that varies in relation to movement of the arm, or any sudden drops in voltage, indicates that the sensor is malfunctioning.

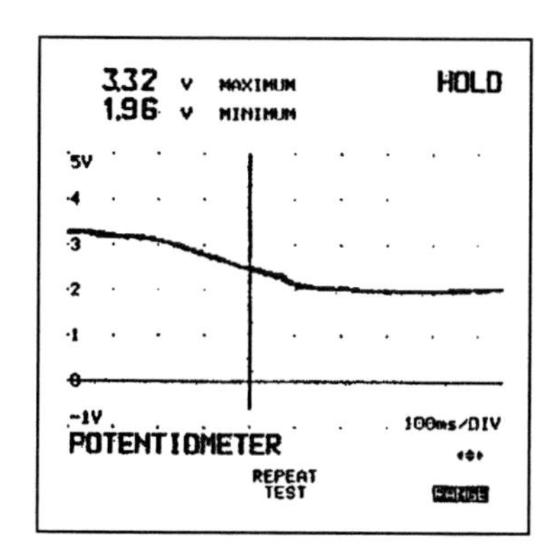

Fig. 7.42 Result display of a voltage test on a ride height sensor

Although a test of this nature will show whether or not the sensor is operating electrically it should be noted that the mechanism that connects the sensor into the suspension system can also be a cause of trouble. Should the sensor's mechanical linkage be worn or damaged, or the sensor be wrongly positioned on the vehicle, the signal that the ECM receives will not accord with its programmed value and this may cause the height control system to malfunction. So, beware, always make a thorough visual check of the system at a very early stage in any diagnostic process.

7.10 Intermittent faults

The self-diagnosis capacity of modern computer controlled systems has provided technicians with an additional source of information that can be of great value in tracking down faults that occur occasionally. The processing power and memory capacity of the on-board computer can be of value in tracing the causes of this type of fault. In addition, many diagnostic scan tools have a data logger capability which is similar to, but of smaller capacity than, the flight recorder on aircraft. Because of this similarity, the data logger function of the scan tool is often known as the 'flight recorder' function.

7.10.1 FLIGHT RECORDER (DATA LOGGER) FUNCTION

The data logger aspect of test equipment capability permits the test equipment to store selected data that the test equipment 'reads' through the serial data diagnostic connector of the ECM. It is particularly useful for aiding the diagnosis of faults, such as an unexpected drop in power that occurs during the acceleration phase. When the test equipment is connected, and proper preparations have been made for a road test, the vehicle is driven by a person who should be accompanied by an assistant to operate the test equipment (for safety reasons). With the test equipment in 'record' mode, the vehicle is driven in an attempt to re-create the default condition and when the 'fault' occurs the test equipment control button is pressed. From this point, data from just before the incident and for a period after is recorded. The stored data can then be played back, on an oscilloscope screen, or printed out later for analysis in the workshop. Figure 7.43 shows an example of live data that was obtained from a test using the Bosch KTS 500 equipment on a Peugot vehicle.

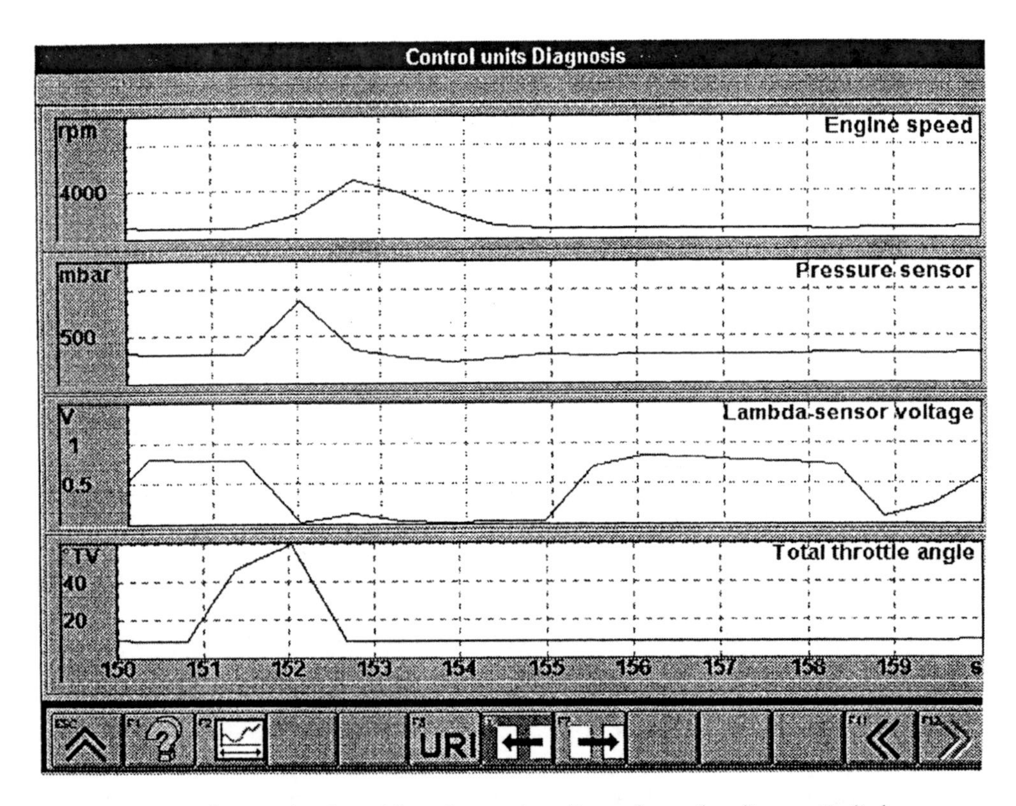

Fig. 7.43 A display of live data as 'read' out from the diagnostic link

In this case the signals all appear to be in order. However, should a defect occur it will show up in displays such as these. Figure 7.44 gives an impression of the type of information that might be seen.

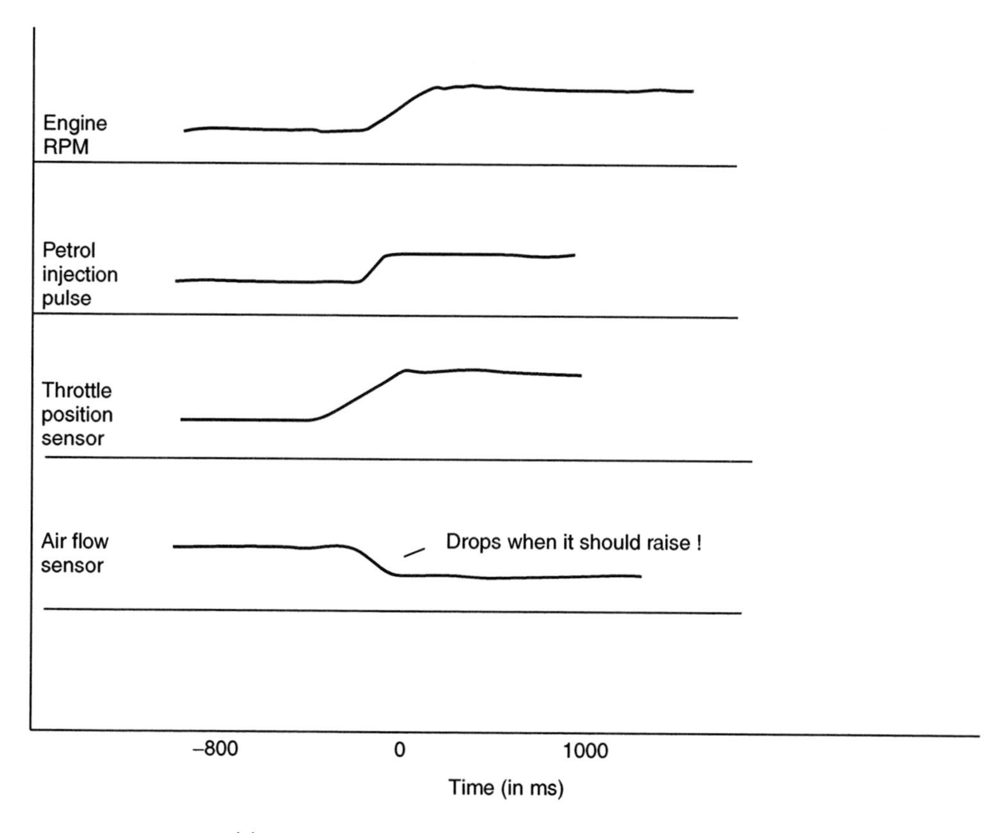

Fig. 7.44 How a probable fault shows up in a live data display

The reasoning here is that the vehicle is accelerating. This is borne out by the increased throttle opening, increased engine speed and increased fuel injection. One would expect these events to be accompanied by an increase in air flow. But the air flow sensor shows a drop in air flow - maybe there is a fault at the air flow sensor!

7.11 Summary

As systems become more sophisticated and the memory space available for diagnostic data increases, it becomes possible for designers to include more diagnostic detail. It is, therefore, quite possible that the on-board computing capacity will enable the precise cause of a fault to be read out. However, it remains the case that the circuit between the sensor or actuator and the computer interface may be the reason why the computer fails to receive, or send, the correct signal.

The ability to check sensors and actuators at their terminals, and then to check the circuits between them and the computer is a valuable asset that is obtained by the use of equipment such as the portable oscilloscope and the digital multimeter. Most garage-type oscilloscopes described in this book have multiple facilities in the sense that they can be used as voltmeters, ohmmeters, frequency meters etc. This makes such equipment a valuable asset to any technician who is undertaking repair and maintenance work on vehicle systems.

At various points I have mentioned that there is an increasing availability of diagnostic data and many publishers produce manuals of fault codes and diagnostic information. This indicates that there is a role for independent repairers in the field of modern technology and, with the advent of European on-board diagnostics, future prospects look bright. I have highlighted the key skills that are necessary for successful diagnosis and I hope that this knowledge will inspire trainees to persevere with study and training in order to acquire these essential skills.

7.12 Review questions (see Appendix 2 for answers)

1. For voltage tests on sensor output the voltmeter is connected:
   (a) in series?
   (b) between the sensor output terminal and the end of the cable that has previously been disconnected?
   (c) between the signal terminal and earth?
   (d) across the vehicle battery terminals?
2. It is useful to be able to test sensor performance at the sensor and at the ECM because:
   (a) it provides a useful check on the wiring between the two points?
   (b) it eliminates the need for intrusive testing?
   (c) the two readings can be averaged to provide a value for the source voltage?
   (d) it is the only way to check voltage in a feedback system?
3. When an exhaust gas oxygen sensor ages:
   (a) the exhaust catalyst stops working?
   (b) the peak-to-peak voltage of the sensor is reduced?
   (c) the peak-to-peak voltage is increased?
   (d) the EGR system stops operating?
4. In a variable reluctance type crank position sensor:
   (a) the air gap has no effect on sensor performance?
   (b) the air gap should be checked if the sensor signal is incorrect?
   (c) the sensor output voltage is not affected by speed of rotation of the crankshaft?
   (d) the voltage signal can only be measured by means of a voltmeter?
5. Cylinder recognition sensors are often fitted to the camshaft:
   (a) because the camshaft turns faster than the crankshaft?
   (b) because they are more accessible in that position?
   (c) because the camshaft rotates once for every two revolutions of the crankshaft?
   (d) in order to operate single-point injection systems?

6. Modern engine management systems often adapt control programs to compensate for wear in components, and after fitting new units the vehicle should be driven for a suitable period to:
   (a) clear the fault code memory?
   (b) allow the ECM to 'learn' a new set of figures?
   (c) allow the alternator to recharge the battery?
   (d) clear unburnt fuel from the exhaust system?
7. A breakout box:
   (a) replaces the ECM for test purposes?
   (b) permits tests to be performed at the ECM without backprobing?
   (c) reads digital data because it is connected to the data bus of the ECM processor?
   (d) needs only one connector to enable it to be used on any make of vehicle?
8. Sensor inputs to the ECM are:
   (a) used only for generating fault codes?
   (b) used to provide data that enables the ECM to know the state of variables, such as engine speed, air flow etc.?
   (c) always analogue signals?
   (d) always frequency signals?

8
Additional technology

This chapter contains additional information about some of the technology that is introduced in the earlier chapters. The treatment of the topics is fairly light but it should be sufficient to provide readers with an introduction that will enable them to embark on further study in their particular area of interest, if they so desire.

8.1 Partial and absolute pressures

Partial pressure of oxygen is mentioned in the context of the exhaust gas oxygen sensors. Dalton discovered that the pressure exerted by a mixture of two gases, in a given space, is equal to the sum of the pressures that each gas would exert if it occupied the same space on its own. Thus, if a mixture of oxygen and nitrogen in a given space, exerts a total pressure of 3 bar and the nitrogen pressure is 2.7 bar, the oxygen pressure will be 0.3 bar. These separate pressures, 2.7 bar for nitrogen and 0.3 bar for oxygen, are known as the partial pressures.

It is the partial pressure of oxygen in exhaust gas that is the variable of interest for the lambda sensor that is used in the emission control system.

Because the majority of pressure gauges are set to record zero pressure in the earth's atmosphere they are, in effect, ignoring the pressure due to the earth's atmosphere. The pressure that gauges record is therefore known as 'gauge pressure' because they only record the pressure above atmospheric pressure. In order to obtain absolute pressure it is necessary to record barometric pressure. If we take an example of atmospheric pressure of 1015 mbar = 1.015 bar and imagine that we have recorded a tyre inflation pressure of 2.1 bar on the gauge, the absolute pressure in the tyre is atmospheric pressure plus gauge pressure which, in this case = 1.015 bar + 2.1 bar = 3.115 bar.

Barometers record absolute pressure and the pressures that are shown on TV weather maps are also absolute pressures.

In automotive control applications the ability to measure absolute pressure is important for at least two reasons.

1. The absolute pressure in the induction manifold, of a throttle controlled engine, is an accurate indicator of the load that the engine is operating under.
2. The speed density method of measuring the mass of air uses the absolute pressure in the manifold, to enable the ECM to estimate accurately the mass of air entering the combustion chambers of the engine.

In broad outline, it is able to do this because the characteristic gas equation gives $pV = mRT$, where p = absolute pressure in the manifold (in bar), V = volume of air entering the engine (in m^3), m = mass of the air entering (in kg), R is the characteristic gas constant for the substance in the manifold (air), and T is the temperature (in degrees Kelvin) of the air in the manifold. The equation can be expressed as $m = pV/RT$, which gives an instantaneous value for mass of air.

8.2 The piezoelectric effect

Some materials that appear in nature, such as quartz, possess the piezoelectric effect. That is to say when an alternating voltage is applied to a suitably sized quartz crystal it will vibrate at the same frequency as the applied voltage. This is because the application of the voltage changes the electrical polarity of the crystal and in the process causes a mechanical strain. This is known as the 'motor effect' and in order to achieve it the accurately sized crystal is placed between metal plates to produce a circuit like the one shown in Fig. 8.1.

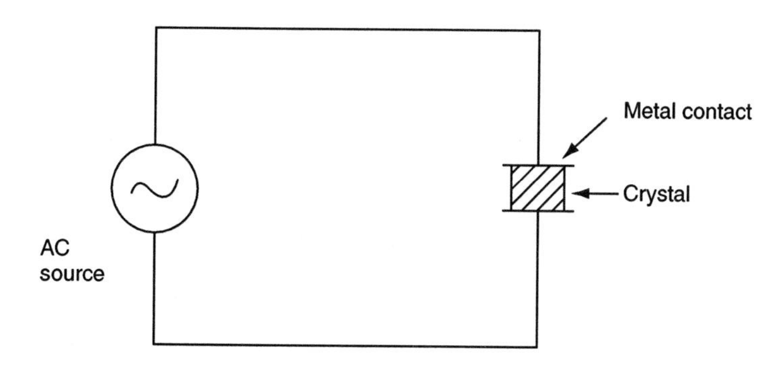

Fig. 8.1 An oscillator based on the piezoelectric effect

When the process is reversed, i.e. by applying force to the crystal, mechanical strain is caused, the electrical polarity is changed, and a voltage is produced between the metal plates. This is known as the 'generator effect' and it is this that is the basis of the knock sensor that is used in many engine management systems (Fig. 8.2).

Natural quartz contains impurities and to overcome this problem quartz is 'grown' in controlled conditions. In addition to quartz and a few other materials, certain ceramics that are oxide alloys of platinum, zirconium and titanium (PZT), also possess piezoelectric properties.

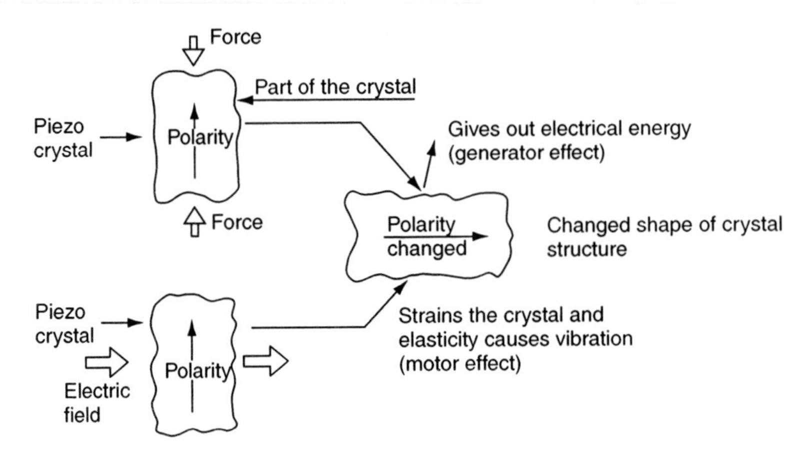

Fig. 8.2 The piezoelectric generator and motor effects

An important property of a piezoelectric material is its coupling coefficient k. This expresses the relation between electrical and mechanical energy. k^2 = mechanical energy output/electrical energy input or electrical energy output/mechanical energy input.

8.3 Liquid crystal displays

Liquid crystal displays (LCDs) are used in many applications, such as laptop computers, pocket calculators, vehicle instrument displays, watches and clocks, and increasingly in diagnostic equipment for automotive systems. The portable oscilloscope that features prominently in this book uses a liquid crystal display and readers may wish to have an insight into the principles of operation of the LCD.

A liquid crystal that is commonly used in LCDs is one that is known as a twisted nematic (TN). This crystal is composed of rod like molecules that are readily polarized by the application of an electric field. When a layer of TN material about 10 μm thick is placed between two transparent electrodes, a capacitor is formed and the application and removal of an electrical potential to the electrodes alters the polarity of the rod-like molecules and this in turn, affects the light transmitting properties of the LCD cell.

In the simple LCD cell shown in Fig. 8.3 the electrode plates are made from indium doped tin oxide (ITO), which is transparent. A thin layer of plastic material covers the inside surface of the electrodes and during the manufacturing process, microscopic grooves are introduced into the plastic by a rubbing process. These grooves are mutually perpendicular, as shown by the sets of parallel lines on the upper and lower sections. The effect of these grooves is to cause the rod-like molecules to rotate smoothly through 90° across the width of the TN layer. The outside of LCD cell is covered with glass and on the surface of the glass is a polarizing filter. The left-hand view shows the electric current removed from the

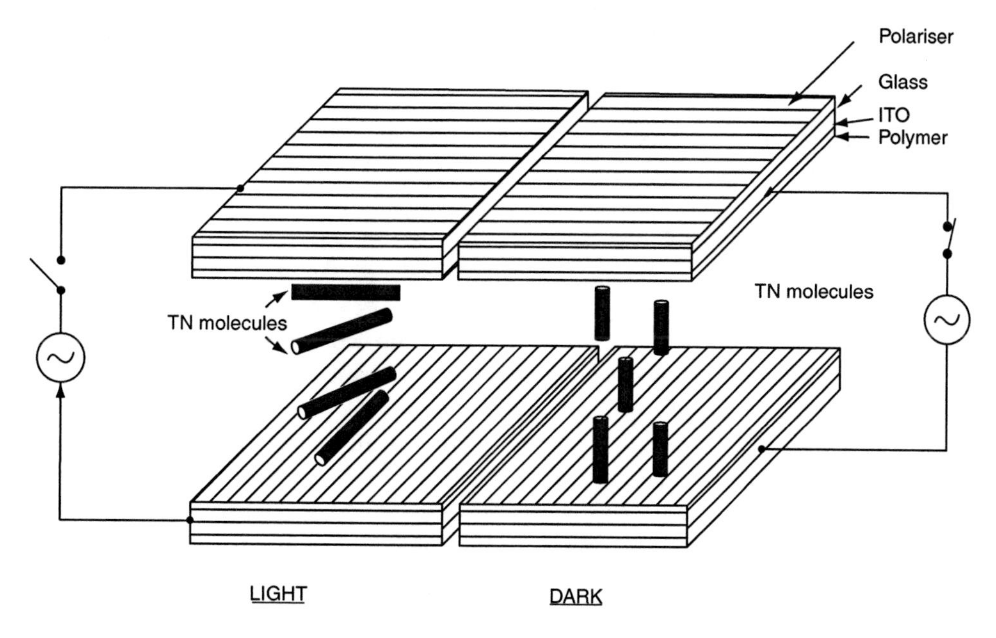

Fig. 8.3 A simplified LCD cell

LCD cell and this allows the TN molecules to guide the light through 90°. If a reflector is placed on one side of the cell, the reflected light makes the cell appear light. Alternatively, back lighting will also shine through the cell. In the right-hand view, the current is switched on. This causes the rod-like molecules to align with the electric field and the passage of light is effectively blocked. This causes the cell to appear dark.

In order to produce a large display, suitable for a computer or oscilloscope screen, it is necessary to have a large number of cells arranged in a matrix form. Figure 8.4 shows the concept.

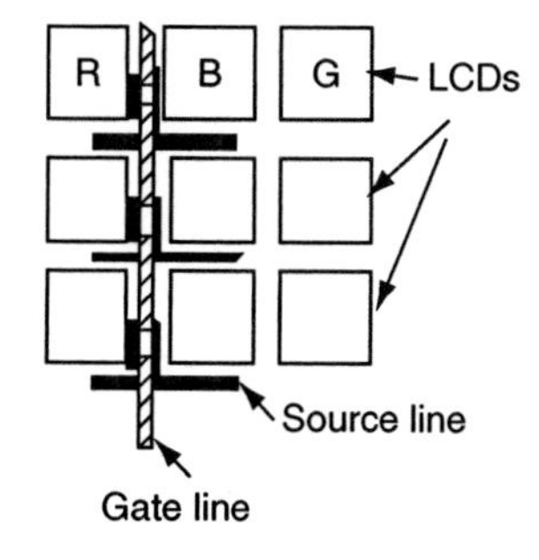

A small part of an LCD screen
Each cell is controlled by a mosfet TFT
via source and gate lines

Fig. 8.4 Arrangement of some cells for the matrix of an oscilloscope screen

The picture elements (pixels) are formed by the LCDs, and by grouping red, blue and green cells as shown, accurate colour images of good definition are produced. Each LCD cell is controlled by a thin film transistor (TFT). By addressing rows and columns these transistors are switched on and off, as required.

8.4 Countering cross-talk

With cables that are conducting data pulses, it is not always possible to separate them by a distance that would prevent 'cross-talk' (one signal interfering with another) and other measures are adopted, for example, screened cables and 'twisted pairs'.

Figure 8.5(a) shows a twisted pair of cables. Capacitive interference from a nearby cable is reduced because the two wires are kept close together and this helps to make the signal-carrying wires less susceptible to capacitive cross-talk. The twisted pair is also used to counter inductive interference from nearby cables. The theory is that the 'interference' effects cancel one another out over the length of the cable.

Figure 8.5(b) shows a screened cable (co-axial). Here the interference is reduced because the interfering current is mainly confined to the outer surface of the screen, where it does not distort the signal in the central conductor. The earthing of the screen is an important factor.

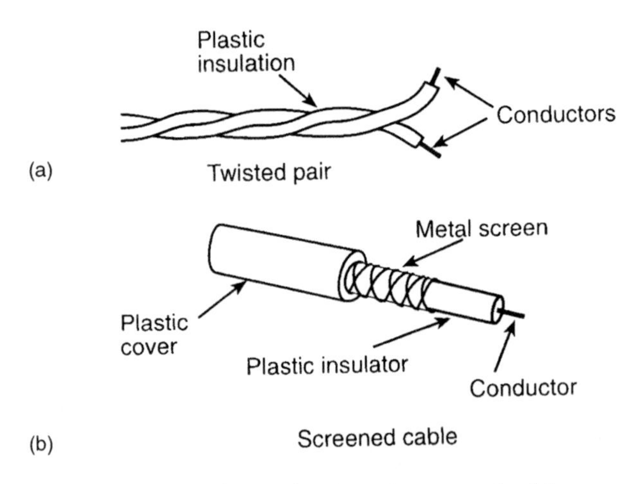

Fig. 8.5 Twisted pair and screened cable

8.5 Logic devices

8.5.1 THE RTL NOR GATE

Figure 8.6 shows how a resistor transistor logic (RTL) gate is built up from an arrangement of resistors and a transistor. There are three inputs: A, B, and C. If one

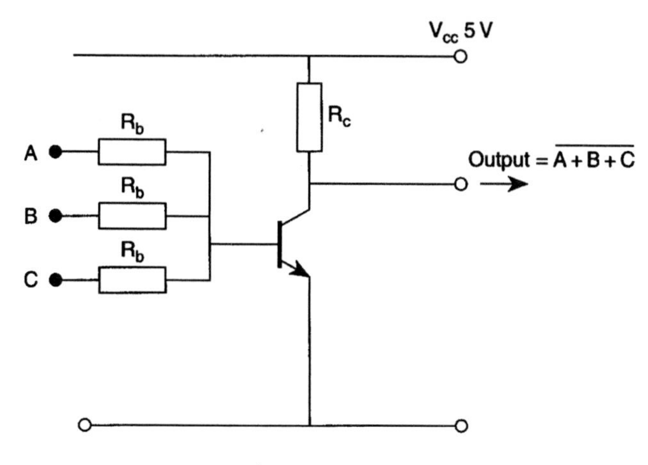

Fig. 8.6 An RTL NOR gate

or more of these inputs is high (logic 1), the output will be low (logic 0). The output is shown as $A + B + C$ with a line, or bar, over the top (the + sign means OR). Thus the $A + B + C$ with the line above means 'not A or B or C' (NOR : NOT OR).

The base resistors R_b have a value that ensures that the base current, even when only one input is high (logic 1), will drive the transistor into saturation to make the output low (logic 0).

8.5.2 TRUTH TABLES

Logic circuits operate on the basis of Boolean logic and terms like, NOT, NOR, NAND etc., derive from Boolean algebra. This need not concern us here, but it is necessary to know that the input-output behaviour of logic devices is expressed in the form of a 'truth table'. The truth table for the NOR gate is given in Fig. 8.7.

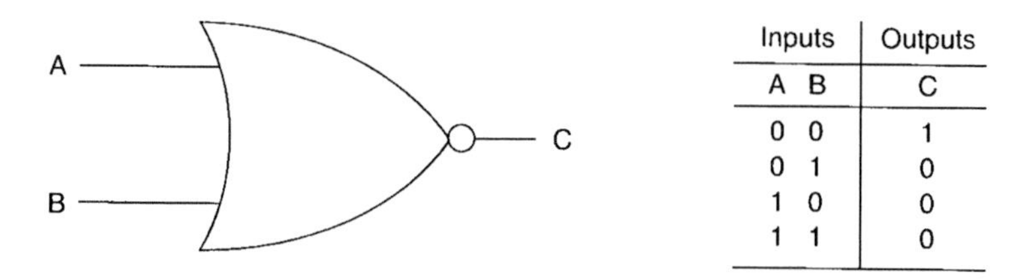

Inputs		Outputs
A	B	C
0	0	1
0	1	0
1	0	0
1	1	0

Fig. 8.7 NOR gate symbol and truth table

In the NOR truth table, when the inputs A and B are both 0 the gate output, C, is 1. The other three input combinations each give an output C=1.

In computing and control systems, a system known as TTL (transistor to transistor logic) is often used. In TTL logic, 0 is a voltage between 0 and 0.8 V and logic 1 is a voltage between 2.0 and 5.0 V. A range of other commonly used logic gates and their truth tables is given in Fig. 8.8.

Type of logic gate	USA symbol	UK symbol	Truth table
AND	A B X	A B & X	Inputs: A B | Outputs: X 0 0 | 0 0 1 | 0 1 0 | 0 1 1 | 1
OR	A B X	A B X	Inputs: A B | Outputs: X 0 0 | 0 0 1 | 1 1 0 | 1 1 1 | 1
(NOT AND) NAND	A B X	A B & X	Inputs: A B | Outputs: X 0 0 | 1 0 1 | 1 1 0 | 1 1 1 | 0
NOT inverter	A X	A X	Inputs: A | Outputs: X 0 | 1 1 | 0
(NOT OR) NOR	A B X	A B X	Inputs: A B | Outputs: X 0 0 | 1 0 1 | 0 1 0 | 0 1 1 | 0

Fig. 8.8 A table of logic gates and symbols

Figure 8.9(a) shows the basic inverter. Here an input of logic 1 becomes an output of logic 0. Figure 8.9(b) shows a feedback connection from the output to the input. This feedback results in an output that oscillates between 0 and 1, at a frequency that is dependent on the propagation delay time of the inverter.

8.5.3 THE SR (SET, RESET) FLIP-FLOP

Just as the basic building blocks of electronics, e.g. transistors, diodes and resistors, can be joined together to make logic gates, so those logic gates can be used as

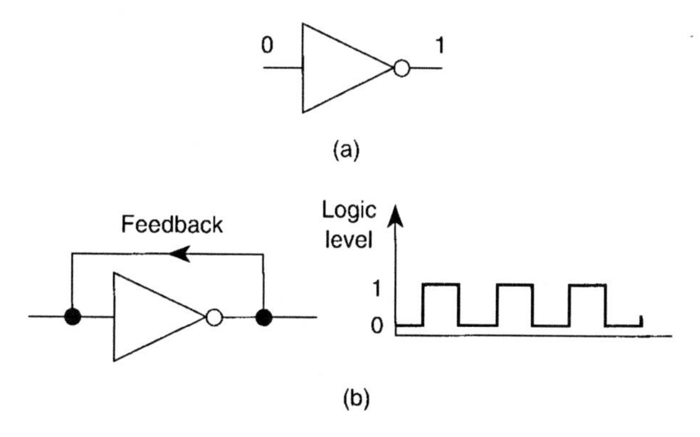

Fig. 8.9 An inverter

basic building blocks to make other logic devices, such as the flip-flop. A flip-flop is a switching device which has a memory. This aspect is important in the operation of sequential switching circuits where the output depends on the present input as well as the past sequence of inputs.

The flip-flop circuit of Fig. 8.10(a) shows a network of two NOR gates connected together. The output of the second gate is fed back to become one of the inputs of the first NOR gate. There are two inputs, *S* and *R*, and two outputs, *Q* and 'not *Q*'. It is usual to show the network 'cross-coupled' as in Fig. 8.10(b).

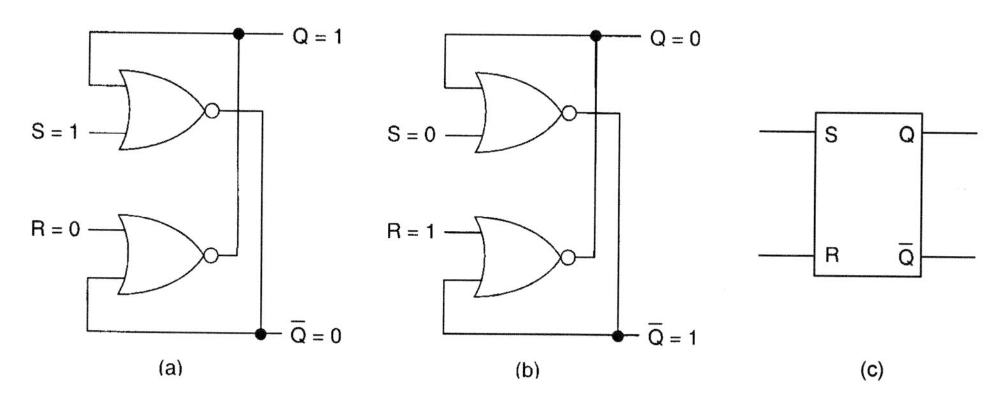

Fig. 8.10 A flip-flop circuit

The network is said to have a memory because the output is dependent on past input sequences as well as present ones. If the inputs are restricted so that *S* and *R* cannot be logic 1 simultaneously, the outputs *Q* and Q' (not *Q*), as shown in the truth table, are always true.

Figure 8.11 shows an *SR* flip-flop in an automatic switching circuit for head lamps.

A number of flip-flops may be used to make other devices, such as registers for holding digital codes, e.g. 1010. (This is a four-bit binary number that represents 10 in ordinary counting.)

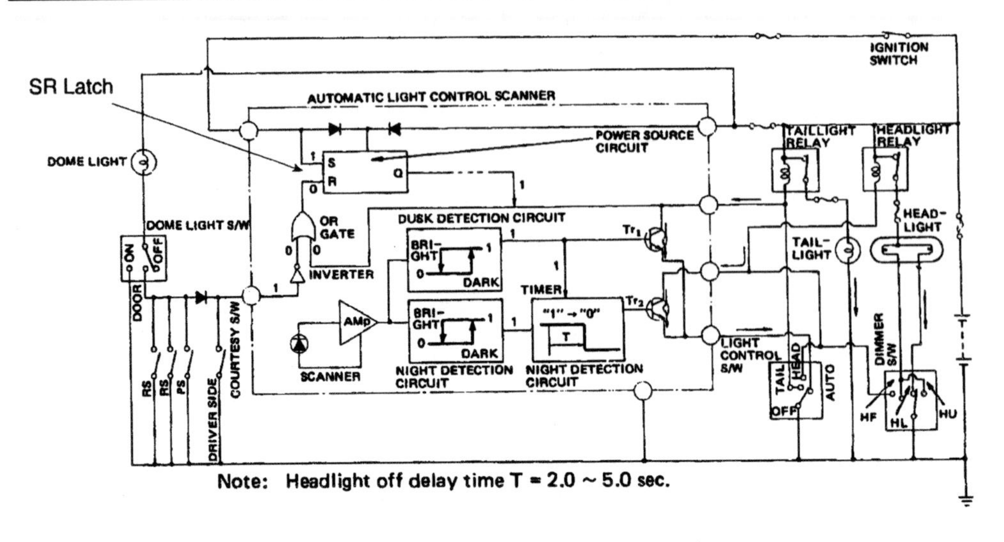

Fig. 8.11 Automatic headlamp circuit (Toyota)

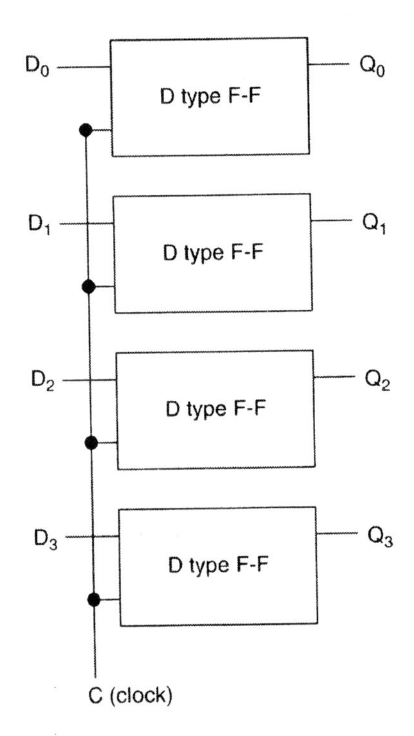

Fig. 8.12 D-type flip-flops used as 4-bit register

In Fig 8.12 the inputs D_0 to D_3 represent data, in the form of logic 0 or 1. When the clock pulses high the data will appear on the respective outputs of the register. A register will hold data until it receives a clock pulse, it will then transfer the data to the outputs which can then be used as inputs to other devices.

Thus, from the basic silicon p-n junction, transistors are made, logic gates are made from transistors, and logic gates are then used to make flip-flops. Flip-flops can then be used to make registers and many other logic circuits.

8.5.4 ANALOGUE TO DIGITAL CONVERSION

In previous chapters there have been many references to interfaces and analogue to digital conversion. A/D conversion is necessary because many sensor signals are of analogue (varying voltage) form. In order for the control computer to function these analogue signals must be converted to binary codes (digital signals). It is therefore appropriate to consider the basics of the design of an A/D converter that could be used at an ECM interface. Conversion from an analogue voltage to a digital code (word) can be done in a number of ways. Figure 8.13 shows one type of A/D converter that is known as a 'flash' converter.

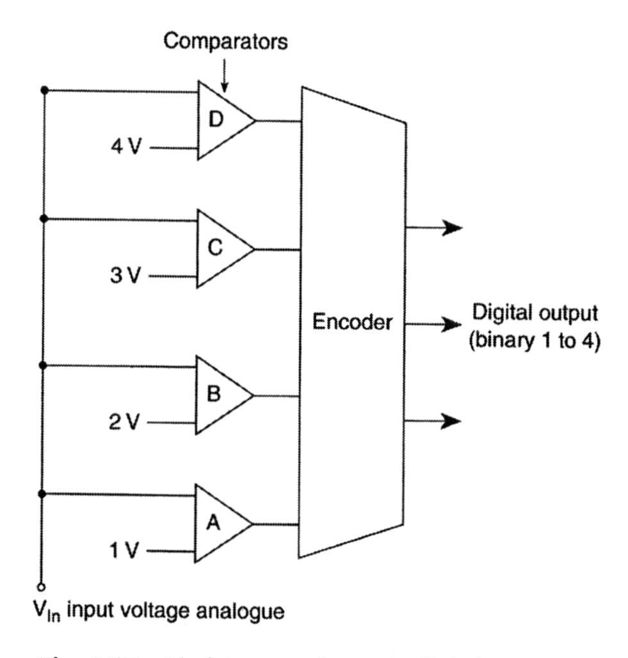

Fig. 8.13 Flash-type analogue to digital converter

The flash converter consists of four comparators and an encoder circuit which takes the comparator outputs and converts them into a binary code. An electronic comparator is a circuit which continuously compares two signals. One of the inputs, at each comparator, is a reference voltage. When the input voltage matches the reference voltage the comparator outputs a logic 1. The reference voltages shown here are 1 V up to 4 V. Table 8.1 shows the input/output performance of the converter.

The encoder contains logic devices (gates etc.) and this enables it to output the binary codes for 1 to 4. These binary codes (0s and 1s) are used by the ECM

Table 8.1 Performance of the flash converter

A/D converter input voltage range	Comparator outputs A B C D	Encoder output
0-1 V	0 0 0 0	0 0 0
1-2 V	1 0 0 0	0 0 1
2-3 V	1 1 0 0	0 1 0
3-4 V	1 1 1 0	0 1 1
4-5 V	1 1 1 1	1 0 0

processor to initiate certain actions. They are moved around the ECM by buses (wires) and consist of very low current electrical pulses. When a binary output command is generated it is normally required to be in analogue form. This requires a digital to analogue converter at the ECU output interface.

8.5.5 DIGITAL TO ANALOGUE CONVERSION

Figure 8.14 shows the basic principle of a digital to analogue converter. The power sources of 8 V, 4 V, 2 V, and 1 V are represented by the small circles. When a binary code is presented at the input, with the most significant bit (MSB) at the 8 V end and the LSB at the 1 V end, the switches are operated electronically. In

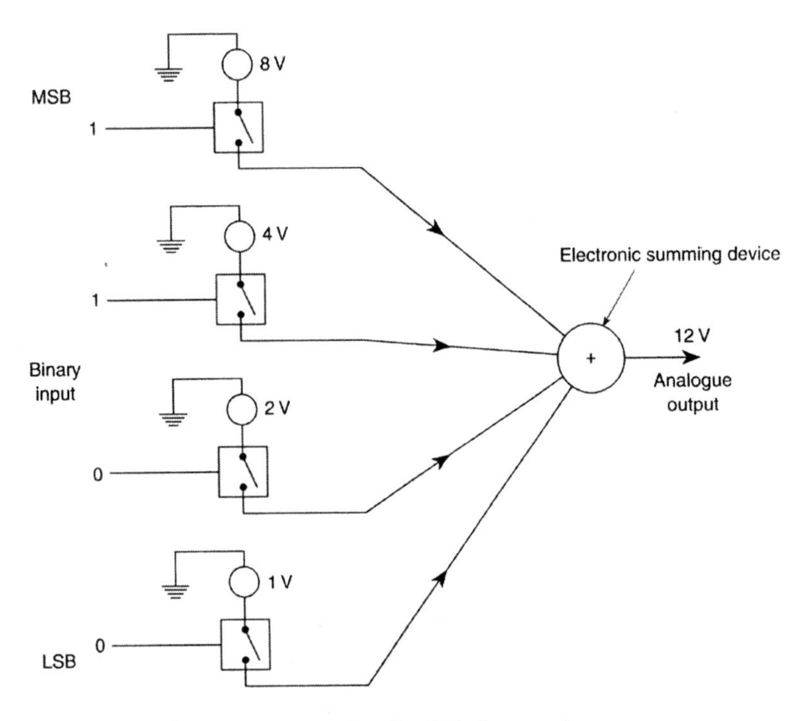

Fig. 8.14 Basic principle of a digital to analogue converter

the diagram binary 1100 (12) is placed at the inputs. This means that the two inputs of 1 will switch their respective voltages of 8 V and 4 V to the output lines, and the electronic summing circuit will add them together to give a 12 V output.

8.6 OBD II

Where it is considered to be pertinent to the topic being covered, aspects of OBD II have been introduced. For example:

- the SAE J 1962 standardized diagnostic connector;
- the downstream oxygen sensor for exhaust catalyst monitoring;
- the make up of fault codes.

Because European on-board diagnostics (EOBD) is likely to become compulsory in the near future and because it will probably embody many of the USA OBD II features, it will be useful to consider some further features of the USA standard. OBD II (on-board diagnostics version 2) applies to cars and light vans that are used in the USA. It affected petrol engined vehicles from model year 1994 and certain aspects of it have applied to diesel engined vehicles from model year 1996. Figure 8.15 shows a Bosch M5 system that is designed for OBD II.

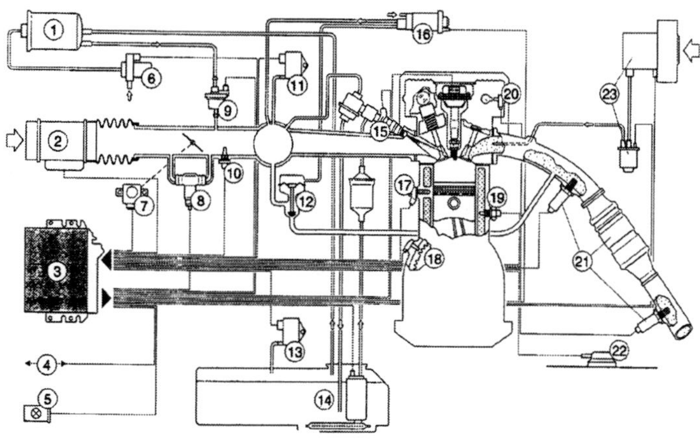

1 *Active-carbon filter*
2 *Air-mass meter*
3 *Control unit*
4 *Diagnostics interface*
5 *Diagnostics light (MIL)*
6 *ACF cut-off valve*
7 *Throttle-valve potentiometer*
8 *Idling actuator*
9 *ACF regeneration valve*
10 *Air-temperature sensor*
11 *Intake-manifold pressure sensor*
12 *Exhaust-gas recirculation (EGR) valve*
13 *Pressure sensor*
14 *Fuel tank*
15 *Injection valve*
16 *Pressure regulator*
17 *Knock sensor*
18 *Engine-speed sensor*
19 *Engine-temperature sensor*
20 *Phase sensor*
21 *Catalytic converter with two lambda sensors*
22 *Vehicle-body acceleration sensor*
23 *Secondary-air pump and secondary-air valve*

Fig. 8.15 A system that meets OBD II requirements (Bosch)

Much of the technology that is included in this system is already in use and it has been covered in earlier chapters. However, there are other aspects of the self-diagnostics capacity of OBD II systems that should be considered here. These are:

- fuel system leakage
- secondary air injection
- freeze frame data.

8.6.1 FUEL SYSTEM LEAKAGE

The ECM must register a fault if a leak equivalent to a hole 1 mm in diameter in the fuel tank ventilation system is detected. Figure 8.16 shows the general principle of the leakage detection system.

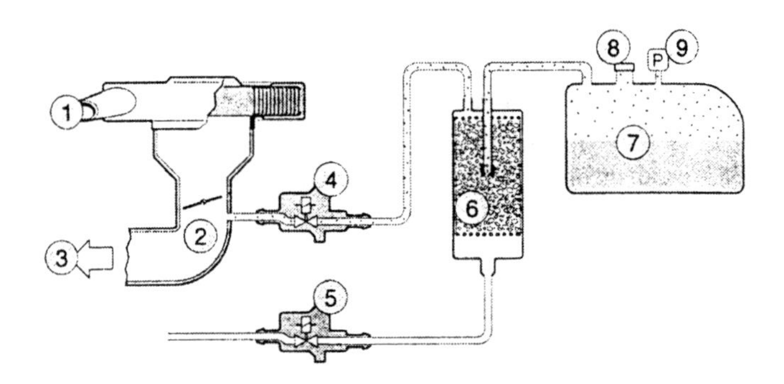

1 *Air filter*
2 *Intake manifold*
3 *Induction via engine*
4 *Active-carbon filter (ACF) regeneration valve*
5 *Active-carbon filter (ACF) shut-off valve*
6 *Active-carbon filter*
7 *Fuel tank*
8 *Tank lid with safety valve*
9 *Pressure sensor*

Fig. 8.16 The elements of a leakage detection system

In addition to the normal evaporative purge control (EVAP) valve (4) there is a valve (5) that the ECM operates to control the supply of fresh air to the carbon canister and a pressure sensor (9) at the petrol tank. When valves (4) and (5) are closed, the fuel evaporation system, from the tank up to these valves, is effectively sealed. The pressure sensor (9) will register a pressure reading that will be read by the ECM. If the EVAP valve is now opened, with the engine running, the manifold vacuum will create a vacuum in the entire fuel evaporation system and the new lower pressure will be recorded by the ECM, via the pressure sensor (9). Valves (4) and (5) are again closed for a period of time during which the ECM monitors the reading from pressure sensor (9). Any significant change in pressure as recorded by the sensor (9) indicates that there is a leak in the evaporation system.

8.6.2 SECONDARY AIR INJECTION

The aim of secondary air injection into the exhaust system is to reduce CO and HC emissions in the period after start-up and during the warm-up phase. The extra oxygen that is introduced assists the catalytic converter with the oxidation process. Figure 8.17 shows a system used by Volvo cars.

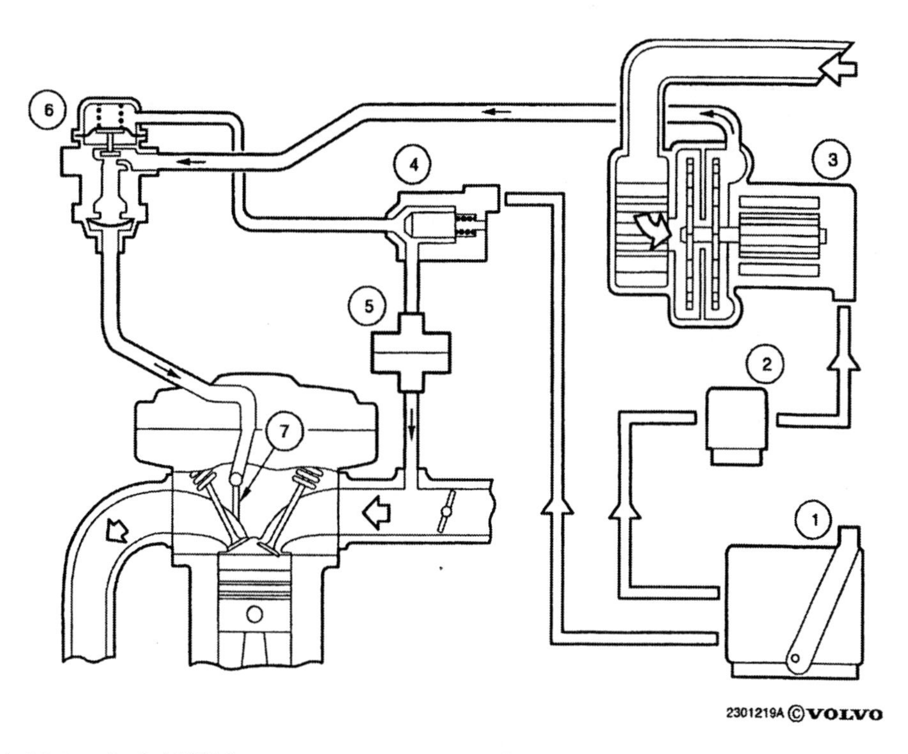

1. Motronic 4.4 ECM
2. Relay
3. Air pump
4. Solenoid valve
5. Check valve
6. Combined air and check valve
7. Air passage

Fig. 8.17 Secondary air injection system

With reference to Fig. 8.17, the solenoid valve (4) is under the control of the ECM, as is the relay (2) that provides the electrical power for the air pump (3). When the ECM starts or cuts off the secondary air, the solenoid valve (4) controls the application of manifold vacuum to the combined air and check valve (6) which opens and closes the valve that passes air to the exhaust injection port at (7). The secondary air flow is monitored by the pre-catalyst oxygen sensor and if the flow rate is not within the limits set by the manufacturer, a fault will be recorded.

The following precis of the self-diagnostic routine for the ECM of the Volvo system gives an insight into the diagnostic power of modern systems.

1. The idling and part load fuel functions are inhibited and the EVAP valve is closed. The ECM checks the signal from the oxygen sensor and if this shows an unchanging maximum value, the air pump is running continuously and the secondary air valve is leaking. This will cause a fault code to be recorded for the pump and valve.
2. The secondary air valve is closed and the air pump is started. The oxygen sensor signal (as determined by the ECM) should remain steady. If the oxygen sensor signal (at the ECM) exceeds a certain value within 6 s, the secondary air valve is leaking and a fault code for this valve will be recorded.
3. The secondary air pump runs continuously and the secondary air valve is opened. In this case the oxygen sensor signals (as assessed by the ECM) should exceed a specified limit within 6 s. If this does not happen, the secondary air pump is not running or the secondary air valve is not opening. This will cause the ECM to record a fault code for the secondary air pump.

8.6.3 FREEZE FRAMES

Whenever the ECM detects the conditions that cause it to record a diagnostic trouble code (fault code), the sensor readings (variables) and the vehicle operating conditions that exist at the time of the occurrence of the fault, are stored in a section of ECM memory. The format in which this information is presented at the diagnostic tool is known as a 'freeze frame'. The information contained in the freeze frame is read out by the scan tool and is of use in diagnosis. Freeze frames can be overwritten if a diagnostic trouble code of higher priority, in the same system, occurs later.

8.6.4 STANDARDIZED FAULT CODES

One of the problems that has confronted independent garages that wish to repair and maintain vehicles equipped with computer controlled systems, has been the lack of information about diagnostic trouble codes (fault codes). To some extent this lack of DTC information is overcome by the companies listed in the Appendix who publish volumes of DTCs for many different types of vehicles. A quick glance through these books will show that there is a lack of uniformity in the codes and yet they are effectively reporting faults on components that are identical (or very similar) on many vehicles.

OBD II overcomes this problem because it stipulates the use of standardized fault codes. It achieves this standardization by making use of SAE standards such as SAE J 1930 and SAE J 2012. The SAE J 1930 standard provides a system for naming the component parts of a computer controlled automotive system and SAE J 2012 details the DTC descriptions. It is understood that the European OBD standards for DTCs are likely to be very similar to the SAE ones and a sample of these is given in the Appendix.

8.7 Computer performance (MIPS)

In earlier chapters I have referred somewhat loosely to performance, power and capacity of computers. This area of computer technology is the subject of debate, but one measure of computer 'performance' that is commonly used is MIPS (millions of instructions per second). This refers to the speed at which the computer processor can deal with instructions. However, some instructions take longer to complete than others and this means that MIPS is just a guide to performance. In the business of computer performance comparison, a more reliable method of comparing computer performance is to use a 'benchmark' program. When such a program is run on different machines it is possible to make an accurate assessment of the comparative performance of the machines.

8.8 Supplementary restraint systems (SRS)

Airbags of the type shown in Fig. 8.18 and seat belt pre-tensioners, such as that shown in Fig. 8.19, are features of a basic supplementary restraint system. In the

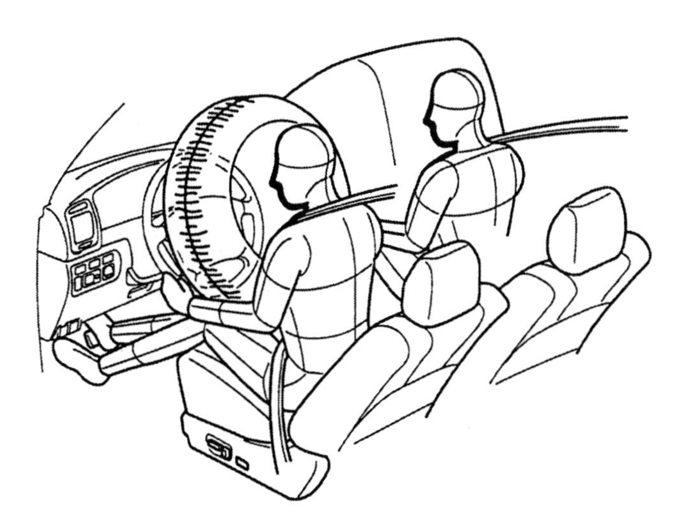

Fig. 8.18 Airbags for driver and front passenger

event of a frontal impact of some severity, the airbags and seat belt pre-tensioners are deployed. The airbags are inflated to protect those provided with them from impact with parts of the vehicle. The seat belt pre-tensioners are made to operate just before the airbags are inflated and they operate by pulling about 70 mm of seat belt onto the inertia reel of the belt. This serves to pull the seat occupant back onto the seat.

The deployment of these supplementary restraint devices is initiated by the action of the collision detection sensing system. A collision detection sensing

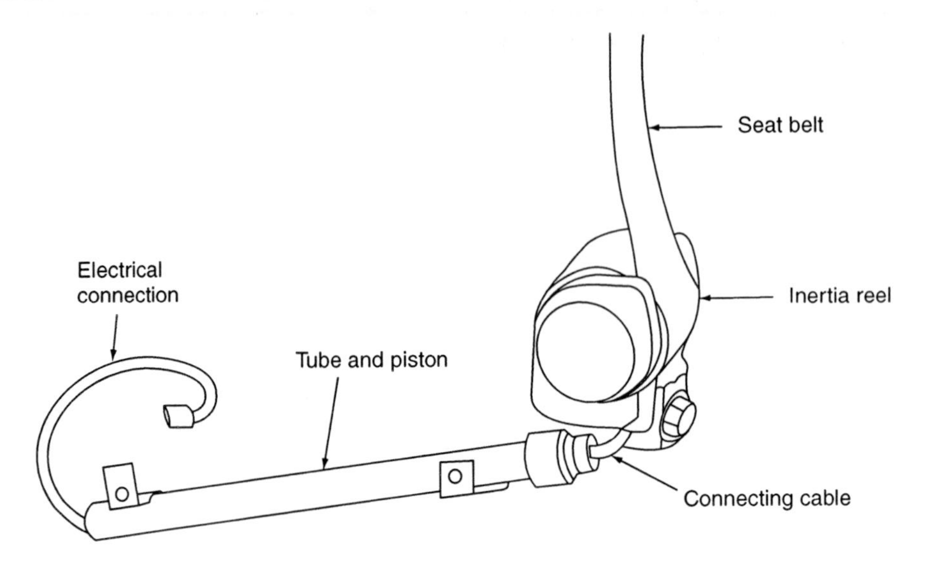

Fig. 8.19 Operation of pre-tensioner

system normally uses signals from two sensors: a 'crash sensor' and a 'safing sensor'. The safing sensor is activated at a lower deceleration than the crash sensor (about 1.5 g less) and both sensors must have been activated in order to trigger the supplementary restraint system. The safing sensor is fitted to reduce the risk of a simple error bringing the airbag into operation. Both of these sensors may be fitted inside the electronic control unit which, in some cases, is known as a diagnostic and control unit (DCU) because it contains the essential self-diagnosis circuits in addition to the circuits that operate the SRS. Figure 8.20 shows the layout of a supplementary restraint system on a Rover Mini.

Airbags are made from a durable lightweight material, such as nylon, and in Europe they have a capacity of approximately 40 litres. The pyrotechnic device that provides the inert gas to inflate the airbag contains a combustion chamber filled with fuel pellets, an electronic igniter and a filter, as shown in Fig. 8.21. Combustion of the fuel pellets produces the supply of nitrogen that inflates the airbag. The plastic cover that retains the folded airbag in place at the center of the steering wheel is designed with built-in break lines. When the airbag is inflated, the plastic cover separates at the break lines and the two flaps open out to permit unhindered inflation of the airbag.

The seat belt pre-tensioners are activated by a similar pyrotechnic device. In this case the gas is released into the cylinder of the pre-tensioner, where it drives a piston along the cylinder. The piston is attached to a strong flexible cable which then rotates the inertia reel of the seat belt by a sufficient amount to 'reel in' the seat belt by approximately 70 mm.

The rotary coupler is a device which is fitted beneath the steering wheel to provide a reliable electrical connection between the rotating steering wheel and airbag, and the static parts of the steering column. The positioning of the rotary coupler is a critical element of the airbag system and it should not be tampered

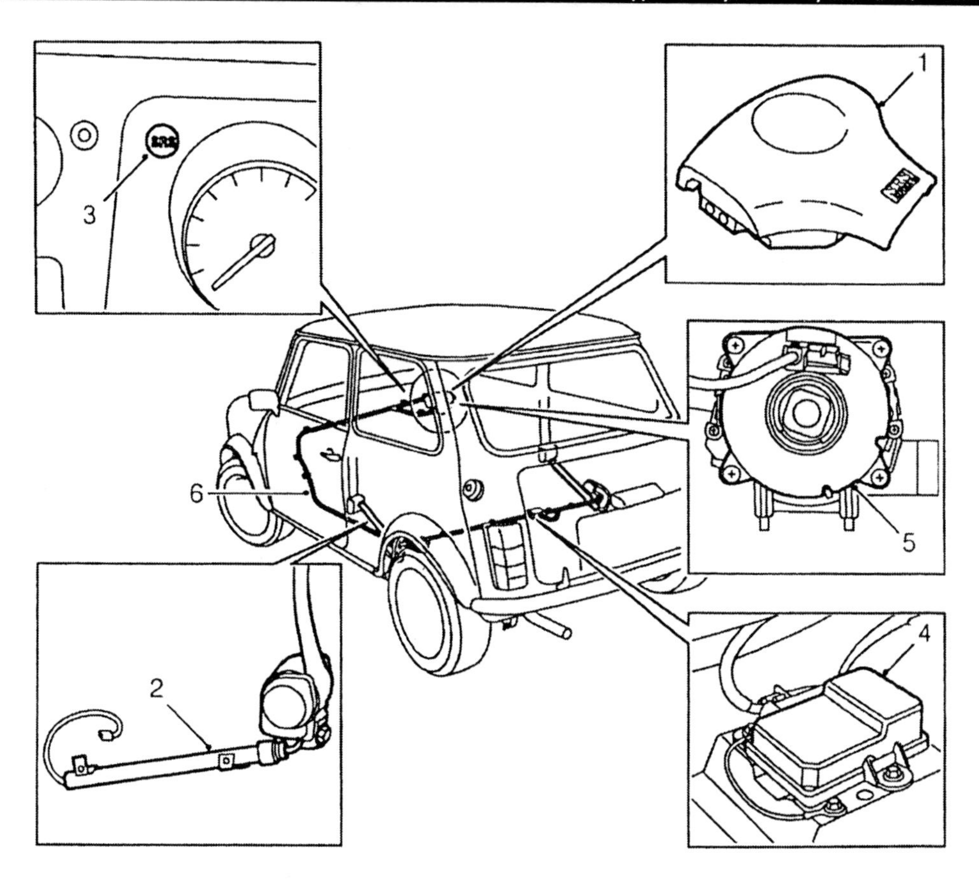

1. Driver's airbag module
2. Seatbelt pre-tensioner
3. SRS warning light
4. Diagnostic and control unit
5. Rotary coupler
6. SRS wiring harness

Fig. 8.20 The elements of a supplementary restraint system (Rover Mini)

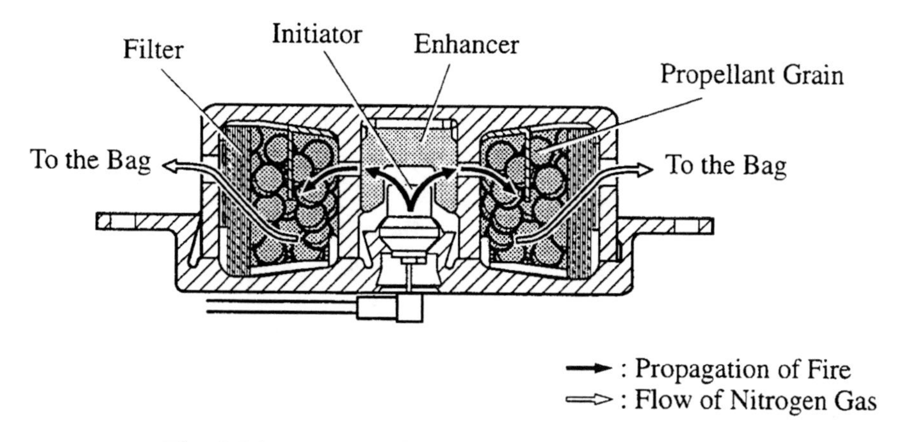

Fig. 8.21 A pyrotechnic device for inflating airbags

with. When working on supplementary restraint systems it is important that a technician is fully acquainted with the system and procedures for working on it.

8.8.1 HANDLING SRS COMPONENTS

The following notes are provided to Rover-trained technicians and they are included here because they contain some valuable advice for all vehicle technicians.

Safety precautions, storage and handling

Airbags and seat belt pre-tensioners are capable of causing serious injury if abused or mishandled. The following precautions must be adhered to:

In a vehicle

- ALWAYS fit genuine new parts when replacing SRS components.
- ALWAYS refer to the relevant workshop manual before commencing work on a supplementary restraint system.
- Remove the ignition key and disconnect both battery leads, earth lead first, and wait 10 min to allow the DCU back-up power circuits to discharge before commencing work on the SRS.
- DO NOT probe SRS components or harness with multi-meter probes unless following a manufacturer's approved diagnostic routine.
- ALWAYS use the manufacturers approved equipment when diagnosing SRS faults.
- Avoid working directly in line with the airbag when connecting or disconnecting multiplug wiring connectors.
- NEVER fit an SRS component which shows signs of damage or you suspect has been abused.

Handling

- ALWAYS carry airbag modules with the cover facing upwards.
- DO NOT carry more than one airbag module at a time.
- DO NOT drop SRS components.
- DO NOT carry airbag modules or seat belt pre-tensioners by their wires.
- DO NOT tamper with, dismantle, attempt to repair or cut any components used in the supplementary restraint system.
- DO NOT immerse SRS components in fluid.
- DO NOT attach anything to the airbag module cover.
- DO NOT transport airbag module or seat belt pre-tensioner in the passenger compartment of a vehicle. ALWAYS use the luggage compartment.

Storage

- ALWAYS keep SRS components dry.
- ALWAYS store a removed airbag module with the cover facing upwards.
- DO NOT allow anything to rest on the airbag module.

- ALWAYS place the airbag module or pre-tensioner in the designated storage area.
- ALWAYS store the airbag module on a flat, secure surface well away from electrical equipment or sources of high temperature.

Airbag modules and pyrotechnic seat belt pre-tensioners are classed as explosive articles and as such must be stored overnight in an approved, secure steel cabinet which is registered with the Local Authority.

Disposal of airbag modules and pyrotechnic seatbelt pre-tensioners
If a vehicle which contains airbags and seat belt pre-tensioners that have not been deployed (activated), is to be scrapped, the airbags and seat belt pre-tensioners must be rendered inoperable by activating them manually, prior to disposal. **This procedure may only be performed in accordance with the manufacturer's instruction manual.**

8.9 The coded ignition key

The engine immobilizer system is a theft deterrent system that is designed to prevent the engine of the vehicle from being started by any other means than the uniquely coded ignition key. The principal parts of a system are shown in Fig. 8.22.

The transponder chip in the handle of the ignition key is an integrated circuit that is programmed to communicate with the antenna (transponder key coil) that surrounds the barrel of the steering lock. When electromagnetic communication is established between the chip in the key and the transponder key coil, the immobilizer section of the ECM recognizes the key code signal that is produced and this effectively 'unlocks' the immobilizer section so that starting can proceed. Vehicle owners are provided with a back-up key that can be used to reset the system should a key be lost.

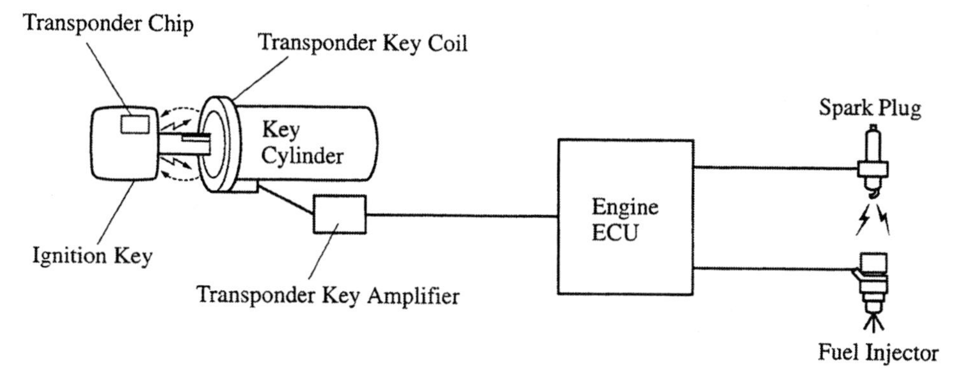

Fig. 8.22 The engine immoblizer system

8.10 Fault tracing

Examples of some other techniques that can be of assistance in fault tracing, when used correctly, include the half-split strategy and heuristics.

Figure 8.23 illustrates the principle of the half-split strategy. This shows how it is possible to use a strategy which limits the number of checks that need to be performed in order to trace a circuit defect. Assume, in the case shown, that the bulb has been checked and it is in order, but when re-inserted in the holder it fails to light. If a voltmeter is placed as shown and it reads battery voltage, it shows that the fault lies between the input to the fuse and the lamp earth. Whilst this may seem very simple, the strategy is quite powerful because it can be used to good effect in more complicated circuits.

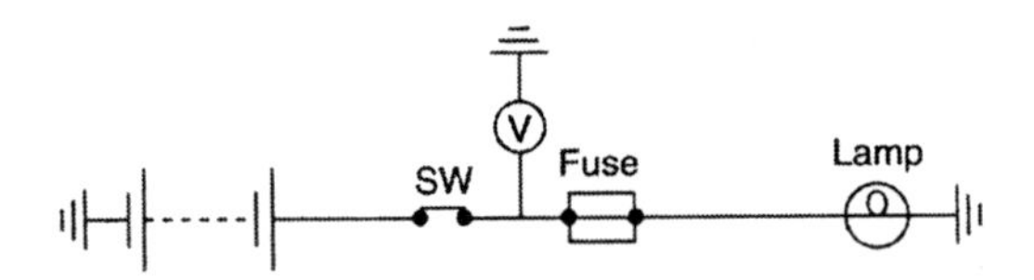

Fig. 8.23 The principle of the half-split method

Another name for heuristics is 'rule of thumb'. As an example assume that there is a vehicle known as the Xmobile. This vehicle is notorious for starting difficulties in damp, misty conditions. It becomes known that drying out HT insulations is often beneficial for producing a start. In the case of a 'call out' for failure to start of one of these vehicles, a sensible use of the 'rule of thumb' would be to take account of the weather conditions and to 'dry out' the HT insulation, before attempting anything more elaborate.

8.11 Precautions when working with computer controlled systems

- Do not perform electric welding on the vehicle without first disconnecting the computer.
- Do not subject the computer to high temperature by placing it in a high bake paint oven.
- Always earth yourself to release the static electricity charge that could damage sensitive components.
- Always use test instruments with a high impedance (internal resistance) to avoid loading electronic devices.
- Do not use chemical cleaning agents on the ECM.
- Avoid using steam cleaning and high pressure water jets in the vicinity of the ECM.
- Do not add units to the vehicle without first assuring yourself that they will not interfere with computer controlled systems.

8.12 Variable capacitance sensor

A variable capacitance type sensor was introduced in Chapter 5 and the following description provides a little more detail about the principles involved.

Figure 8.24(a) and (b) shows how altering the distance between the capacitor plates changes the value of the capacitance. When a variable capacitance is

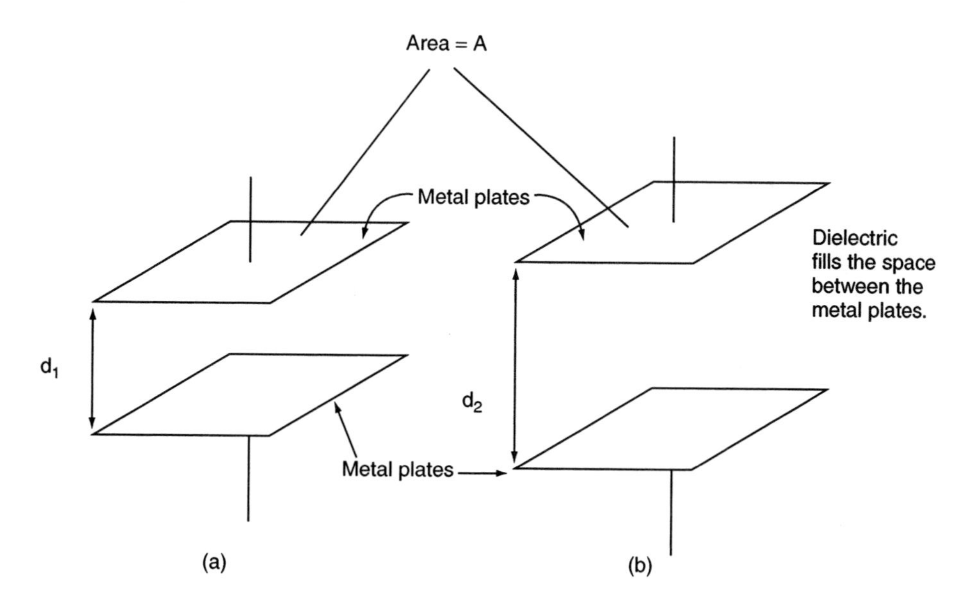

The substance between the metal plates is known as the dielectric. A property of the dielectric is its permittivity e.

The capacitance $C = \frac{e\,A}{d}$. If e & A are kept constant and d is changed, the valve of C also changes. For example,

if d_1 (a) is 5 units and e x A = 50 units, $C = \frac{50}{5} = 10$, if as in (b), d_2 is increased to 10 units $C = \frac{50}{10} = 5$ units.

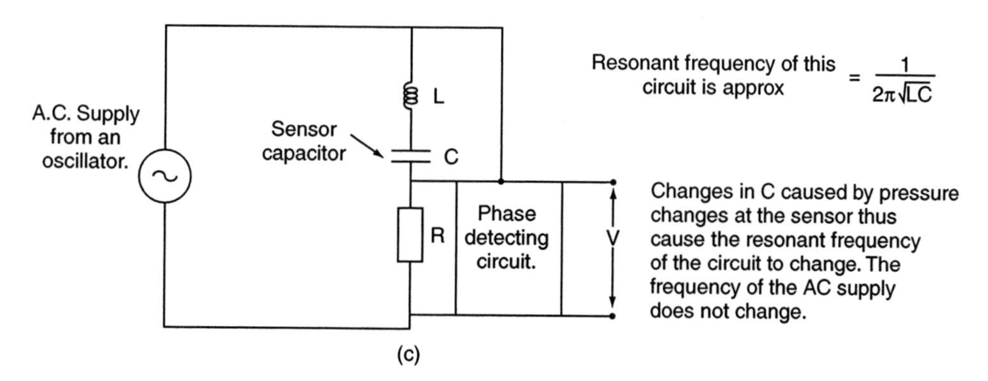

Fig. 8.24 An RLC circuit and the principle of the variable-capacitance sensor

introduced into a circuit of the form shown in Fig. 8.24(c) it has an effect on the frequency at which the voltage across the capacitor and the voltage across the inductance (coil) cancel each other out, electrically. The frequency at which this happens is known as the resonant frequency of the circuit. When the circuit is resonating, the voltage across the resistor R is in phase with the a.c. supply. Because the resonant frequency will change as the pressure at the sensor causes the capacitance value C to change, the terminal voltage V at the phase detector circuit, is related to the resonant frequency and manifold pressure.

8.13 Optoelectronics

For electronics purposes, light is often considered to consist of discrete elements called photons. Photons possess energy and in certain materials and under suitable conditions photon energy can be transferred to and from electrons to change their electronic behaviour. One way in which the change in behaviour of electrons affects a material is to change its conductivity and this is made use of in photodiodes, light sensitive sensors, etc. Another way that it is made use of is to produce a voltage and this is the basis of a fuel cell.

Figure 8.25 shows the basic principle of an optocoupler. This is a device which forms the basis of sensors that are used for position sensing, as in

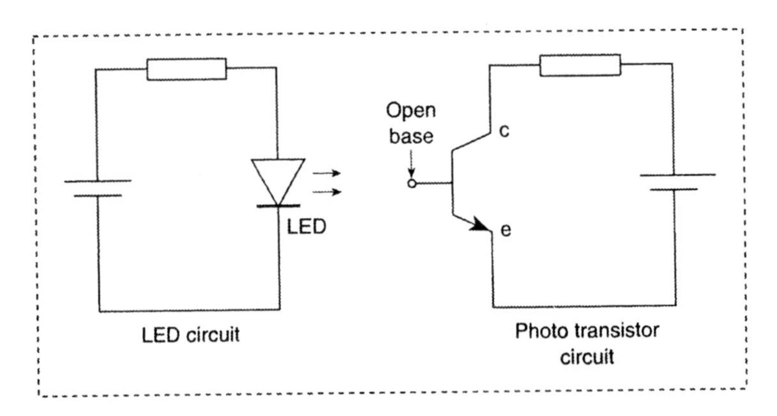

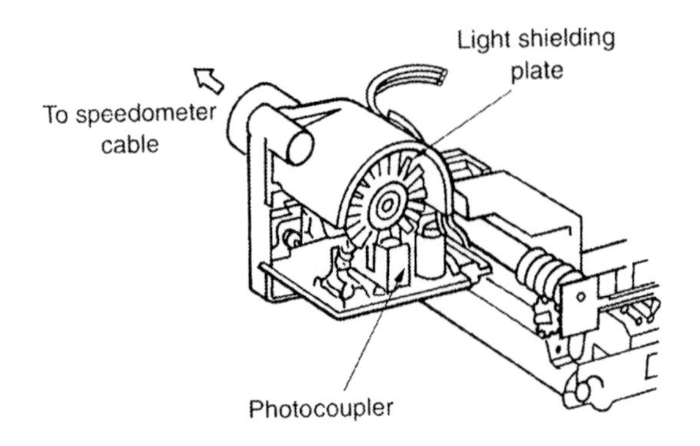

Fig. 8.25 An optocoupler

the steering position sensor, or a vehicle speed sensor. Light from the LED is directed onto the phototransistor and this affects the output from the transistor. When the light from the LED is interrupted by an opaque object the transistor output changes. The interruption of the light from the LED is achieved by the rotation, between the LED and the transistor, of a disc with holes in it.

8.14 Review questions (see Appendix 2 for answers)

1. The piezoelectric generator effect is used in:
   (a) frequency-type MAP sensors?
   (b) knock sensors?
   (c) air flow sensors?
   (d) sparking plug insulators?
2. Manifold absolute pressure in a petrol engine is an accurate indicator of:
   (a) engine speed?
   (b) a misfiring sparking plug?
   (c) engine temperature?
   (d) engine load and a factor in determining air flow in speed density systems?
3. OBD II is mainly concerned with:
   (a) emissions related problems?
   (b) vehicles fitted with traction control systems?
   (c) making it easier for owner drivers to service their own vehicles?
   (d) reducing the fuel consumption of vehicles?
4. If barometric pressure is 1.017 bar and a pressure gauge reads 5.2 bar the absolute pressure is:
   (a) 1.017 bar?
   (b) 5.29 bar?
   (c) 6.217 bar?
   (d) zero?
5. At the input interface of an ECM, an analogue to digital converter may be used to:
   (a) reduce the voltage of the input?
   (b) convert a voltage reading into a binary-coded word?
   (c) slow down the speed of input so that the ECM processor has time to work on it?
   (d) convert a.c to d.c?
6. A twisted nematic material is used:
   (a) to reduce cross-talk between cables?
   (b) in liquid crystal displays?
   (c) as an anti-knock additive in petrol?
   (d) as a sealant to prevent moisture getting into an ECM?
7. A carbon canister is used:
   (a) in evaporative emissions control systems?
   (b) to provide a reserve supply of fuel?

(c) to allow fuel vapor to be introduced into the exhaust system?
(d) to improve the cetane rating of diesel fuel?

8. In an AND gate:
   (a) the output is 1 when the two inputs are different?
   (b) the output is 1 when both inputs are the same?
   (c) the output is 1 when both inputs are 1?
   (d) the output is 1 when one input is 1 and the other input is zero?

Appendix

A.1 Companies who supply equipment and diagnostic data

Robert Bosch Ltd, PO Box 98, Broadwater Park, Denham, Uxbridge UB9 5HJ. (The code reading and diagnostic functions of the Bosch KTS300 instrument have now (2001) been incorporated into the KTS500 instruments.) FKI Crypton Ltd, Bristol Road, Bridgewater, Somerset TA9 6BX. Garage Equipment Association, 2-3 Church Walk, Daventry, Northants NN11 4BL. Gunson Ltd, Acorn House, Coppen Road, Dagenham, Essex RM8 1NU. Haynes Publishing, Sparkford Road, Yeovil, Somerset BA22 7JJ. Lucas Aftermarket Operations, Test Equipment Division, Unit 9, Monkspath Business Park, Highlands Road, Solihull, West Midlands, B90 4AX. Omitec Instrumentation, Hopton Industrial Estate, London Road, Devizes SN10 2EU. Snap-on Tools, Palmer House,154 Cross Street, Sale, Cheshire M33 1AQ.

A.2 Answers to review questions

Chapter 1
Question 1. Answer (b)
Question 2. Answer (b)
Question 3. Answer (b)
Question 4. Answer (b)
Question 5. Answer (b)
Question 6. Answer (a)
Question 7. Answer (c)
Question 8. Answer (a)

Chapter 2
Question 1. Answer (c)
Question 2. Answer (a)
Question 3. Answer (a)
Question 4. Answer (b)
Question 5. Answer (a)
Question 6. Answer (c)
Question 7. Answer (b)
Question 8. Answer (a)

Chapter 3
Question 1. Answer (a)
Question 2. Answer (a)
Question 3. Answer (b)
Question 4. Answer (c)
Question 5. Answer (b)
Question 6. Answer (b)
Question 7. Answer (c)
Question 8. Answer (a)

Chapter 4
Question 1. Answer (b)
Question 2. Answer (a)
Question 3. Answer (b)
Question 4. Answer (b)
Question 5. Answer (a)
Question 6. Answer (c)
Question 7. Answer (b)
Question 8. Answer (b)

Chapter 5
Question 1. Answer (d)
Question 2. Answer (c)
Question 3. Answer (a)
Question 4. Answer (a)
Question 5. Answer (a)
Question 6. Answer (b)
Question 7. Answer (a)
Question 8. Answer (c)

Chapter 6
Question 1. Answer (b)
Question 2. Answer (b)
Question 3. Answer (b)
Question 4. Answer (a)
Question 5. Answer (c)
Question 6. Answer (c)
Question 7. Answer (b)
Question 8. Answer (b)

Chapter 7
Question 1. Answer (c)
Question 2. Answer (a)
Question 3. Answer (b)
Question 4. Answer (b)
Question 5. Answer (c)
Question 6. Answer (b)
Question 7. Answer (b)
Question 8. Answer (b)

Chapter 8
Question 1. Answer (b)
Question 2. Answer (d)
Question 3. Answer (a)
Question 4. Answer (c)
Question 5. Answer (b)
Question 6. Answer (b)
Question 7. Answer (a)
Question 8. Answer (c)

A.3 OBD II standard fault codes

A principal aim of this book is to show readers that there are significant amounts of technology that are common to all computer controlled automotive systems. In reaching this conclusion it is natural to enquire why, when there is so much common ground, there are so many different approaches to diagnostics. It has long been acknowledged that the 'profusion' of diagnostic trouble codes, access to data, different interfaces between scan tools and ECMs, and other details, have made the work of service technicians more difficult than it perhaps needs to be. This acknowledgement of 'difficulties' at the user end of a vehicle's life cycle has resulted in various standards being incorporated into OBD II, and similar standards are likely to feature in EOBD. The standard diagnostic plug (SAE J 1962) and the format of DTCs (SAE J 2012) have been covered in this book.

The following list of standard diagnostic trouble codes is derived from SAE J 2012, by courtesy of Robert Bosch Ltd. It is important to note that these codes apply to the USA, although the European ones are expected to be similar. They are provided here to show the principle and to demonstrate the apparent advantages of standardization. When dealing with diagnostic trouble codes (fault codes) on an actual vehicle it is imperative that the method of obtaining the codes and the interpretation of them is performed in accordance with the vehicle manufacturer's instructions.

The meaning of these standard codes is described in Chapter 3, section 3.2.2.

MALFUNCTION CODE	FUNCTION
P0291	INJECTOR CIRCUIT LOW - CYLINDER 11
P0292	INJECTOR CIRCUIT HIGH - CYLINDER 11
P0293	CYLINDER 11 CONTRIBUTION/BALANCE FAULT
P0294	INJECTOR CIRCUIT LOW - CYLINDER 12
P0295	INJECTOR CIRCUIT HIGH - CYLINDER 12
P0296	CYLINDER 12 CONTRIBUTION/BALANCE FAULT
P03XX	**Ignition System or Misfire**
P0300	RANDOM/MULTIPLE CYLINDER MISFIRE DETECTED
P0301	CYLINDER 1 MISFIRE DETECTED
P0302	CYLINDER 2 MISFIRE DETECTED
P0303	CYLINDER 3 MISFIRE DETECTED
P0304	CYLINDER 4 MISFIRE DETECTED
P0305	CYLINDER 5 MISFIRE DETECTED
P0306	CYLINDER 6 MISFIRE DETECTED
P0307	CYLINDER 7 MISFIRE DETECTED
P0308	CYLINDER 8 MISFIRE DETECTED
P0309	CYLINDER 9 MISFIRE DETECTED
P0310	CYLINDER 10 MISFIRE DETECTED
P0311	CYLINDER 11 MISFIRE DETECTED
P0312	CYLINDER 12 MISFIRE DETECTED
P0320	IGNITION/DISTRIBUTOR ENGINE SPEED INPUT CIRCUIT MALFUNCTION
P0321	IGNITION/DISTRIBUTOR ENGINE SPEED INPUT CIRCUIT RANGE/PERFORMANCE
P0322	IGNITION/DISTRIBUTOR ENGINE SPEED INPUT CIRCUIT NO SIGNAL
P0323	IGNITION/DISTRIBUTOR ENGINE SPEED INPUT CIRCUIT INTERMITTENT
P0325	KNOCK SENSOR 1 CIRCUIT MALFUNCTION (BANK 1 OR SINGLE SENSOR)
P0326	KNOCK SENSOR 1 CIRCUIT RANGE/PERFORMANCE (BANK 1 OR SINGLE SENSOR)
P0327	KNOCK SENSOR 1 CIRCUIT LOW INPUT (BANK 1 OR SINGLE SENSOR)
P0328	KNOCK SENSOR 1 CIRCUIT HIGH INPUT (BANK 1 OR SINGLE SENSOR)
P0329	KNOCK SENSOR 1 CIRCUIT INPUT INTERMITTENT (BANK 1 OR SINGLE SENSOR)

MALFUNCTION CODE	FUNCTION
P0330	KNOCK SENSOR 2 CIRCUIT MALFUNCTION (BANK 2)
P0331	KNOCK SENSOR 2 CIRCUIT RANGE/PERFORMANCE (BANK 2)
P0332	KNOCK SENSOR 2 CIRCUIT LOW INPUT (BANK 2)
P0333	KNOCK SENSOR 2 CIRCUIT HIGH INPUT (BANK 2)
P0334	KNOCK SENSOR 2 CIRCUIT INPUT INTERMITTENT (BANK 2 OR SINGLE SENSOR)
P0335	CRANKSHAFT POSITION SENSOR CIRCUIT MALFUNCTION
P0336	CRANKSHAFT POSITION SENSOR CIRCUIT RANGE/PERFORMANCE
P0337	CRANKSHAFT POSITION SENSOR CIRCUIT LOW INPUT
P0338	CRANKSHAFT POSITION SENSOR CIRCUIT HIGH INPUT
P0339	CRANKSHAFT POSITION SENSOR CIRCUIT INTERMITTENT
P0340	CAMSHAFT POSITION SENSOR CIRCUIT MALFUNCTION
P0341	CAMSHAFT POSITION SENSOR CIRCUIT RANGE/PERFORMANCE
P0342	CAMSHAFT POSITION SENSOR CIRCUIT LOW INPUT
P0343	CAMSHAFT POSITION SENSOR CIRCUIT HIGH INPUT
P0344	CAMSHAFT POSITION SENSOR CIRCUIT INTERMITTENT
P0350	IGNITION COIL PRIMARY/SECONDARY CIRCUIT MALFUNCTION
P0351	IGNITION COIL A PRIMARY/SECONDARY CIRCUIT MALFUNCTION
P0352	IGNITION COIL B PRIMARY/SECONDARY CIRCUIT MALFUNCTION
P0353	IGNITION COIL C PRIMARY/SECONDARY CIRCUIT MALFUNCTION
P0354	IGNITION COIL D PRIMARY/SECONDARY CIRCUIT MALFUNCTION
P0355	IGNITION COIL E PRIMARY/SECONDARY CIRCUIT MALFUNCTION
P0356	IGNITION COIL F PRIMARY/SECONDARY CIRCUIT MALFUNCTION

MALFUNCTION CODE	FUNCTION
P0357	IGNITION COIL G PRIMARY/SECONDARY CIRCUIT MALFUNCTION
P0358	IGNITION COIL H PRIMARY/SECONDARY CIRCUIT MALFUNCTION
P0359	IGNITION COIL I PRIMARY/SECONDARY CIRCUIT MALFUNCTION
P0360	IGNITION COIL J PRIMARY/SECONDARY CIRCUIT MALFUNCTION
P0361	IGNITION COIL K PRIMARY/SECONDARY CIRCUIT MALFUNCTION
P0362	IGNITION COIL L PRIMARY/SECONDARY CIRCUIT MALFUNCTION
P0370	TIMING REFERENCE SIGNAL "A" HIGH RESPONSE MALFUNCTION
P0371	TOO MANY HIGH RESOLUTION SIGNAL "A" PULSES
P0372	TOO FEW HIGH RESOLUTION SIGNAL "A" PULSES
P0373	INTERMITTENT/ERRATIC HIGH RESOLUTION SIGNAL "A" PULSES
P0374	NO HIGH RESOLUTION SIGNAL "A" PULSES
P0375	TIMING REFERENCE SIGNAL "B" HIGH RESPONSE MALFUNCTION
P0376	TOO MANY HIGH RESOLUTION SIGNAL "B" PULSES
P0377	TOO FEW HIGH RESOLUTION SIGNAL "B" PULSES
P0378	INTERMITTENT/ERRATIC HIGH RESOLUTION SIGNAL "B" PULSES
P0379	NO HIGH RESOLUTION SIGNAL "B" PULSES
P0380	GLOW PLUG/HEATER CIRCUIT MALFUNCTION
P0381	GLOW PLUG/HEATER INDICATOR CIRCUIT MALFUNCTION
P04XX	**Auxiliary Emission Controls**
P0400	EXHAUST GAS RECIRCULATION FLOW MALFUNCTION
P0401	EXHAUST GAS RECIRCULATION FLOW INSUFFICIENT DETECTED
P0402	EXHAUST GAS RECIRCULATION FLOW EXCESSIVE DETECTED
P0403	EXHAUST GAS RECIRCULATION CIRCUIT MALFUNCTION
P0404	EXHAUST GAS RECIRCULATION CIRCUIT RANGE/PERFORMANCE

MALFUNCTION CODE	FUNCTION
P0405	EXHAUST GAS RECIRCULATION SENSOR A CIRCUIT LOW
P0406	EXHAUST GAS RECIRCULATION SENSOR A CIRCUIT HIGH
P0407	EXHAUST GAS RECIRCULATION SENSOR B CIRCUIT LOW
P0408	EXHAUST GAS RECIRCULATION SENSOR B CIRCUIT HIGH
P0410	SECONDARY AIR INJECTION SYSTEM MALFUNCTION
P0411	SECONDARY AIR INJECTION SYSTEM INCORRECT FLOW DETECTED
P0412	SECONDARY AIR INJECTION SYSTEM SWITCHING VALVE A CIRCUIT MALFUNCTION
P0413	SECONDARY AIR INJECTION SYSTEM SWITCHING VALVE A CIRCUIT OPEN
P0414	SECONDARY AIR INJECTION SYSTEM SWITCHING VALVE A CIRCUIT SHORTED
P0415	SECONDARY AIR INJECTION SYSTEM SWITCHING VALVE B CIRCUIT MALFUNCTION
P0416	SECONDARY AIR INJECTION SYSTEM SWITCHING VALVE B CIRCUIT OPEN
P0417	SECONDARY AIR INJECTION SYSTEM SWITCHING VALVE B CIRCUIT SHORTED
P0420	CATALYST SYSTEM EFFICIENCY BELOW THRESHOLD (BANK 1)
P0421	WARM UP CTALYST EFFICIENCY BELOW THRESHOLD (BANK 1)
P0422	MAIL CATALYST EFFICIENCY BELOW THRESHOLD (BANK 1)
P0423	HEATED CATALYST EFFICIENCY BELOW THRESHOLD (BANK 1)
P0424	HEATED CATALYST TEMPERATURE BELOW THRESHOLD (BANK 1)
P0430	CATALYST SYSTEM EFFICIENCY BELOW THRESHOLD (BANK 2)
P0431	WARM UP CATALYST EFFICIENCY BELOW THRESHOLD (BANK 2)
P0432	MAIL CATALYST EFFICIENCY BELOW THRESHOLD (BANK 2)

MALFUNCTION CODE	FUNCTION
P0433	HEATED CATALYST EFFICIENCY BELOW THRESHOLD (BANK 2)
P0434	HEATED CATALYST TEMPERATURE BELOW THRESHOLD (BANK 2)
P0440	EVAPORATIVE EMISSION CONTROL SYSTEM MALFUNCTION
P0441	EVAPORATIVE EMISSION CONTROL SYSTEM INCORRECT PURGE FLOW
P0442	EVAPORATIVE EMISSION CONTROL SYSTEM LEAK DETECTED (small leak)
P0443	EVAPORATIVE EMISSION CONTROL SYSTEM PURGE CONTROL VALVE CIRCUIT MALFUNCTION
P0444	EVAPORATIVE EMISSION CONTROL SYSTEM PURGE CONTROL VALVE CIRCUIT OPEN
P0445	EVAPORATIVE EMISSION CONTROL SYSTEM PURGE CONTROL VALVE CIRCUIT SHORTED
P0446	EVAPORATIVE EMISSION CONTROL SYSTEM VENT CONTROL MALFUNCTION
P0447	EVAPORATIVE EMISSION CONTROL SYSTEM VENT CONTROL OPEN
P0448	EVAPORATIVE EMISSION CONTROL SYSTEM VENT CONTROL SHORTED
P0450	EVAPORATIVE EMISSION CONTROL SYSTEM PRESSURE SENSOR MALFUNCTION
P0451	EVAPORATIVE EMISSION CONTROL SYSTEM PRESSURE SENSOR RANGE/PERFORMANCE
P0452	EVAPORATIVE EMISSION CONTROL SYSTEM PRESSURE SENSOR LOW INPUT
P0453	EVAPORATIVE EMISSION CONTROL SYSTEM PRESSURE SENSOR HIGH INPUT
P0454	EVAPORATIVE EMISSION CONTROL SYSTEM PRESSURE SENSOR INTERMITTENT
P0455	EVAPORATIVE EMISSION CONTROL SYSTEM LEAK DETECTED (gross leak)
P0460	FUEL LEVEL SENSOR CIRCUIT MALFUNCTION
P0461	FUEL LEVEL SENSOR CIRCUIT RANGE/ PERFORMANCE
P0462	FUEL LEVEL SENSOR CIRCUIT LOW INPUT

MALFUNCTION CODE	FUNCTION
P0463	FUEL LEVEL SENSOR CIRCUIT HIGH INPUT
P0464	FUEL LEVEL SENSOR CIRCUIT INTERMITTENT
P0465	PURGE FLOW SENSOR CIRCUIT MALFUNCTION
P0466	PURGE FLOW SENSOR CIRCUIT RANGE/PERFORMANCE
P0467	PURGE FLOW SENSOR CIRCUIT LOW INPUT
P0468	PURGE FLOW SENSOR CIRCUIT HIGH INPUT
P0469	PURGE FLOW SENSOR CIRCUIT INTERMITTENT
P0470	EXHAUST PRESSURE SENSOR MALFUNCTION
P0471	EXHAUST PRESSURE SENSOR RANGE/PERFORMANCE
P0472	EXHAUST PRESSURE SENSOR LOW
P0473	EXHAUST PRESSURE SENSOR HIGH
P0474	EXHAUST PRESSURE SENSOR INTERMITTENT
P0475	EXHAUST PRESSURE CONTROL VALVE MALFUNCTION
P0476	EXHAUST PRESSURE CONTROL VALVE RANGE/PERFORMANCE
P0477	EXHAUST PRESSURE CONTROL VALVE LOW
P0478	EXHAUST PRESSURE CONTROL VALVE HIGH
P0479	EXHAUST PRESSURE CONTROL VALVE INTERMITTENT
P05XX	**Vehicle Speed Control and Idle Control System**
P0500	VEHICLE SPEED SENSOR MALFUNCTION
P0501	VEHICLE SPEED SENSOR RANGE/PERFORMANCE
P0502	VEHICLE SPEED SENSOR LOW INPUT
P0503	VEHICLE SPEED SENSOR INTERMITTENT/ERRATIC/HIGH
P0505	IDLE CONTROL SYSTEM MALFUNCTION
P0506	IDLE CONTROL SYSTEM RPM LOWER THAN EXPECTED
P0507	IDLE CONTROL SYSTEM RPM HIGHER THAN EXPECTED
P0510	CLOSED THROTTLE POSITION SWITCH MALFUNCTION
P0530	A/C REFRIGERANT PRESSURE SENSOR CIRCUIT MALFUNCTION
P0531	A/C REFRIGERANT PRESSURE SENSOR CIRCUIT RANGE/PERFORMANCE
P0532	A/C REFRIGERANT PRESSURE SENSOR CIRCUIT LOW INPUT

MALFUNCTION CODE	FUNCTION
P0533	A/C REFRIGERANT PRESSURE SENSOR CIRCUIT HIGH INPUT
P0534	A/C REFRIGERANT CHARGE LOSS
P0550	POWER STEERING PRESSURE SENSOR CIRCUIT MALFUNCTION
P0551	POWER STEERING PRESSURE SENSOR CIRCUIT RANGE/PERFORMANCE
P0552	POWER STEERING PRESSURE SENSOR CIRCUIT LOW INPUT
P0553	POWER STEERING PRESSURE SENSOR CIRCUIT HIGH INPUT
P0554	POWER STEERING PRESSURE SENSOR CIRCUIT INTERMITTENT
P0560	SYSTEM VOLTAGE MALFUNCTION
P0561	SYSTEM VOLTAGE UNSTABLE
P0562	SYSTEM VOLTAGE LOW
P0563	SYSTEM VOLTAGE HIGH
P0565	CRUISE ON SIGNAL MALFUNCTION
P0566	CRUISE OFF SIGNAL MALFUNCTION
P0567	CRUISE RESUME SIGNAL MALFUNCTION
P0568	CRUISE SET SIGNAL MALFUNCTION
P0569	CRUISE COAST SIGNAL MALFUNCTION
P0570	CRUISE ACCEL SIGNAL MALFUNCTION
P0571	CRUISE/BRAKE SWITCH "A" CIRCUIT MALFUNCTION
P0572	CRUISE/BRAKE SWITCH "A" CIRCUIT LOW
P0573	CRUISE/BRAKE SWITCH "A" CIRCUIT HIGH
P06XX	**Computer and Output Circuits**
P0600	S ERIAL COMMUNICATION LINK MALFUNCTION
P0601	INTERNAL CONTROL MODULE MEMORY CHECK SUM ERROR
P0602	CONTROL MODULE PROGRAMMING ERROR
P0603	INTERNAL CONTROL MODULE KEEP ALIVE MEMORY ERROR
P0604	INTERNAL CONTROL MODULE RANDOM ACCESS MEMORY (RAM) ERROR
P0605	INTERNAL CONTROL MODULE ROM TEST ERROR
P0606	PCM PROCESSOR FAULT

MALFUNCTION CODE	FUNCTION
P07XX	**Transmission**
P0700	TRANSMISSION CONTROL SYSTEM MALFUNCTION
P0701	TRANSMISSION CONTROL SYSTEM RANGE/PERFORMANCE
P0702	TRANSMISSION CONTROL SYSTEM ELECTRICAL
P0703	TORQUE CONVERTER/BRAKE SWITCH "B" CIRCUIT MALFUNCTION
P0704	CLUTCH SWITCH INPUT CIRCUIT MALFUNCTION
P0705	TRANSMISSION RANGE SENSOR CIRCUIT MALFUNCTION (PRNDL INPUT)
P0706	TRANSMISSION RANGE SENSOR CIRCUIT RANGE/PERFORMANCE
P0707	TRANSMISSION RANGE SENSOR CIRCUIT LOW INPUT
P0708	TRANSMISSION RANGE SENSOR CIRCUIT HIGH INPUT
P0709	TRANSMISSION RANGE SENSOR CIRCUIT INTERMITTENT
P0710	TRANSMISSION FLUID TEMPERATURE SENSOR CIRCUIT MALFUNCTION
P0711	TRANSMISSION FLUID TEMPERATURE SENSOR CIRCUIT RANGE/PERFORMANCE
P0712	TRANSMISSION FLUID TEMPERATURE SENSOR CIRCUIT LOW INPUT
P0713	TRANSMISSION FLUID TEMPERATURE SENSOR CIRCUIT HIGH INPUT
P0714	TRANSMISSION FLUID TEMPERATURE SENSOR CIRCUIT INTERMITTENT
P0715	INPUT TURBINE/SPEED SENSOR CIRCUIT MALFUNCTION
P0716	INPUT TURBINE/SPEED SENSOR CIRCUIT RANGE/PERFORMANCE
P0717	INPUT TURBINE/SPEED SENSOR CIRCUIT NO SIGNAL
P0718	INPUT TURBINE/SPEED SENSOR CIRCUIT INTERMITTENT
P0719	TORQUE CONVERTER/BRAKE SWITCH "B" CIRCUIT LOW
P0720	OUTPUT SPEED SENSOR CIRCUIT MALFUNCTION
P0721	OUTPUT SPEED SENSOR CIRCUIT RANGE/PERFORMANCE
P0722	OUTPUT SPEED SENSOR CIRCUIT NO SIGNAL
P0723	OUTPUT SPEED SENSOR CIRCUIT INTERMITTENT

MALFUNCTION CODE	FUNCTION
P0724	TORQUE CONVERTER/BRAKE SWITCH "B" CIRCUIT HIGH
P0725	ENGINE SPEED INPUT CIRCUIT MALFUNCTION
P0726	ENGINE SPEED INPUT CIRCUIT RANGE/PERFORMANCE
P0727	ENGINE SPEED INPUT CIRCUIT NO SIGNAL
P0728	ENGINE SPEED INPUT CIRCUIT INTERMITTENT
P0730	INCORRECT GEAR RATIO
P0731	GEAR 1 INCORRECT RATIO
P0732	GEAR 2 INCORRECT RATIO
P0733	GEAR 3 INCORRECT RATIO
P0734	GEAR 4 INCORRECT RATIO
P0735	GEAR 5 INCORRECT RATIO
P0736	REVERSE INCORRECT RATIO
P0740	TORQUE CONVERTER CLUTCH CIRCUIT MALFUNCTION
P0741	TORQUE CONVERTER CLUTCH CIRCUIT PERFORMANCE OR STUCK OFF
P0742	TORQUE CONVERTER CLUTCH CIRCUIT STUCK ON
P0743	TORQUE CONVERTER CLUTCH CIRCUIT ELECTRICAL
P0744	TORQUE CONVERTER CLUTCH CIRCUIT INTERMITTENT
P0745	PRESSURE CONTROL SOLENOID MALFUNCTION
P0746	PRESSURE CONTROL SOLENOID PERFORMANCE OR STUCK OFF
P0747	PRESSURE CONTROL SOLENOID STUCK ON
P0748	PRESSURE CONTROL SOLENOID ELECTRICAL
P0749	PRESSURE CONTROL SOLENOID INTERMITTENT
P0750	SHIFT SOLENOID A MALFUNCTION
P0751	SHIFT SOLENOID A PERFORMANCE OR STUCK OFF
P0752	SHIFT SOLENOID A STUCK ON
P0753	SHIFT SOLENOID A ELECTRICAL
P0754	SHIFT SOLENOID A INTERMITTENT
P0755	SHIFT SOLENOID B MALFUNCTION
P0756	SHIFT SOLENOID B PERFORMANCE OR STUCK OFF
P0757	SHIFT SOLENOID B STUCK ON
P0758	SHIFT SOLENOID B ELECTRICAL
P0759	SHIFT SOLENOID B INTERMITTENT

MALFUNCTION CODE	FUNCTION
P0760	SHIFT SOLENOID C MALFUNCTION
P0761	SHIFT SOLENOID C PERFORMANCE OR STUCK OFF
P0762	SHIFT SOLENOID C STUCK ON
P0763	SHIFT SOLENOID C ELECTRICAL
P0764	SHIFT SOLENOID C INTERMITTENT

Index